Microbial life on Façades

Wolfgang Karl Hofbauer · Georg Gärtner

Microbial life on Façades

 Springer Spektrum

Wolfgang Karl Hofbauer
Institutsteil Holzkirchen
Fraunhofer-Institut für Bauphysik
Valley, Germany

Georg Gärtner
Institut für Botanik
Universität Innsbruck
Innsbruck, Austria

ISBN 978-3-662-54831-8   ISBN 978-3-662-54833-2   (eBook)
https://doi.org/10.1007/978-3-662-54833-2

Responsible Editor: Stefanie Wolf
This Springer Spektrum imprint is published by the registered company Springer-Verlag GmbH, DE part of Springer Nature.
The registered company address is: Heidelberger Platz 3, 14197 Berlin, Germany

Preface

Microbial life on building façades is a worldwide known phenomenon and has often led to controversial discussions between scientists, architects and industry. On the one hand growth of microorganisms may be regarded as a decorative element on historical buildings, on the other hand it may represent stages of ruination and corrosion of building surfaces and is therefore recognized as a nuisance that should be eliminated. To understand microbial growth on façades comprehensive knowledge of its diversity, dynamics and ecology is required. This can help us to develop effective management and mitigation strategies. The diversity of organisms on buildings is of much broader variety than reflected in most discussions in which it is reduced to algal components only. Besides algae also fungi, lichens, mosses and other groups even of higher plant and animal life forms are members of the associations of pioneer organisms on such extreme habitats like façades. Numerous authors have discussed countermeasures against growth on buildings despite the vague knowledge of the biological background. Still an uncertainty remains regarding the composition of different types of growth and especially the very beginning of microbial colonization on man-made structures. Since 2001 when W. Hofbauer was appointed as scientific collaborator at the Fraunhofer Institute for Building Physics (IBP) in Holzkirchen, he started studying aeroterrestric algae and fungi on different external building components. During various projects into the beginning of microbial growth – initial colonization – specimens were exposed at the institute's outdoor area. Isolation and cultivation of microorganisms for taxonomic analysis as well as ecological experimental work resulted in W. Hofbauer's doctoral thesis (supervised by G. Gärtner) and are the main component of this book. G. Gärtner, who is an expert specialized in aeroterrestrial algae and their cultivation, participated in the taxonomical identification and provided support with cultures for comparison and literature.

Together with the descriptions of morphological and ecological data of the analysed organisms, the book provides information on chemical and physical processes of façades as well as notes on different methods for preventing growth on external building components. A dichotomous key for identifying micobial layers on building surfaces, a glossary of technical and biological terms and a broad list of references are also included.

Thanks are due to Werner Kofler for preparing scanning electron microscope (SEM) pictures, to the Fraunhofer IBP for support and to the Deutsche Bundesstiftung Umwelt (DBU) for funding a part of the investigations. Special thanks go to Maya Stoyneva for valuable discussions and to the editor Stephanie Wolf for her great patience.

The authors hope that the book will be of interest not only to specialists in aeroterrestrial organisms, but also provide technicians and building engineers with an essential reference to manage a biological phenomenon, which has been and is a part of our life since centuries.

Holzkirchen	W. K. Hofbauer
Innsbruck	G. Gärtner
January 2020	

Acknowledgments

Part of the presented investigations was performed in a project funded by the Deutsche Bundesstiftung Umwelt (DBU, AZ 17974), industry partners and industry associations (Hofbauer et al. 2006). W. Hofbauer thanks the Fraunhofer Institute for Building Physics (IBP) for the possibility of conducting applied scientific work in many successful projects. We acknowledge the important hints from industry partners which are involved in further scientific collaborations.

The authors are much obliged to the Springer Verlag Heidelberg, Germany, for the interest in the project and to Stefanie Wolf and co-workers for valuable help, comments and support during preparation of the manuscript.

Both authors are deeply grateful to their families for their support, understanding and patience.

Contents

Introduction

Façades Colonized by Aerophytic Microorganisms

Since man-made buildings exist on earth, specialized microorganisms have found habitats on them, developing in different ways from positively accepted appearances as natural beauty to negatively conceived effects of deteriorating surfaces up to the destruction of artists' and architects' craft. In recent times, a lot of scientific research has been done, dealing with various aspects of microbial growth on man-made structures. Only a small minority of the enormous amount of publications dealt comprehensively with the whole spectrum of occurring organisms on modern building surfaces and almost none focused on the very beginning of the microbial colonization, the *primary colonization*, as we call it.

Façades as Functional Part of a Building

The façade of a building as the outermost layer of the construction has important protective functions, e.g., against driving rain or freezing conditions. But these are not the only functions of a façade, and it usually is also an element of decoration and design. Regarding the design and style of façades, there are many regional differences which are typical constituents of the flair of villages in different landscapes and economic zones. Usually, there are not only differences between geographical regions but also between rural and suburban or urban areas. The state of the art of house construction and the regional design are under continuous development, and the used materials and technologies are permanently optimized.

In our time, legal regulations have interfered in the design process substantially via, e.g., the German thermal insulation ordinance ("Wärmeschutzverordnung"). Producers

© Springer-Verlag GmbH Deutschland, ein Teil von Springer Nature 2021
W. K. Hofbauer and G. Gärtner, *Microbial life on Façades*,
https://doi.org/10.1007/978-3-662-54833-2_1

as well as manufacturers have reacted accordingly in an adaptation of their products and techniques. In order to protect buildings as much as possible against loss of warmth and energy, different highly performing systems have been developed. Examples are among others: external thermal insulation compound systems (ETICS), facings with air space, insulated bricks and insulating mortars—all these may additionally be composed of different materials.

The Environment Façade

Physical structures, substrate chemistry and environmental conditions define a façade as an extreme environment. It can be colonized by organisms which endure the often harsh conditions. In view of the fact that approximately 99% of the whole biomass of all ecosystems on land and in water is produced by photosynthetic organisms (Larcher 2001; Raven et al. 2005), it seems feasible to have a close look on plant forms which are a main part of the initial growth on buildings as a base for further biological succession.

To what extent components of building coatings may support growth of organisms and how the pioneer phase proceeds has scarcely been investigated and is still discussed controversially. Not only algae are capable of colonizing the outer surface of buildings, as a rule complex biocoenoses consisting of various organisms (algae incl. Cyanoprokaryota, fungi, animals, etc.) are established.

Development of lichens and mosses on monuments or ancient buildings is a well-known and accepted phenomenon. The composition of microbial growth interfering with materials/substrates (mainly degradative but also protective) has been analyzed and documented mostly for historic buildings, monuments and in nature (e.g., growth on stone) (e.g., Richardson 1975; Krumbein and Jens 1981; Del Monte et al. 1987; Galun 1988; Jones 1988; Sabbioni and Zappia 1991; Nimis et al. 1992; Caneva 1993; Piervittori and Laccisaglia 1993; Crispim and Gaylarde 2005; Khobragade et al. 2006, Darienko et al. 2013).

Potential destructive influences of microbial growth on building surfaces are of vital importance for the assessment of the situation and for the design of countermeasures. In general, a visible microbiological colonization of a façade within the first few years after construction is seen as problematic and discussed under different points of view (e.g., Richardson 1975; Caneva 1993; Bagda et al. 1999; Saiz-Jimenez 1997; Künzel 2000; Künzel and Sedlbauer 2001; Sedlbauer and Krus 2001; Sedlbauer 2002; Hladik 2003; Rindi and Guiry 2004; Hofbauer et al. 2005a, b, c, 2006; Crispim et al. 2006). Recently, the growth of Cyanoprokaryota, algae, fungi and lichens on external walls has increased dramatically. This is also due to increased insulation and the thereby reduced drying potential of wet external walls. Surfaces of insulated walls, e.g., of ETICS which are loaded by dew or driving rain, may remain wet for a longer period of time which favors biological growth (Künzel and Sedlbauer 2001; Sedlbauer 2002). Different studies indicated that a change in the quality of outside air (e.g., less SO_2 content) enhances the growth of certain aerophytic microorganisms (Hawksworth et al. 1973; Bates et al. 1990, 1996, 2001; Farmer et al. 1991, 1992; Gilbert 1992; Künzel 2000; Hauck et al. 2001, 2002; Hauck 2003, 2005; Schnug et al. 2004). An increasing eutrophication of the atmosphere (e.g., through increase of nitrogen compounds and hydrocarbons) and also climatic processes ("global change") are additionally recognized as factors which may favor the colonization of external building surfaces (Leathy and Colwell 1990; Cerniglia 1993; Pitcairn and Fowler 1995; Pitcairn et al. 2006; Ortega-Calvo and Saiz-Jimenez 1996; Saiz-Jimenez 1995, 1997; Leith et al. 1999, 2001; Mitchell et al. 2004; Raven et al. 2005). Frahm (2008) stated that the eutrophication of the atmosphere is also connected to the catalysts used in car engines. Microbial growth on building surfaces, which is relevant for damage cases, usually consists of different organisms. Not one single form (one alga or one fungus) alone is responsible for the perceived damage, but microbial growth is almost always caused by different organisms. Avoidance or reduction of unwanted growth is demanded, not only in a commercial view but also in a sustainable use of materials. Substantial growth on façades demands early and expensive renovation measures. In the long term, material damage or optic defacement and therefore greater use of materials cannot be ruled out. Added biocidal substances or chemical cures according to evident experiences have a certain time of action but do not last in the long term. Furthermore, if washed out, they may harm the environment. Common regulations within the EC result in a remarkable limitation in the choice of available biocides.

Within the scope of this book, general results regarding the diversity and ecophysiological parameters of biological growth on external building parts are presented.

Because of the rapidly and permanently changing conditions of temperature and moisture, building surfaces must be regarded as extreme environments. The constituents of biological crusts occurring in such environments are well equipped to face harsh conditions. They can withstand, e.g., extreme temperatures and other adverse influences. A filamentous soil crust alga (*Zygnema* sp.) was demonstrated to be insensitive to experimental UV (from 280 nm upwards) exposure (Holzinger et al. 2009).

History of Aerobiology in Respect of Research on Man-Made Surfaces

The history of the aerobiology is connected both to the development of microscopic instruments and techniques as well as to the establishment of laboratory cultures, as shown by Sitte et al. (2002). In the nineteenth century, Ferdinand Cohn, the founder of bacteriology, was able to keep *Haematococcus* (Chlorophyceae-Volvocales) in his laboratory in Breslau for a certain time. He named this process "cultivation" (Cohn 1850). The Russian plant physiologist Famintzin used for the first time Knop's solution for the cultivation of algae (Famintzin 1871). This culture medium with some inorganic compounds was developed by Knop for research on vascular plants in 1865 and is still in use today (Preisig and Andersen 2004). Aerobiology, as the study of aerophytic microorganisms, was mainly influenced by the classical research of Louis Pasteur and Robert Koch (Deichfelder 1985) at the end of the nineteenth century, confirming the distribution of microorganisms by air. Further important steps in the research on aerophytic microorganisms were the foundation of culture collections, as was done by Chodat (e.g., 1913, 1928) and Pringsheim (1924), and the technology of maintaining isolated microorganisms, especially algae (Pringsheim 1954, Preisig and Andersen 2004). A specialized collection of aerophytic (soil, airborne and lichen) algae was established in Innsbruck based partially on the collections made by W. Vischer in Basel, Switzerland (Gärtner 2004).

Aerophytic cryptogams, with emphasis on aerophytic algae, were investigated by Puymaly (1924) in France. Aerophytic green algal layers and their components were studied by Brand (Brand and Stockmayer 1925) and aerophytic biocoenoses on rock and cryptogamic epiphytic coenoses including algal associations by Barkman (1969). For taxonomic studies of eukaryotic aeroterrestrial algae and phycobionts, see Ettl and Gärtner (1995, 2014).

The Fraunhofer Institute for Building Physics (IBP, founded 1929 as an institute for technical physics) in Holzkirchen/Bavaria is specialized in the investigations of building materials and constructions, and for many years the colonization of materials and constructions with microorganisms has been intensively studied.

Our Study on Aerophytic Organisms on Building Surfaces

In an interdisciplinary study (see also Hofbauer et al. 2006), we undertook many different investigations in connection with the initial biological succession on modern building surfaces. The applied research work was mostly carried out in the years 2002–2007. Apart from newly constructed specimens, exposed at three different study sites in Germany (Holzkirchen, Heggen/Finnentrop, Ernsthofen/Oberramstadt), also old specimens exposed for 10 years and many additional growth situations (Sect. 2.1) as well as background concentrations were measured. Qualitative and quantitative microbiological analyses as well as continuous observations of newly built specimens at different outdoor weathering stations were the main emphasis of the investigations (Sect. 2.2–2.4). Furthermore in cooperation with colleagues, also data of the structural–physical characteristics of the different building materials were gained (Sect. 4.2). Our work presents the first comprehensive investigation of the initial succession on modern building surfaces. Occurring organisms were differentiated as far as possible (especially algae, Cyanoprokaryota, fungi, bryophytes and lichens) and compiled with chemical–physical measurements and data from the literature (Chap. 3). The most important species of the initial succession were documented. In the course of our investigations, more than 220 different taxa were identified as part of the initial succession (see Sect. 4.1). The biggest part of the diversity was provided by algae and cyanoprokaryota (ca. 85 species), followed by fungi (ca. 80 species). The remaining taxa were allocated to further groups of organisms. Remarkably also bryopsida (ca. 12 species) and lichens (ca. 15 species) contributed to the initial growth. Fungi were discovered to be prominent in the initial phase of surface colonization on modern building surfaces, especially intensely pigmented forms like melanogenous fungi and Coelomycetes. Algae and Cyanoprokaryota were present after a certain lag phase. According to the given microclimatic conditions on the surfaces, especially the water availability mainly through dew or by high relative humidity, eukaryotic algae were favored and only few Cyanoprokaryota occurred. Surface growth was mostly dominated by aerophytic green algae (especially Trebouxiophyceae) and strongly pigmented fungi (dematiaceous fungi and Coelomycetes) on new specimens after 2 and 3 years' exposure, in additional growth situations and on old specimens. Overall, a correspondence existed between the different measurements regarding microclimate, local climate, material characteristics and the identified organisms. Growth was more intense on materials which stayed moist for a longer time or which contained more nutrients. Growth diminished on materials which were characterized by strong chalking. It became obvious that the contribution of nutrients to the surrounding environment plays an important function in initial succession. A further important result was that microbial growth develops in cycles. During the dry seasons (summer, winter), surface growth stagnated or decreased; whereas in the cool and damp seasons especially in autumn/early winter, a distinct increase of growth development happened. Furthermore, we demonstrated that driving rain is an important factor in the propagation and settlement of microorganisms involved in the initial succession on outer building surfaces. For all identified organisms, detailed data regarding taxonomy and physiology are provided (Chap. 3). The gained data are an important base for future work on initial succession on façades and can also be used to design effective countermeasures against "unwanted microbial growth". Target organisms for potential chemical measures were better defined. An ample culture collection of microorganisms relevant to building parts has been established. In the course of our study, more than 400 isolates were acquired and integrated into the collection. Our study clearly demonstrates that the control of moisture on building surfaces is of crucial importance. This can be reached by different measures and strategies. As the important organisms were determined also their needs regarding moisture, temperature and further influence factors were defined. The observed algae are known to be active in a range of relative air humidity of 68–100%; the fungi start at ca. 73.3% relative humidity. The upper temperature limits lay at 57 °C (active) or even at 100 °C (dormant stage). The lowest temperature for physiological activity of the identified organisms is ca. −15 °C. In Chap. 3 also the pH limits, tolerated salt concentrations, substrate specifications, etc., of the different observed forms are given. As examples, *Trentepohlia iolithus* (Ulvophyceae, Chlorophyta) and *Sarcogyne regularis* (Lecanoromycetidae, Ascolichenes) showed a preference for mineral surfaces (both species showed up in the second year of outdoor exposure or on older surfaces only). The gained data also offer an option for extended and enhanced accuracy of mathematical models in the assessment of damage tolerance/resilience of new materials and constructions regarding microbial growth.

Tools and Methods

For the experimental investigation and the analysis of *primary growth* (=initial growth) on building surfaces as conducted in the study by the authors, different methods were applied. In this chapter, we describe our investigation methods as well as some further or different approaches to assess microorganisms on modern building structures. Furthermore, testing methods for susceptibility of building products against microbial growth are discussed.

Specimens–Weathering Exposure–Locations

Specimens: To investigate the settlement and development (succession) of growth on surfaces of different ETICS, a special kind of specimen was created and exposed at different outdoor locations in Germany (Hofbauer et al. 2006). The structure of the specimens comprised the complete layer sequence of a real ETICS (from top to bottom: paint—top plaster—reinforcement and ground plaster—insulation layer). In all used materials, film conservation was excluded. Additionally, specimens were sealed on the side by a double layer of an epoxy coating. To reduce side effects by run-off water or stagnant water, the top of the specimens was slanted and at the lower end a drip edge was attached. The dimension of the specimens was ca. 35 × 30 cm. We did not provide the wall building material (e.g., concrete or brick) but glued the insulation layer onto a tile which functioned as a strengthening part and attached the specimens onto a supporting construction. Thus, the specimens were subjected to the influence of the outdoor climate from all directions and did not get any protection of a back wall. Real insulated ETICS walls may get tiny proportions of warmth from the room behind but rough calculations revealed that this effect could be neglected. The general composition of the specimens is illustrated in Fig. 2.1.

© Springer-Verlag GmbH Deutschland, ein Teil von Springer Nature 2021
W. K. Hofbauer and G. Gärtner, *Microbial life on Façades,*
https://doi.org/10.1007/978-3-662-54833-2_2

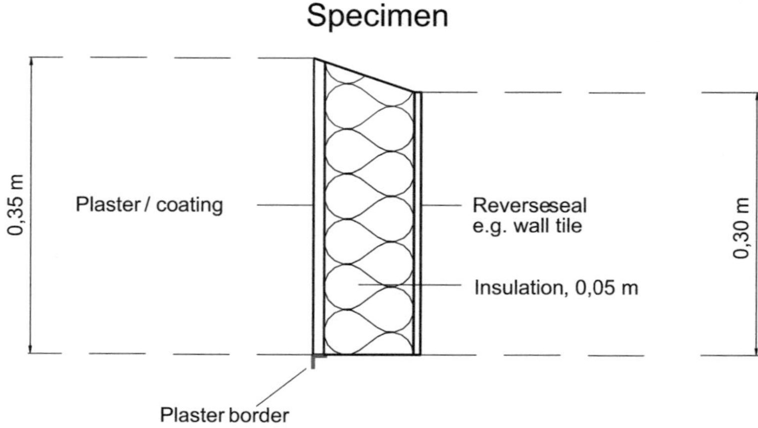

Fig. 2.1 Schematic structure of a specimen (Hofbauer et al. 2006; Hofbauer 2007)

Overall, the conditions on the surface of the specimens were slightly more challenging for the material itself and slightly more favorable for microbial growth than real wall constructions. Therefore, the results are reliable in respect of the long-term resistance of surface materials against growth. Altogether, 15 different variants were investigated comprising three different material groups (Table 2.1 for variants exposed in Valley/Holzkirchen).

For each variant, 18–21 replicates were produced, altogether about 300 specimens were involved in the investigations. In Valley/Germany, 10 specimens for each variant were exposed supplemented by extra specimens of some variants with "model character" which had attached PT100 thermoelements on the top layer to measure surface temperature and gain additional information on dew point shortfall. Since the attachment of thermoelements meant some disturbance of the surface, these specimens were excluded from the biological assessment. At additional locations (see below), four specimens of each variant again supplemented by "model specimens" with thermoelements were exposed. Specimens were attached to the supporting structure in approximately breast height (for better assessment of the surface) facing exactly west, and their angle was absolutely vertical. The outdoor exposure started in June 2002 (Fig. 2.2). In order to protect specimens against excess bird influence, a thread was attached above the specimens, as birds tend to sit on projecting structures in the field and leave droppings. Materials were varied in order to gain information about different factors which potentially influence the establishment of microbial growth. Generally, the surface was created in the typical way of a structure plaster with a structure kernel size of 2 mm, in one variant this was only 1 mm and one variant had a so-called scratched surface. Some variants had additional paint as finish, and some were without. For paints, also color and hydrophobicity were varied. Further, we had variants with super hydrophobicity and with infrared (IR) effect. Specimens in Holzkirchen were assessed (biological surface development) in a monthly or two monthly cycle, respectively.

Table 2.1 Overview of "new" variants exposed in Valley/Holzkirchen (main location) and in Heggen/Finnentrop and Ernsthofen/Oberrramstadt (exception: variants NV7–NV9 were only exposed in Holzkirchen)

Term	Plaster system	Top coat/color
NV1	Mineral thin	Dispersion silicate paint 1, purple, reduced hydrophoby
NV2	Mineral thin	Dispersion silicate paint 2, white, reduced hydrophoby
NV3	Mineral thin	Silicone resin paint 1, white
NV4	Mineral thin	Silicate paint 2
NV5	Mineral thick	Silicate paint 1, white
NV6	Mineral thin	Dispersion silicate paint 1, white
NV7	Mineral thin	Colored paint 1, gray
NV8	Mineral thin	Colored paint 1, gray, additional IR effect
NV9	Mineral thin	Silicone resin paint 3, white, super hydrophoby
NV10	Mineral thin	–
NV11	Mineral thin, structure kernels only 1 mm	–
NV12	Mineral thick, scratched surface	–
NV13	Silicone resin	–
NV14	Silicone resin	Silicone resin paint 1, white
NV15	Silicone resin	Silicone resin paint 2, white, super hydrophoby

Fig. 2.2 Exposition of the specimens at the main location in Valley/Holzkirchen shortly after the start. Above the specimens, bird protecting threads can be recognized (Hofbauer et al. 2006; Hofbauer 2007)

"Old" specimens: In addition to the newly built specimens, a double set of older specimens (already exposed to the local climate in Valley at an angle of 60° for 10 years [since 1992]) which remained of an earlier project were used. The structure of the "old" specimens was similar to the newly produced specimens apart from the following two differences: They were not slanted on top and did not have a drip edge. Some of the "old" specimens originally had film conservation, but it was only known which ones were equipped with it not the composition of the biocides. These older specimens were very useful in giving information on successional stages of growth development, and they also allowed us to evaluate and optimize methods of assessment since they were available from the beginning of the study. Eleven different variants of "old" specimens were chosen and biologically assessed (Table 2.2).

Whole wall constructions: For comparison and practical issues, a choice of different system compositions was installed on a whole building wall at the institute, again facing west (Fig. 2.3). The chosen system compositions (Table 2.3) mostly comprised the "model systems," and they were equipped with temperature sensors. Biological assessment was performed in areas not disturbed by the installed temperature sensors.

Additional locations: In order to have some variation in weather condition, additional locations for weathering exposure were chosen: Sauerland–Heggen at Finnentrop (outdoor area of company Weber and Broutin, Fig. 2.4) and Odenwald–Bergstraße–Ernsthofen at Oberramstadt (outdoor area of company Deutsche Amphibolin Werke, Fig. 2.5). Both

Table 2.2 Overview of the composition of "old" specimens

Term	Plaster system	Binders	Biocidal equipment
OV1	Synthetic resin plaster	Styrolacrylate Vinyl ester-terpolymer	+
OV2	Synthetic resin plaster	Styrolacrylate Vinylacetat/ethylene/vinyl chloride-copolymer	+
OV3	Mineral thin	Calcium silicate	–
OV4	Mineral thin	Lime–cement plaster	–
OV5	Mineral thin	Lime–cement plaster	–
OV6	Mineral thin	Lime-cement plaster	–
OV7	Mineral thin	Lime-cement plaster	–
OV8	Mineral thick	White lime hydrate white cement	–
OV9	Silicone resin plaster	Silicon resin emulsion KH dispersion	+
OV10	Silicate plaster	Potassium water glass Styrolacrylate	–
OV11	Silicate plaster	Potassium water glass Styrolacrylate	–

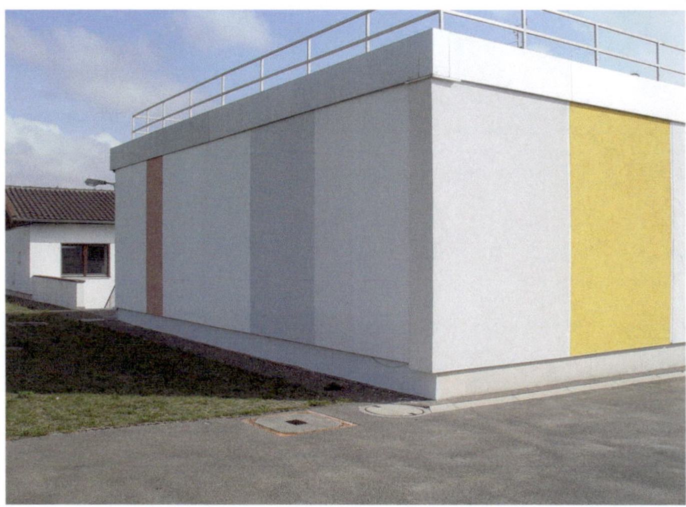

Fig. 2.3 View of the whole wall construction with attached ETICS facing west (Hofbauer et al. 2006; Hofbauer 2007)

Table 2.3 Variant specimens mounted as whole wall constructions with the same material types as used for new specimens

Term	Plaster system	Paint/color
NV1	Mineral thin	Dispersion silicate paint 1, purple, reduced hydrophoby
NV2	Mineral thin	Dispersion silicate paint 2, white, reduced hydophoby
NV3	Mineral thin	Silicone resin paint 1, white
NV4	Mineral thin	Silicate paint 2
NV5	Mineral thick	Silicate paint 1, white
NV6	Mineral thin	Dispersion silicate paint 1, white
NV7	Mineral thin	Colored paint 1, gray
NV8	Mineral thin	Colored paint 1, gray, additionally IR effect
NV9	Mineral thin	Silicone resin paint 3, white, super hydrophoby
NV10	Mineral thin	–

additional locations have a different climate compared to Valley/Holzkirchen with a tendency to be warmer and with more relative humidity. In contrast to the site in Holzkirchen, the additional locations are situated in depressions not far from running water. Therefore, some differences in the surface growth development were expected. Simultaneous to the installation of the specimens at the main exposure site also the exposure at the additional locations was started. The frequency of the surface assessment at the additional locations was for logistical reasons set at quarterly intervals.

Fig. 2.4 Exposition of the specimens in Heggen/Finnentrop (Hofbauer 2007)

Fig. 2.5 Exposition of the specimens in Ernsthofen/Oberramstadt (Hofbauer 2007)

Analysis of Microbial Growth

A major goal of our investigations was to systematically assess the composition and time course of the biological processes at the newly produced surfaces. Apart from taxonomical/systematical microbial analyses, different microscopic techniques (e.g., light and electron microscopy) were conducted.

A quantitative germ count was performed at defined intervals (½, 1 and 2 years). Special emphasis was laid on the qualitative composition and taxonomic background of the developing initial succession. This was also documented in defined intervals (½, 1,

2 and 3 years) and done for all three locations. Additionally, the qualitative assessment was compared to further analyses of established growth on building structures and on the older specimens. Visual growth development and pattern of initial growth were documented by a newly developed assessment method including a rating scale (Hofbauer et al. 2003) in monthly intervals.

Quantitative Biological Analysis

Visual Assessment

A special method was created on the base of synsystematic vegetation assessment to record the precise start of visible growth and developmental patterns (Hofbauer et al. 2003). Therefore, the specimens or part of wall areas were scanned, and the developed growth was rated by a special scale (Table 2.4). To avoid side effects, the upper

Table 2.4 Visual rating scale for assessing growth intensity on building surfaces. (Hofbauer et al. 2006; Hofbauer 2007; Künzel et al. 2011)

Rating figure	Rating symbol	Microbiological growth covering examined area on an average of
10	+++++	**Very intense**, almost the whole area is overgrown, >87.5–100% of the total area covered by clear surface growth or soiling
9	++++(+)	Surface growth/soiling **between intense and very intense**, >75–87.5% of the total area covered by surface growth or soiling
8	++++	**Intense** surface growth/soiling, >62.5–75% of the total area overgrown or soiled
7	+++(+)	Surface growth/soiling **between clear and intense,** >50–62.5% of the total area overgrown or soiled
6	+++	Surface growth/soiling **clear,** >37.5–50 % of the total area overgrown or soiled
5	++(+)	Surface growth/soiling **between moderate and clear,** numerous dots or marks, overgrown area >25–37.5%
4	++	**Moderate** surface growth, numerous dots or sparse spots, >5–25% of the total area affected by surface growth and soiling
3	+(+)	**Low** surface growth, some dots or marks, in all less than 5%
2	+	**Very low** surface growth, two to few single spots (clearly visible but less than 5%)
1	(+)	**Minimal** surface growth, 1–3 small single spots
0	-	No visually recognisable surface growth (0%)

4 cm of the specimens (slant) and the lower most 2 cm (plaster rim) were excluded. Differentiation of the scale is finer in the lower ranges in order to assess small differences in starting growth patterns, whereas higher ranges of the scale are oriented at the covered surface. This is in contrast to ASTM D3719-00 (2000) which only assesses growth on a linear scale based on the covered area. Emerging growth thus was regularly scanned on a monthly and bi-monthly (Valley/Holzkirchen) and half-year (Oberramstadt/ Ernsthofen; Finnentrop/Heggen) basis, respectively.

In Figs. 2.6, 2.7 and 2.8, examples for the rating are given. In addition, growth patterns were assessed and documented by photography and by the aid of a template foil, which was used to transcribe the pattern on corresponding sketches (Fig. 2.9), in order to observe the development of the growth pattern in detail. Complementary to the regular assessment of the specimens, the analogous assessment was also done for the trial walls.

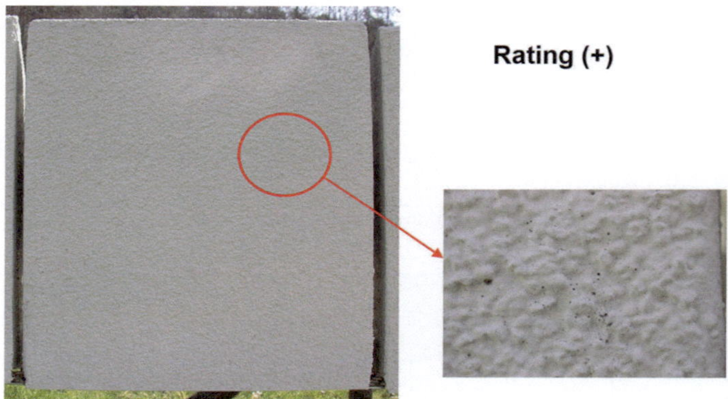

Fig. 2.6 Rating scale level + with detail of observed growth (Hofbauer et al. 2006; Hofbauer 2007)

Fig. 2.7 Rating scale level +(+) with detail of observed growth (Hofbauer et al. 2006; Hofbauer 2007)

Fig. 2.8 Rating scale level +++(+) with detail of observed growth (Hofbauer et al. 2006; Hofbauer 2007)

Sample Collection and Preparation

At first, a set of "old" specimens (exposed for 10 years) was collected. Regarding the newly exposed specimens, a time scheme for collecting samples was established: Samples were taken after ½, 1 and 2 years of exposure. Specimens were taken into the laboratory in a dry state (not wetted by precipitation) for further processing. Of a representative area of the surface of a subsample, the size of 5×5 cm^2 was cut out in the thickness of the whole plaster and paint layers using a stone cutting saw. This was performed under conditions to avoid additional contamination and cross-contamination. The blade of the stone cutting saw was cleaned and disinfected with 70% ethanol or isopropanol before each specimen. For transport and/or temporary storage, the cutout samples were immediately transferred into fresh and sterile ziplock bags. Remaining parts of the old specimens were transferred back to the outdoor site as backup.

Further processing of the samples was done in a microbial laboratory under sterile conditions. Storage of samples, if necessary, took place under dark, dry and cool (room temperature) conditions.

Isolation of Microorganisms

Isolation of microorganisms was performed analogous to Berner et al. (1997) or Sterflinger and Prillinger (2001), respectively. Samples were cautiously manually grinded in an enamel mortar under laminar flow. This yielded a homogenous sample powder with grain size ≤0.1–0.5 mm and ensured the greatest possible recuperation of viable colony-forming units (CFU) under the aspect of more or less difference of hardness and elasticity of the sample material.

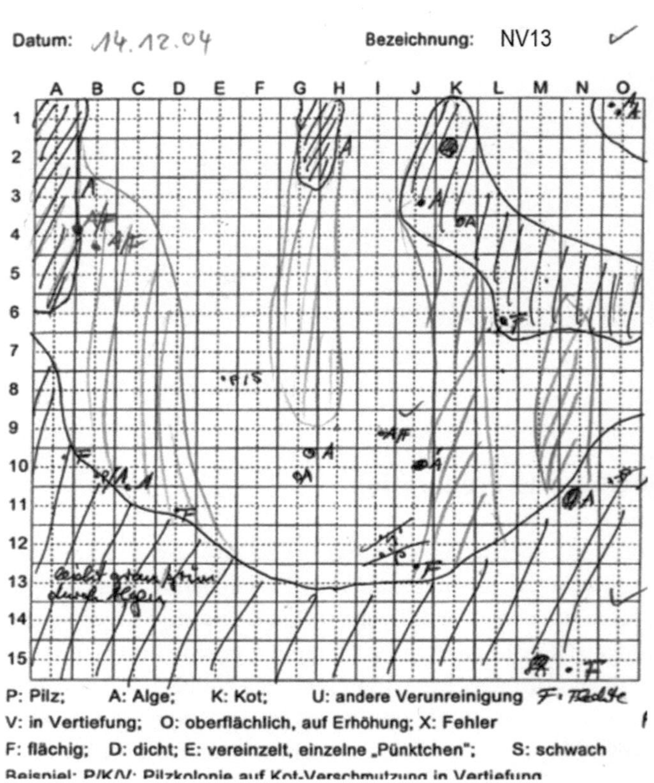

Fig. 2.9 Raster sketch of developing algal (A) and fungal (F) growth (Hofbauer et al. 2006; Hofbauer 2007)

Part of the gained homogenized powder (approx. 1 g) was given into 10 ml dilution solution (0.9% NaCl, 0.01% TWEEN 80) and shaken for 1 h at 300 rpm. A combination of physiological salt solution with the mild detergence TWEEN 80 has proven to be good in different microbial investigations and was applied, so that propagation units of microorganisms adhering to the substrate were removed in a mechanically and physiologically gentle way (e.g., Wollum 1982; Craig et al. 1987; Perrissol et al. 1993; Ehrlich 2002; Labuda et al. 2003; Ranilla and Carro 2003; Çelen and Kiliç 2004; Kłyszejko et al. 2005). After shaking, the powder suspension was allowed to settle for a few minutes. Working in a laminar flow, 100 µl of each diaspore-containing supernatant was plated onto Petri dishes containing different culture media by using Trigalsky spatula. In addition, a dilution series was performed according to the "most-probable-number-method" (MPN-method; Alexander 1982), employing three dilution steps (1:10 each, always diluting 1 ml of the former suspension with 9 ml of dilution medium). After that the different dilutions were plated, resulting in a "diluted plating" (Skinner et al. 1952; Alexander 1982; Hoekstra et al. 2002; Samson et al. 2002). Further to the dilution series, a weighted amount of the sample powder (ca. 0.5 g up to ca. 1 g) was applied in "direct plating" (Skinner et al. 1952; Hoekstra et al. 2002; Samson et al. 2002). Direct plating was employed in order to cultivate organisms that appeared in low numbers on the investigated surfaces. In the direct plating trials, primarily algae-medium and a low nutrient medium were used because it was assumed that in the initial phase especially algae and other oligophilous microorganisms would be present at the investigated surfaces in low germ numbers. Each batch (dilution and medium) was done in triplicate. The inoculated Petri dishes were placed into culture cabinets (Binder Company), under a diurnal regime of 12/12 h light/dark and 25 °C (light) and 16 °C (dark), respectively. For irradiation, special plant culture light tubes (Osram Company) with an illuminance of ca. 1200 W/m^2 were chosen, according to previous experience in the cultivation of aerophytic algae (Ettl and Gärtner 1995; Gärtner 1996). In order to provide an additional selective effect for organisms that may survive on façade surfaces prone to light, not only the culture batches for algae but also the batches for fungi and bacteria were kept under daily variation of irradiation.

In the culture experiments, the following media were used: malt extract peptone agar (MEA) for mesophilic fungi (Booth 1971a; Samson et al. 2002; Domsch et al. 2007), Czapek-Dox agar (CD) undiluted for mesophilic fungi and diluted 1:50 for oligophilic/ oligotrophic fungi and bacteria (Domsch et al. 2007), dichlorane-glycerin agar (DG18) for xerotolerant fungi (King et al. 1979, 1986; Hocking and Pitt 1980; Samson et al. 1992, 2002; Frändberg and Olsen 1999), plate count agar (PCA) for aerobic bacteria (Domsch et al. 2007; Samson et al. 2002) and Bold's basal medium (BBM) for algae (modified according to Bischoff and Bold 1963; Ettl and Gärtner 1995).

The Petri dishes were regularly controlled for several weeks because part of the investigated organisms was frugal or slow growing. For quantitative analyses, the emerging colonies (colony-forming units, CFU) were observed and counted after 10 and 90 days, respectively. In the total germ counts according to the MPN method (Alexander 1982), we distinguished between bacteria, fungi and algae (in single cases also for further organism groups like e.g. ferns). Finally, the total germ numbers were extrapolated to 1 g sample powder.

Assessment of Starter Germ Load of Materials

Of each material/composition, reserve samples were generated. The system composition (plaster and paint) was identical to specimens used in weathering experiments. Part of the reserve samples was used for assessment of the starter germ load immediately after completion of the hardening of the material while stored at room temperature. This was undertaken to clarify the following questions: (1) If the specimens were loaded with germs of microorganisms right from the start, (2) what kind of microorganisms were present and (3) if microorganisms were present which caused damage later on. Reserve samples were called 0-probes, because they resembled the condition at time zero of the weathering exposure. Sample collection and sample preparation as well as the culture experiments were done as described previously. Cultures were assessed quantitatively and qualitatively. Aerobic bacteria were assessed as a whole group only. In principle, it was expected that starter germ loads of fresh finished coatings would be very low, because present germs usually are killed during the production of the material (e.g., rather high alkalinity of plaster in preparation, temperature evolvement, etc.). For data, see Hofbauer et al. (2006).

Culture-Based Taxonomic Analysis

Cultures on selective media with subsequent isolation of pure cultures allow an identification of the occurring microorganisms. Imperative for a taxonomic treatment are comparison and reference cultures, because most microorganisms cannot be compared with herbarium material like, e.g., vascular plants. Drying usually destroys important diagnostic characteristics of microorganisms. Furthermore, knowledge and investigation of different developmental stages, which can only be achieved in parallel with cultures, are often necessary for a successful differentiation (Gärtner 1986; Ettl and Gärtner 1995). At the IBP, a culture collection for the most important species known to occur on building surfaces has been built up, mainly based on proprietary investigations. This collection is still maintained and expanded. It comprises the only comprehensive culture collection which is specialized on building relevant microorganisms so far (Fig. 2.10; see also Hofbauer et al. 2003).

Taxonomy

Colonizers of building surfaces stem from almost all today accepted major groups of organisms and comprise representatives of prokaryota and eukaryota. In a more narrow sense, organisms of the regnums of bacterial life forms (incl. Cyanoprokaryota), fungi, plants, protists, etc., can be found.

In Chap. 3, basal groups are arranged mainly according to Ettl and Gärtner (1995, 2014), Hoek van den et al. (1995), Esser (2000), Komárek and Anagnostidis (1998), Graham et al. (2009), Komárek and Anagnostidis (2005) and Raven et al. (2005). For detailed questions regarding certain forms and species, specialized current literature was used, which is mentioned in the discussion of the respective forms. As much as possible, we tried to follow the natural system. The basis for the identification of the different taxa is a comprehensive taxonomic concept with emphasis on morphological

Fig. 2.10 Part of the culture collection of building relevant microorganisms at the IBP (Hofbauer 2007; Hofbauer et al. 2003, 2006; www.ibp.fraunhofer.de: Forschung im Fokus: Mikroorganismen)

characteristics (Bubrick et al.1984; Komárek and Fott 1983; Gärtner 1984, 1985; Komárek and Anagnostidis 1998; Graham et al. 2009). Additionally, synoptic works were used for the identification of diverse form groups such as Domsch et al. (2007), Ettl and Gärtner (1995, 2014), Bergey's Manual of Systematic Bacteriology (Boone and Castenholz 2001), Komárek and Anagnostidis (1998), Komárek and Anagnostidis (2005) and AlgaeBase (Guiry and Guiry 2019) as well as specialized literature which is specified at the respective groups. Technical equipment at hand was powerful reflected light and transmission microscopes employed in video, polarization and fluorescence microscopy. Magnifications up to 1000- to 1250-fold (immersions objectives) facilitated light microscopic analyses of cell morphology. For highlighting cellular structures different staining methods were used for microscopic preparation (see Ettl and Gärtner 1995), e.g., Lugol's solution, carmine acetic acid, methylene blue, Gram staining, sudan red and others. Microscopic investigations were accompanied by extensive photographic documentation. In some cases, also scanning electron microscope (SEM) techniques were used. Emphasis on determination of taxonomy was laid on photosynthetic active organisms (including algae, Cyanoprokaryota, bryophytes, ferns and lichens) and fungi. Bacteria were mainly assessed as a whole group (aerobic). Results of recent genetic–taxonomic research were also considered (e.g., Buchheim et al. 1990, 1996, 2001; Huss and Sogin 1990; Kantz et al. 1990; Buchheim and Chapman 1991; Lewis et al. 1992; Wilcox et al. 1992; Surek et al. 1994; Friedl 1995; Melkonian and Surek 1995; Bhattacharya et al. 1996; Nakayama et al. 1996; Booton et al. 1998; Chapman et al. 1998; An et al. 1999; Nedelcu et al. 2000; Turmel et al. 2002; Krienitz et al. 2003; Lewis and McCourt

2004; Pombert et al. 2004, etc.). Naming of the detailed taxonomic units was aligned to the International Code of Nomenclature for algae, fungi and plants (Shenzhen Code; Turland et al. 2018).

Pre-culture Investigation

In a pre-culture investigation, the surface of the specimens was scanned for traces of growth first by naked eye. The overall impression of the surface condition was noted (weathering, macroscopic discernible growth, etc.).

In the second step, probes of supposed growth or dirt/graying were investigated using scanning and transmission light microscopes. A detailed investigation of already visible established growth leads to subsequent cultures on selective media and helped to interpret the results. It cannot be ruled out that by chance forms developed in culture whose diaspores only accidentally had fallen on the surface of the specimens. Pre-culture analysis helped to assess the proportions that were measured in culture. Therefore, we tried to identify as much forms as possible in pre-culture investigation. This was only achieved in a limited way because many organisms showed reduced or altered features due to the changing and partly extreme outdoor conditions at the surfaces. It is well known that not all organisms that colonize building surfaces can be cultivated with standard media, as, e.g., lichens and some bryophytes. These organisms were identified mainly on original material from the surfaces; if enough differentiating characteristics were found, otherwise they were grouped within the next higher rank.

Sample Collection and Preparation

Collection of samples and preparation was the same as for the quantitative analysis which is described in Sect. 2.3.1. Raw cultures of the quantitative analysis after ½, 1 and 2 years of exposure in Holzkirchen were the base not only for the germ counts but also for the differentiation of the occurring microorganisms. Therefore, subsequent pure cultures of the isolates were established.

After 3 years of exposure in Valley/Holzkirchen, an additional survey was performed in a simplified procedure: With a lancet needle, the surface of the specimens was scratched at randomly selected points and the gained material was investigated microscopically, and the rest was spread evenly on Petri dishes with selective media for algae and fungi. In this way, raw cultures for further differentiation were gained. These cultures allowed to assess coarse abundance classes (dominant, subdominant, accessory/scattered).

Determination of the occurring organisms was conducted in the same way for the additional locations Ernsthofen/Oberramstadt and Heggen/Finnentrop. Furthermore, another set of samples of "additional variants/surfaces" (Table 2.5) was also investigated based on scratch samples and subsequent cultures.

Table 2.5 Overview of additional variants and surfaces

Symbol	Location	Substrate/building material/construction	Age (in years)
AV1	Ernsthofen	Specimen, ETICS,	ca. 10
AV2	Oberramstadt	Specimen, ETICS,	ca. 2
AV3	Oberramstadt	Specimen, ETICS,	ca. 3
AV4	Oberramstadt	Specimen, ETICS,	ca. 3
AV5	Holzkirchen, IBP	Wall base, ETICS, shady	Unknown (>10)
AV6	Holzkirchen, IBP	Concrete slab, horizontal	Unknown (>10)
AV7	Holzkirchen, IBP	Specimen: sandstone, cut ("Asterix")	ca. 20
AV8	Innsbruck	Concrete base of lantern pole	Unknown (>10)
AV9	Innsbruck	Algal crust on pavement	Several years
AV10	Kufstein fortress	Base of travertine built wall	Unknown (>10)
AV11	Innsbruck	Plaster surface, massive construction, green stain	Unknown (>10)
AV12	Pfaffenhofen	Dark spots at ETICS	Unknown (>10)
AV13	Berlin	Flat roof	Several years
AV14	Innsbruck	Algal crust at fair-faced concrete	Unknown (>10)
AV15	Kufstein/Zell	Concrete surface	Unknown (>10)
AV16	Stuttgart	ETICS, green stain	Unknown (>10)
AV17	Gladbeck	Specimen, coated façade facing slab	Unknown (>10)
AV18	Nürnberg	ETICS, green stain	Unknown (>10)
AV19	Wesel	Special type of concrete, pre-fabricated construction part	Several years
AV20	Alzenau	Special type of concrete, pre-fabricated construction part	Several years
AV21	Grevesmühle	ETICS, green stain	Unknown (>10)
AV22	Wismar	Timber surface, green stain	Unknown (>10)
AV23	Holzkirchen, IBP	Algal crust, window sill	Several years
AV24	München	Travertine, zoological garden	Unknown (>10)
AV25	Schleswig	ETICS, green stain	Unknown (>10)

Establishment of Pure Cultures

In general, isolation of microorganisms was performed as described above, analogous to Berner et al. (1997) or Sterflinger and Prillinger (2001), respectively. To establish pure cultures, one or two cleaning steps were needed during which the respective forms were transferred to further culture or selective media. If necessary, also micromanipulator techniques were performed by use of specially produced thin glass tips or threads (cleaning by pulling over an agar surface) or by very thin pipettes (cleaning by sucking up of single cells), as described by Ettl and Gärtner (1995).

Background Load

Microorganisms growing at specimens and building parts reach these by different means of transport. It has previously been shown that, e.g., algae may spread over land by animal vectors (Migula 1888; Wille 1923; Strøm 1926; Foged 1953; Schlichting 1960; Luther 1963; Maguire 1963; Parsons et al. 1966; Proctor 1966; Roscher 1967; Milliger and Schlichting 1968; Schlichting et al. 1978; Schwabe and Hickel 1978; Kristiansen 1996), which applies mostly for limnic species. Therefore, this means may be quite important for building surfaces as well. For aerophytic algae, distribution through air has been demonstrated (Overeem 1937; Messikommer 1943; Schlichting 1961; Brown et al. 1964; Geissler and Gerloff 1965; Saxena 1983; Rosas et al. 1989; Kristiansen 1996; Marshall and Chalmers 1997; Tormo et al. 2001; Sharma et al. 2006), but so far no connection to building part surfaces was established.

Quantitative and qualitative assessments of airborne diaspores are achieved by different methods, e.g., air germ measurements, sedimentation, etc. Especially for bacteria, fungi and pollen, standard methods and protocols are available (e.g., Hirst et al. 1967; Hoekstra et al. 2002). As a reference for colonization and succession on outer surfaces of building parts, only viable distribution units are relevant. Therefore, it is not required to reveal the complete composition of the background concentration including dead or nonviable biogenic particles, as it is necessary in, e.g., allergologic questions (Burge et al. 1977; Bhati and Gaur 1979).

Air Germ Measurement

To assess the background concentration of fungal spores in the immediate surroundings of the specimens at different times, air germ measurements using a filtration method were performed (Jensen et al. 1998; Hoekstra et al. 2002). A defined air current (volume, velocity) was drawn through a filter with a mesh size small enough that all distribution units (spores) were collected. Very useful were gelatine filters because this material reduces the collection stress at the distribution units of microorganisms. After collection, the loaded gelatine filters were put on appropriate selective media and subsequently the growing microorganism colonies were counted and determined. If the spore load was expected to be rather high, gelatine filters were dissolved easily and the spore concentration was diluted. We used the MD18 apparatus of Sartorius Company with adjustments of 50 l/min. In a single measurement, always 50 l of air were sampled. For the assessment, selective media MEA, DG18, PCA and BBM were employed. Resulting germ counts were calculated for 1 m^3 air volume (results not shown).

Sedimentation

As a complementary assessment of concentration and composition of distribution units of microorganisms in the surrounding air of the specimens, open Petri dishes containing BBM (according to Brown et al. 1964) were exposed. As a good reference in relation to the specimens, some Petri dishes were attached vertically at the average height of the specimens (approx. 1.2 m above ground) by double-sided adhesion film (Figs. 2.11, 2.12). Additionally, some Petri dishes were attached horizontally at the same height. After different sedimentation times, the Petri dishes were closed and incubated (12/12 h light–dark cycle, at 16 °C).

Germ Load of Precipitation

Occasionally, algae are transported as a result of aerosol causing processes by appropriate air currents over great distances (Hamilton and Lenton 1998). Regarding this, a trap for driving rain was employed to get an impression about the concentration of algal distribution units which are transported via rainfall incidences. The driving rain trap was disinfected and cleaned after every emptying using 70% ethanol or isopropanol. The gained rainwater was either spread on BBM (100 µl in triplicate) or 100 ml were incubated with an aliquot of liquid BBM at 12/12 h light–dark cycle at 16 °C (enrichment culture); whereas direct plated cultures were also assessed quantitatively, the enrichment

Fig. 2.11 Petri dishes for sedimentation measurements were attached vertically and horizontally at a height of approx. 1.2 m above ground, the same average height of the specimens (Hofbauer 2007)

Fig. 2.12 Detail of a
vertically attached Petri dish
mounted by use of short strips
of double-sided adhesion film
(Hofbauer 2007)

cultures were mainly used to assess algal diversity. As different species behaved differently in the enrichment cultures, they could not be used to assess the original samples. Further, the inoculated Petri dishes were also investigated for fungi (quantitative, qualitative). Bacterial growth was not assessed because it was assumed that bacteria already had multiplied substantially during the stay in the driving rain trap.

Ecophysiological Data

In addition to the taxonomic criteria given in the description of the different forms in Chap. 3, ecophysiological and geographic data are provided for the discussed forms using data from literature and own observations. Especially, connections to the situation at building parts are emphasized.

The following important parameters were noted, if available:

- temperature limits for growth (upper and lower limit, optimum),
- water balance (moisture demands),
- pH value (range),
- osmotic concentration/salinity,
- nutrients (substrate specificity, metabolism, etc.),
- distribution geography (hitherto known occurrences).

Advanced Methodologies for Taxonomic Analysis

Molecular Genetic Methods—Genetic Barcoding

Different state-of-the-art molecular approaches were used to reveal the diversity of biological crusts in man-made surfaces with varying success (Lamenti et al. 2002). Denaturing gradient gel electrophoresis (DGGE) (Rölleke et al. 1996; Zanardini et al. 2002; Gonzalez and Saiz-Jimenez 2004) is a method which does not need sequencing of DNA, but breaks DNA up into short and medium length sections which then are separated and used as fingerprinting technique. In a different approach, in situ hybridization targeted probes can be used for the identification of certain organisms (Sterflinger and Hain 1999).

Very promising is the so-called genetic barcoding. Genetic barcoding is a method of species identification using short sections of DNA of a specific gene or genes. By comparison with a reference library of such DNA sequences, an individual sequence can be used to uniquely identify an organism to species level. These "barcodes" are sometimes used to identify unknown species, parts of an organism or simply to catalog as many taxa as possible and compare them with traditional taxonomy to determine species boundaries (https://ibol.org/).

Different gene regions are used to identify the different organismal groups using barcoding. Many existing protocols have been implemented by the Consortium for the Barcode of Life (CBOL) which represents the leading international authority concerning this revolutionary technique. With this technique, the ITS region (eukaryotes) or the 16S RNA code (prokaryotes) is used as basic barcode for differentiation. In many organism groups, these basic barcodes already allow a sufficient differentiation on species level. If this cannot be achieved by the use of the basic barcode, additional taxon-specific sequences are utilized. We previously applied the barcoding approach successfully in the differentiation of moss samples of the critical genus, *Schistidium*, isolated from masonry (Hofbauer et al. 2014, 2016). In this work, even probes of moss plants, which are very tiny and therefore lack some characteristics which are necessary for a morphological species identification, were assigned by barcoding to certain species groups.

Recently, another epoch-making breakthrough was made. By applying next-generation sequencing techniques, it is now possible to investigate environmental samples which contain different organisms and to isolate barcode sequences of several organisms simultaneously, so-called metabarcoding. Several approaches of metabarcoding have been developed and used, e.g., for land plants and fungi (Hollingsworth et al. 2009; Stielow et al 2015), and it is expected that soon also aerophytic algae may be covered.

In the investigation of systematic–taxonomic contexts, comparative analyses of genetic sequences are particularly useful and necessary, although a genetic analysis can hardly provide a differentiation between dead and living material. It is often not possible to discriminate between traces of organisms which were transported to the surface by chance and real growth; we therefore recommend an accompaniment by suitable culture methods. It is increasingly noticed that for in detail analysis several genetic–biochemical characters have to be used together with morphological characters (Soltis et al. 1999; Gontcharov et al. 2004). Various genetic methods (PCR analysis, Gene-sondes, etc.) have previously been used in the characterization of surface crusts (e.g., Laiz et al. 2003). As the species which grow on building surfaces are only known in part, genetic analyses may only provide partial results. Science is still at the beginning of the systematic assessment of organisms which occur on the outer surfaces of building parts (Rindi 2007). Concerning aerophytic algae specialists' opinion is that only approximately 20% of all occurring species are known at present (Friedl 2002). Taxonomically characterized pure cultures as established in our study may provide basic material for further genetic research on building part-relevant organisms.

Advanced Microscopic and Spectrometric Techniques

In addition to the discussed methods, advanced microscopic and spectrometric techniques can be employed as are, e.g., confocal microscopy/spectrometry, fluorescence techniques or image analysis.

Particularly, in crust buildups, where also formation of precipitates is involved, confocal microscopy is very useful for monitoring the colonization process and for revealing the three-dimensional structure of the crust community (Roldán et al. 2002, 2004a, b). Using this method, structures are shown in very great detail, although it is not yet possible to assess the taxonomic affiliation of the observed organisms in depth.

In order to assess the abundance and physiological state of the photosynthetic organisms thriving on building surfaces, pulse–amplitude–modulated fluorescence (PAM) was used. If only photosynthetic active organisms and their physiology were of question, this would be the method of choice.

Employing modern digital microscopic and photographic techniques, image analysis of building surfaces allows to assess growth pattern and intensity (Thornbush and Viles 2003, 2007). A chlorophyll extract measurement calibrated approach has already been developed to quantify algal growth on concrete (Alum et al. 2009). Other applications detect starting corrosion processes, e.g., in metals, but do not allow quantification (Angelini et al. 2010). Because of the very different nature of potential surface organisms, recent methods can only record certain organisms, but not yet resolve the whole picture. With the development of further sensor and image analysis capacity, the actual problems will be solved in the future.

Testing Methods for Microorganism Susceptibility of Building Products

For producers and customers, the resistance of a building coating against biological colonization is an important question in respect of the buying motive. Another concept/term used to express resistance/susceptibility against "infection" is the so-called bioreceptivity (Guillitte 1995; D´Orazio et al. 2014; Coutinho 2015). We take the analogous terms as synonyms. Many building products are tested for their susceptibility/resistance against unwanted growth of microorganisms according to various regulations. Examples for testing regulations concerning the durability against colonization by aerophytic microorganisms are ISO 2810, ASTM D 3456, ASTM D 3719, ASTM D 5589, etc.

Most of the mentioned regulations are more or less capable of giving a kind of comparison between different products. Apart from outdoor testing, they are often not sufficient for all products under all potential situations because the natural environment is much too complex. Some research groups developed advanced testing procedures that include more complexity (e.g., Barberousse et al. 2006). By an interdisciplinary research group, an accelerated testing setup was created which is based on real climate data (Hofbauer et al. 2006; Künzel et al. 2011).

Aerophytic Organisms Colonizing Façades: Diversity, Taxonomy and Ecophysiology

Table A.1 to Table A.11 (see Appendix) show the results of the taxonomic analyses of the surfaces of "new" and "old" specimens, "additional" variants and surfaces as well as ambient air and driving rain probes. Almost species which was detected and identified in our study is presented in this chapter with its characteristics and often a photography. Taxonomic information about class, order, family and genus as well as ecophysiological data were compiled from the literature.

Life Form Prokaryota

Prokaryota are defined by special cellular characters. Karyoplasm typically is not segregated from other cell plasma by a special membrane. Cell sap vacuoles are lacking. Respiratory and photosynthetic processes, if present, are bound to the plasma membrane system and not accommodated in independent organelles. Ribosomes of the 70S type are distributed within the plasma. There is no endoplasmic reticulum established. Plasma currents and linked processes (e.g., phagocytosis, formation of pseudopodia, etc.) and flagella of the $9+2$ type are unknown (Krieg 2001). To date, some 5,000 species of bacteria (incl. Cyanoprokaryota) have been described, although different authors state many more species (Brenner et al. 2001).

Regnum Eubacteria–Subregnum Glycobacteria–Division Cyanoprokaryota

Cyanoprokaryota (Komárek and Anagnostidis 1998, 2005) are also called Cyanobacteria (Castenholz 2001a, b; Garrity and Holt 2001) or Chloroxybacteria (Graham et al. 2009).

© Springer-Verlag GmbH Deutschland, ein Teil von Springer Nature 2021
W. K. Hofbauer and G. Gärtner, *Microbial life on Façades*,
https://doi.org/10.1007/978-3-662-54833-2_3

Typical prokaryotic organized forms (" blue green algae") are able to perform oxygenic photosynthesis like eukaryotic algae and higher plants. The widely used expression "blue-green algae" will subsequently be applied to this group of organisms. The oldest known fossil forms are presumably about 3.5 billion years old, and therefore, blue-green algae are regarded as the oldest known plant growth forms (Schopf 1993; Graham et al. 2009). According to the current understanding, the plastids of all eukaryotic plants evolved from cyanoprokaryoid ancestors via endosymbiosis (Hoek, van den et al. 1995; Lee 2008; Graham et al. 2009). The cell wall of the Cyanoprokaryota contains polysaccharide murein, like that of gram-negative bacteria. The photosynthetic apparatus contains apart from chlorophyll a the accessory pigments phycocyanine und phycoerythrine (phycobiliproteids). In rare cases, the phycobiliproteids are substituted by chlorophyll b, as is the case in the plastids of the eukaryotic Chlorophyceae, or additionally chlorophyll d is developed as in the plastids of the eukaryotic Rhodophyceae (Castenholz 2001). Some species are able to conduct gaseous nitrogen (N_2) fixation and therefore enhance nutrients in their direct environment (Esser 2000). Cyanoprokaryota are mainly organized coccal, capsal and trichal (Esser 2000). In course of their very long ontogeny, Cyanoprokaryota have colonized various habitats. Apart from marine and limnic forms, numerous species are present in soil and on/in rock. Some species also thrive in extreme niches like thermal water or rocks in deserts; some are adapted to high solar radiation, while others prefer low radiation. Cyanoprokaryota are important members of lichen symbiosis (Tschermak-Woess 1988a). Classical studies based on morphological criteria were performed by Geitler (1932) and by Bourrelly (1970). With molecular genetic research, the discussion about species delimitation is still controversial (Castenholz 2001a, b). We followed the combined morphological and molecular genetic investigations conducted by Anagnostidis and Komárek (1985, 1988, 1990), Komárek and Anagnostidis (1986, 1989, 1995, 1998, 2005) and Komárek (1994, 2013). The Cyanoprokaryota flora in initial colonization of modern building systems (especially of WDVS) is relatively poor. A richer diversity in species is found in older colonization stages or in cases regularly influenced by liquid water (dropping water, basal areas of buildings, etc.). Different studies documented colonization of buildings by blue greens, e.g., Saiz-Jimenez et al. (1990), Ortega-Calvo et al. (1991, 1993, 1995), Palmer and Hirsch (1991), Ariño and Saiz-Jimenez (1996), Gaylarde and Gaylarde (2000), Lamenti et al. (2000), Ortega-Morales et al. (2000), Crispim and Gaylarde (2005) and Sánchez-Antón and Asencio Martínez (2007).

Order Chroococcales sensu Komárek et Anagnostidis

Family **Synechococcaceae** Komárek et Anagnostidis, subfamily **Aphanothecoideae** Komárek et Anagnostidis

Genus *Cyanothece* Komárek

This genus is characterized by relatively big ellipsoid to slightly cylindrical cells occurring single or in dual groups. The length/width ratio of the cells is about 2:1. Reproduction runs by binary fission and formation of two identical daughter cells. Sometimes a thin homogenous slime sheath covers the cells. The typical granulation of the cell content is based on irregular arrangement of thylakoids (Komárek 1976; Komárek and Anagnostidis 1998; Komárek and Cepák 1998). Komárek and Anagnostidis (2005) proposed to include the genus provisorily in the family Cyanobacteriaceae. Discussion is still ongoing (see, e.g., Rudi et al. 1997; Porta et al. 2000; Turner et al. 2001), but we followed Komárek and Anagnostidis (1998). The genus *Cyanothece* is also designated as form-genus V (Rippka et al. 2001a) of subsection I (Herdmann et al. 2001a) of Cyanobacteria (Castenholz 2001c). The presumably cosmopolitan genus occurs mainly in freshwater and marine biotopes (Komárek and Anagnostidis 1998).

Cyanothece aeruginosa (Nägeli) Komárek (Figs. 3.1 and 3.4) (Syn.: *Synechococcus aeruginosus* Nägeli; *Coccochloris aeruginosa* [Nägeli] Drouet et Daily).

Type of organization: single cell or two cell aggregates. Single cells were oval egg-shaped (8.6) 10.3–12 µm broad, (14.3) 14.9–20 µm long. Often cells in division were

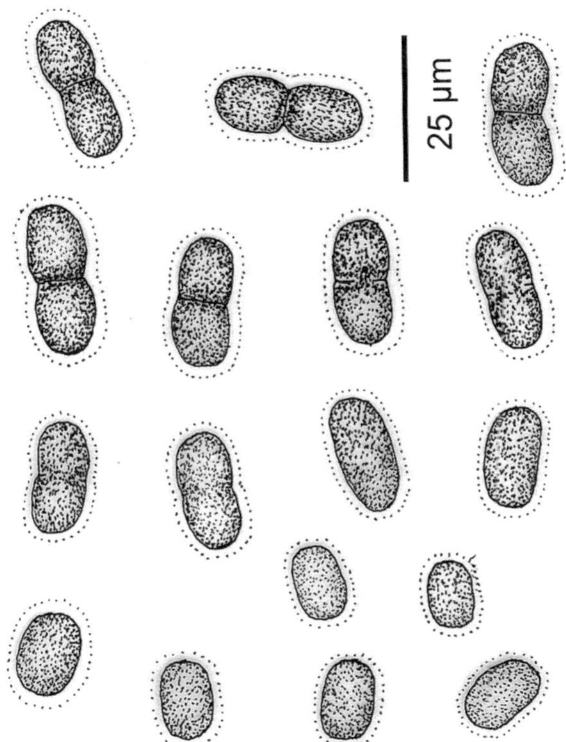

Fig. 3.1 *Cyanothece aeruginosa*, different stages of division and single cells of an original sample. A thin colorless mucilaginous layer around the cells is characteristic and forms a bright frame around the cells. (Hofbauer 2007)

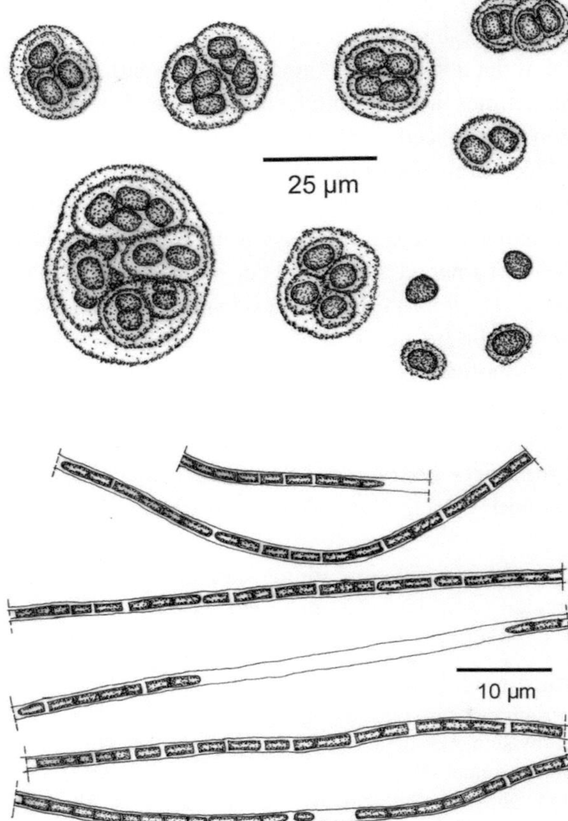

Fig. 3.2 *Gloeocapsa sanguinea*, differently developed colonies. Bottom right: four permanent cells with and without shrunken and densified jelly. (Hofbauer 2007)

Fig. 3.3 *Planktolyngbya limnetica,* different trichomes. (Hofbauer 2007)

found; these stages were up to 21.7 µm long. Division of cells was only by binary fission; after division, the new daughter cells were connected for a time. These cell duplets were 27.4–29.7 µm long. The cells of the investigated form were often covered with thin, homogenous and colorless slime which was visible as a bright margin by light microscopy. Cell content was slightly reticulate or keritomized (=irregular arrangement of thylakoids), and a fine granulation was visible within the turquoise-colored matrix. Additionally, regular dark granules were seen. Dimensions and cell morphology coincided with Komárek and Anagnostidis (1998). The results given here were based on free-living material.

Cyanothece aeruginosa has a known worldwide distribution in temperate and circumpolar climate zones. According to Komárek and Anagnostidis (1998), *C. aeruginosa* is a species living mainly limnic and on wet rocks, but is also found in mires (e.g., Nováková 2002; Rauch and Schagerl 2003) as a member of the soil algal flora, in desert soils and subaeric habitats (e.g., Fehér 1936; Vinatzer 1975; Broady 1989a; Mataloni et al. 2000; Hawkes and Flechtner 2002; Uher and Kováčik 2002). The species was also described

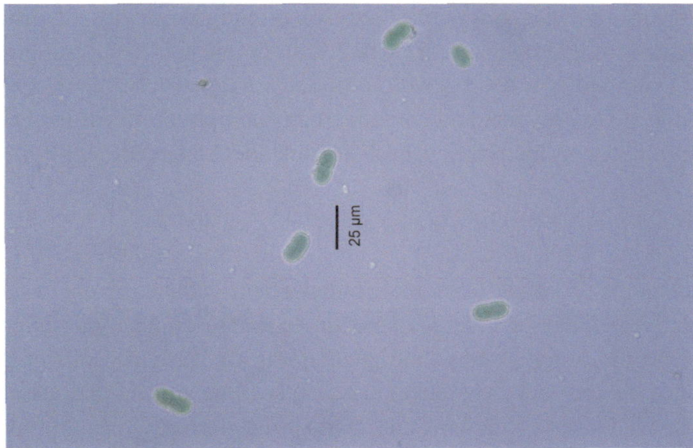

Fig. 3.4 *Cyanothece aeruginosa*, microscopic images of fresh material showing single cells and different stages of division (Hofbauer et al. 2003). The occurrence of dual cell aggregates is characteristic

living in caves as a member of the "Lampenflora" (Smith and Olson 2007). As colonizer of building materials, it was first documented by Hofbauer et al. (2003).

Regarding the ecophysiological parameters, the pH values of habitats are given at 5.5 (Fehér 1936) and 5.75 (Vinatzer 1975), between 4.1 and 5.2 (Nováková 2002) and<7 (Komárek and Anagnostidis 1998). A soil water content of 25% (weight percentage) is necessary for the species according to Fehér (1936).

Some marine isolates were shown to assimilate N_2 from the air (e.g., Spiller and Shanmugam 1987; Soper and Reddy 1991; Reddy et al. 1993; Bergman et al. 1997; Cólon-López et al. 1997; Meunier et al. 1995, 1997, 1998; Schneegurt et al. 1994, 2000). It still has to be verified, if the discussed form on building surfaces is capable of Nitrogen fixation.

Family **Microcystaceae** Elenkin

Genus ***Gloeocapsa*** Kützing

Cells often form macroscopic colonies surrounded by stratified mucilage. Single cells are mainly globular or slightly irregular. Reproduction is by cell division in two daughter cells in two successive planes, the original mucilaginous sheath is enlarged, and the daughter cells form their individual sheath. Mucilage is often colored. In different species, nanocytes are formed; sometimes, *Aphanocapsa*-stages (diffluent homogenous mucilage) and/or resting stages (resting cells) are observed (Komárek and Anagnostidis 1998). The genus *Gloeocapsa* is also designated as form-genus VIII (Herdman et al. 2001b) of subsection I (Herdman et al. 2001a) of Cyanobacteria (Castenholz 2001c).

Different forms of *Gloeocapsa* may survive prolonged periods of drought and are especially adapted to light exposure (Lewin 2006). In one strain, special pigments which protect against ultraviolet radiation were detected (Garcia-Pichel et al. 1993). The cosmopolitan genus is very diverse and contains many aerophytic forms as well as edaphic, limnic and thermal water inhabiting taxa (Komárek and Anagnostidis 1998).

Gloeocapsa sanguinea (Agardh) Kützing (Figs. 3.2, 3.5 and 3.6)

Type of organization is usually colonial, independent cells were enclosed by a stratified mucilage. Aggregates of colonies sometimes macroscopically visible. Mainly single colonies with up to eight compound generations are formed. Cells of grayish-green color, embedded in red mucilage envelopes. Cells more or less globular, before division slightly oval. Observed dimensions in material derived from colonized masonry were (4) 4.9–6.9 (7.4) μm, sometimes slightly bigger than the type. Resting cells up to 8.6×9.7 μm with stable walls and an inner homogenous brownish mass were formed, surrounded by a compact gelatinous layer. In some colonies, the surface of the outermost gelatinous envelope appeared rough, which resembled *Gloeocapsa novacekii* Komarek et Anagnostidis documented from wet rocks only in the Czech Republic. Perhaps *G. novacekii* should be united with *G. sanguinea,* as originally proposed by Geitler (1932). Our investigated strains fell in the variability of *G. sanguinea.*

Gloeocapsa sanguinea is known worldwide from mountainous regions. According to Komárek and Anagnostidis (1998), it lives on surfaces of non-calcareous rocks. Jaag (1945) found the species on all types of rocks, including artificial stone. Barberousse et al. (2006) described *G. sanguinea* on façades in France. Hofbauer observed the

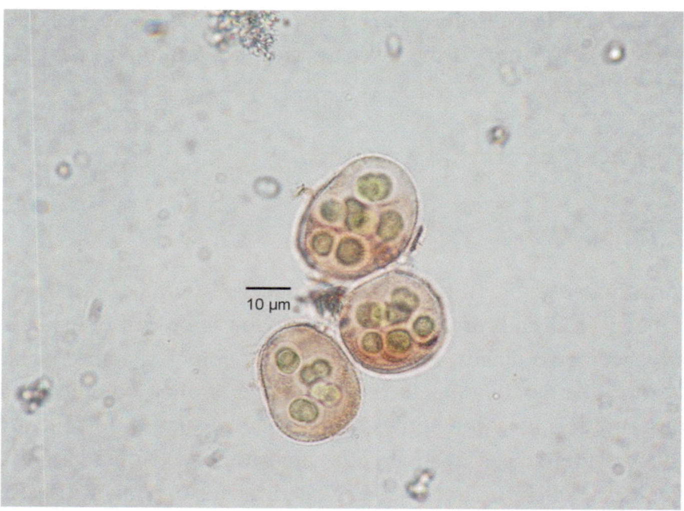

Fig. 3.5 *Gloeocapsa sanguinea,* different colonies

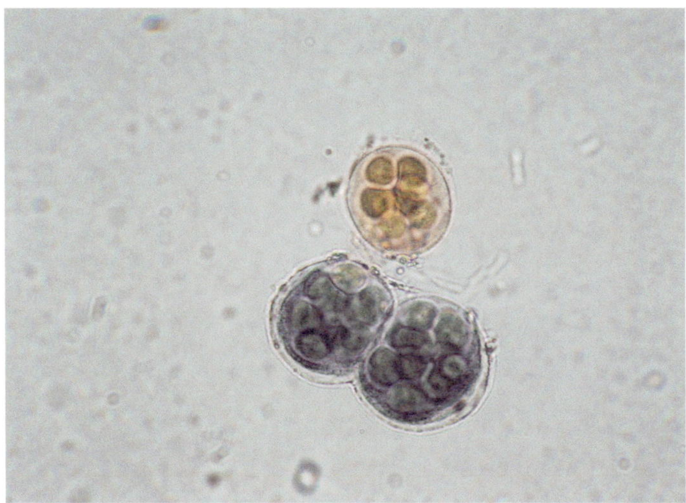

Fig. 3.6 *Gloeocapsa sanguinea.* Top: colony typical ruby-colored. Bottom: colonies with color changed to blue-violet according to higher pH value as described by Jaag (1945). (Hofbauer et al. 2006; Hofbauer 2007)

species in association with other microorganisms mainly on habitats where water is occasionally running down the surface ("Fahnenbildung"), but it did not appear in places with dew or dripping water or in splash water zones. *G. sanguinea* is also known as phycobiont of different lichens, e.g., *Pyrenopsis*- and *Synalissa*-species (Jaag 1945; Büdel and Henssen 1988; Tschermak-Woess 1988). Jaag (1945) discussed the deterioration of stones by *G. sanguinea* and suggested this could only happen in lichenized stages. It is also discussed that the alga alone could alter the surface like other Cyanoprokaryota which actively enter the surface of stones (Komárek and Anagnostidis 1998). So far, neither a culture of lichen symbiosis with *G. sanguinea* nor an effect of biodeterioration has been experimentally proven.

Regarding environmental factors, it is remarkable that the substrate pH lies between 5.24 and 6.09 according to Fehér (1936) and the moisture of substrate between 20.1 and 37.8% (weight %).

Jaag (1945) reported a change in the color of the mucilage envelopes according to pH from reddish to blue violet to blue (Fig. 3.6). The critical value for the change in color is pH 6.5, and forms with red mucilage envelopes are only found at a pH<6.5. Investigations of Hofbauer produced a remarkable appearance of single colonies which were partially or totally blue violet.

Gloeocapsa atrata Kützing (Fig. 3.7)

Gloeocapsa atrata sometimes formed complexes of colonies with colorless mucilage envelopes, similar to a form of *Gloeocapsopsis*. The typical formation of colonies resembled *Gloeocapsa* (lamellate mucilage envelopes, mode of cell division). Single

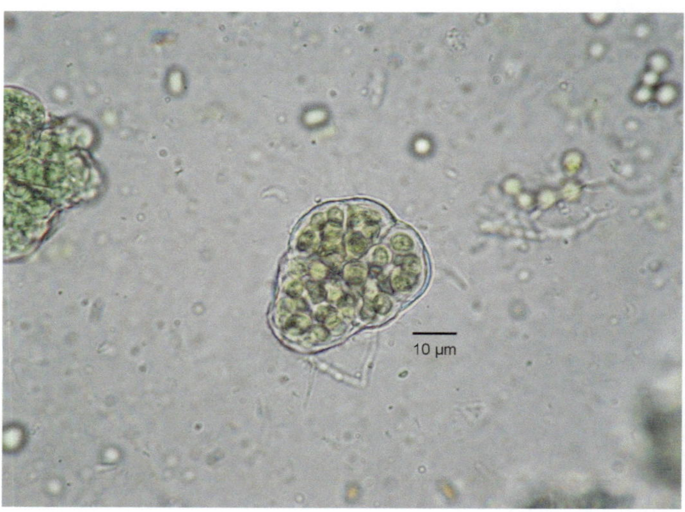

Fig. 3.7 *Gloeocapsa atrata*, a small compact aggregate of colonies with stratified but colorless mucilage envelopes

cells measured (3.7) 4–5.7 μm and were subspherical; resting cells were brownish, up to 8 μm, and were seldom found in culture. In culture, colonies were blue green to grayish green; in older stages, the color changed to yellow green.

Gloeocapsa atrata is a subaerophytic species, widely distributed in the temperate zone (Komárek and Anagnostidis 1998). This species of *Gloeocapsa* was isolated from the surface of concrete and was also documented on artificial substrates (walls) by Komárek and Anagnostidis (1998).

One more species of the genus *Gloeocapsa*, ***Gloeocapsa violacea*** (Corda) Rabenhorst, was isolated by Hofbauer from a stone wall (Sill gorge) near Innsbruck (Hofbauer et al. 2003). Barberousse et al. (2006) documented *G. violacea* on façades in France.

During our study, additional taxa of *Gloeocapsa* were observed but could not be assigned to a species.

Family **Chroococcaceae** Nägeli

Genus ***Chroococcus*** Nägeli

The genus *Chroococcus* develops colonies of cells in mucilage envelopes, but in contrast to *Gloeocapsa* with lesser cells per colony. Cell division runs in planes in right angles to each other, and daughter cells do not always grow to the size of mother cells until the next division takes place. Single cells are globular but after division semiglobular cells appear. The genus *Chroococcus* is also designated as form-genus II Nägeli (Rippka et al. 2001b) of subsection I (Herdman et al. 2001a) of Cyanobacteria (Castenholz 2001c); however, Komárek and Anagnostidis (1998) used different features to describe the genus. Most taxa of this cosmopolitan genus are limnic and some also aerophytic.

Chroococcus pallidus Nägeli (Fig. 3.8)

According to the dimensions (cell diameter 8 µm) and typical morphology several observations of older succession stages at masonry were determined to be close to *Chroococcus pallidus*. Prolonged cultivation failed.

The species has previously been documented on historical buildings and monuments (Kováčik 2000).

Genus *Gloeocapsopsis* Geitler

The genus is characterized by colonies of compact, irregularly arranged cells or cell groups with gelatinous envelopes. The mucilage is typically colored. In contrast to *Gloeocapsa*, the intensity of the coloration increases from the outer margin to the inner areas of the aggregates. Cells are subspherical, in complex colonies also irregular, and cell division runs in different planes. During reproduction, single cells or groups of cells are formed by rupture of the gelatinous envelope of the mother cells. Sometimes development of nanocytes and resting cells with firm and intensely colored mucilage are

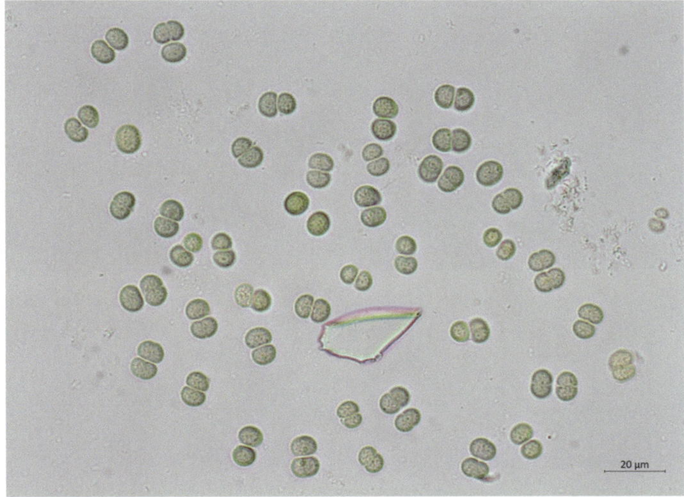

Fig. 3.8 Cells and coenobia of *Chroococcus pallidus* from culture living material

observed. The taxon was not included in the work by Boone et al. (2001). In the presumably cosmopolitan genus, aerophytic forms are prevailing (Komárek and Anagnostidis 1998).

Gloeocapsopsis pleurocapsoides (Nováček) Komárek et Anagnostidis (Fig. 3.9)

The investigated isolates were characterized by compact aggregates of cells, each cell separated by a thin mucilage envelope. The mainly ovoid to spherical or cubic aggregates were separated from other colonies by thicker mucilage, but also boxes appeared in the box systems. The dimensions of the oval to irregular cells were variable between ca. 4–12 µm in diameter. Cells in the marginal parts of colonies were olive green, in the central regions more blue green (Fig. 3.9). A diagnostic feature was the inner structure of the cells: a colorless central part differs from the chromatoplasma of the periphery. Older cells were sometimes transferred into resting cells with a firm cell wall or mucilage, respectively. Although identification according to morphological criteria was correct, we observed a relatively colorless mucilage envelope of the colonies (difference to the type!). Sometimes, in older and central regions, only the mucilage appeared yellow brownish. Due to the colorless mucilage, there were similarities to *Cyanosarcina parthenonensis* Anagnostidis in Anagnostidis et Pantazidou (*Myxosarcina concinna* Printz sensu Anagnostidis et al. [1983]), but *C. parthenonensis* forms sarcinoid (cubic) aggregates of cells with smaller dimensions. The species was first documented on marble at the Parthenon in Athens (Anagnostidis et al. 1983), later also found on roman walls and

Fig. 3.9 *Gloeocapsopsis pleurocapsoides,* colonies of the marginal area of a cluster with olive-green cells

façades in France (Ariño and Saiz-Jimenez 1996; Barberousse et al. 2006) and on historical granite buildings in Spain (Rifon-Lastra and Noguerol-Seoane 2002). Similarities also exist to *Chroococcidiopsis kashayi* Friedmann (differentiation in central and chromatoplasma) which shows more regular arrangement of single cells and less mucilage. The taxon needs revision according to Komárek et Anagnostidis (1998). It was detected near caves in Israel (Friedmann 1961a, b) and also on monuments and buildings in Spain (Asencio and Aboal 1996; Uher et al. 2005a) and caves in Italy (Abdelahad 1985).

Similar forms like ours were documented by other authors but with different taxonomic designations: e.g., *Chroococcidiopsis* sp. on buildings in Brazil (Crispim et al. 2006), on marble statues in Florence (Lamenti et al. 2000) and historical fountains in Spain (Peraza Zurita et al. 2005). *Chroococcidiopsis umbratilis* Dor, Carl et Barlinger, was demonstrated on monuments in Bratislava (Uher et al. 2005b), *Myxosarcina* spp. on historical Maya buildings in Mexico (Ortega-Morales et al. 2000) and *Cyanosarcina chroococcoides* (Geitler) Kováčik (sub *Myxosarcina chroococcoides* Geitler, originally a metaphytic form of bogs and mires) on monuments in Spain (Noguerol-Seoane and Rifón-Lastra 1997a, b). A revision of these strains to clarify their taxonomic position should be performed by cultures and molecular techniques.

The discussed form was only isolated once from a specimen exposed in Ernsthofen/Oberrramstadt in Germany during the first year of exposure (Hofbauer 2007).

A peculiar further form of *Gloeocapsopsis* sp. with orange mucilage envelopes was documented by Hofbauer et al. (2003) on roof tiles of the outdoor experimental laboratory of the Fraunhofer Institute for Building Physics in Holzkirchen.

Order **Oscillatoriales** Elenkin sensu Komárek et Anagnostidis

Family **Pseudanabaenaceae** Anagnostidis et Komárek, subfamily **Leptolyngbyoideae** Anagnostidis et Komárek

Genus *Leptolyngbya* Anagnostidis and Komárek

The genus *Leptolyngbya* Anagnostidis and Komárek is characterized by trichomes with sheath, cells with parietal concentric arranged thylakoids and non-planktonic life (Anagnostidis and Komárek 1988). The trichomes are normally in groups, mats and layers. The genus *Leptolyngbya* is designated as form-genus V Anagnostidis and Komárek (Castenholz et al. 2001b) of subsection III (Castenholz et al. 2001a) of Cyanobacteria (Castenholz 2001c). This genus is known to occur in all possible biotopes (Komárek and Anagnostidis 2005).

Leptolyngbya foveolarum (Rabenhorst) Anagnostidis et Komárek (Fig. 3.10)

Cells were mainly isodiametric (1.5–2.5 μm diameter), constricted at cross-walls. In young trichomes, the mucilaginous sheath was tender and hyaline, in older trichomes distinct and thicker. Colonies and mats showed a blue-green or black-green color.

Leptolyngbya foveolarum was documented on monuments in Spain (Noguerol-Seoane and Rifón-Lastra 1997a,b) and on marble of the Acropolis in Athens (Anagnostidis 1983). In addition to its development, in some aerophytic associations on artificial surfaces, we isolated this taxon even from air (by use of a spore sedimentation technique).

The pH of the substrate ranges from 6.35–8.4 according to Fehér (1936), from 7.8–8.2 according to John (1942).

Leptolyngbya notata (Schmidle) Anagnostidis et Komárek (Fig. 3.11)

Trichomes were thin, irregularly curved and densely aggregated but not attached to each other. Sheaths were thin but clearly visible. Cells of the investigated strain were between 0.8–1.3 μm broad and 2–3 times longer. Typical features were one or two small granules in the cell lumen (the thylakoids were slightly colored at the margin of the cell) and a tendency to false branching of trichomes, especially in old cultures.

Genus ***Planktolyngbya*** Anagnostidis and Komárek

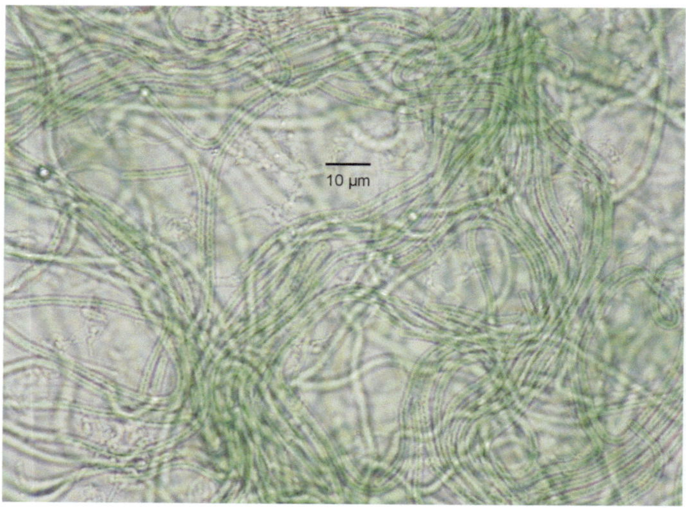

Fig. 3.10 Mass of trichomes of *Leptolyngbya foveolarum* in a 5 weeks old culture. The sheaths of the trichomes are fine and less visible

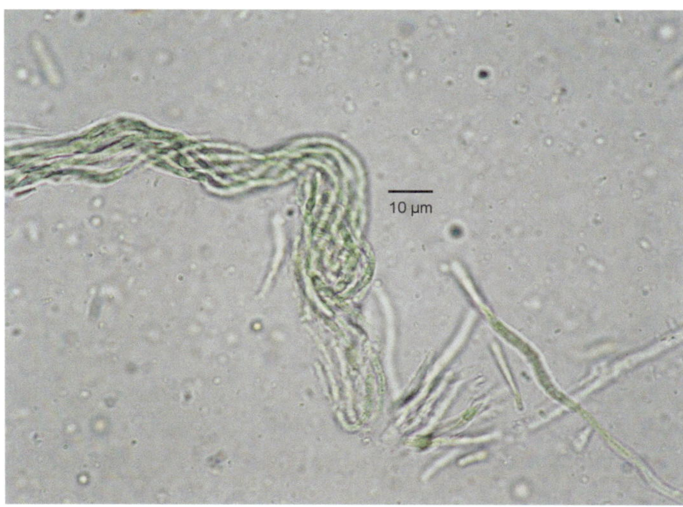

Fig. 3.11 *Leptolyngbya notata*, raw culture. In well-growing cultures, trichomes often intermingle and even pleat

The genus *Planktolyngbya (Lyngbya* subg. *Limneticae* Forti) was transferred from the genus *Lyngbya* to the family Pseudanabaenaceae by morphological and ecological features (Anagnostidis and Komárek 1988). *Lyngbya* shows irregular arranged thylakoids, while in Pseudanabaenaceae, thylakoids are arranged in parallel layers in the external areas of the cells. According to Komárek and Anagnostidis (2005), the probably cosmopolitan genus is found almost only planktic in freshwater and salty waters; the occurrence in other biotopes is quite exceptional.

Planktolyngbya limnetica (Lemmermann) Komárková-Legnerová et Cronberg (Fig. 3.3)

The investigated isolate showed delicate trichomes of light blue-green color (see also Hofbauer 2003) without heterocysts and resting cells. Trichomes were thin (about 2 μm diameter), surrounded by a delicate but visible mucilage. Cells of trichomes were 1.5 μm broad, 3.5 μm long, intercalary cells barrel shaped, apical cells rounded or slightly cylindrical. Cell content was divided in a colored marginal area and a pale central area (feature of Pseudanabaenaceae!). Trichomes on agar developed straight or slightly curved. A morphologically related form is the terrestrial *Leptolyngbya notata* (Schmidle) Anagnostidis et Komárek which forms additionally false branching and has often granules in the cell matrix (see above).

According to Komárek and Anagnostidis (2005), the genera of subfamily Leptolyngbyoideae are mainly determined by their ecology and the genus *Planktolyngbya* should be excluded from aeroterrestric habitats. On the other hand, the taxon was documented for aerophytic habitats like monuments in Spain (Noguerol-Seoane and Rifón-Lastra 1997a,b).

Family **Phormidiaceae** Anagnostidis et Komárek, subfamily **Phormidioideae** Anagnostidis et Komárek

Genus *Phormidium* Kützing

Phormidium Kützing includes many species in different habitats. Trichomes have cylindrical cells without aerotopes. Thylakoids in the cells are radial and longitudinal, gelatinous sheaths are obligatory or facultative, and there is only one trichome per sheath. Trichomes are motile; reproduction occurs with motile hormogonia. Boone et al. (2001) did not follow Castenholz et al. (2001a) with their division of subsection III of Cyanobacteria (Castenholz 2001c) and tansfer *Phormidium*-taxa to other form-genera. The cosmopolitan genus is very diverse and occurs in many different biotopes, marine, freshwater and aerophytic (Komárek and Anagnostidis 2005).

Phormidium autumnale (Agardh) Trevisan ex Gomont (Fig. 3.12)

Filaments were straight or slightly curved with a thin distinct gelatinous sheath. Cells at cross-walls were usually not constricted and were as long as broad (or also shorter,

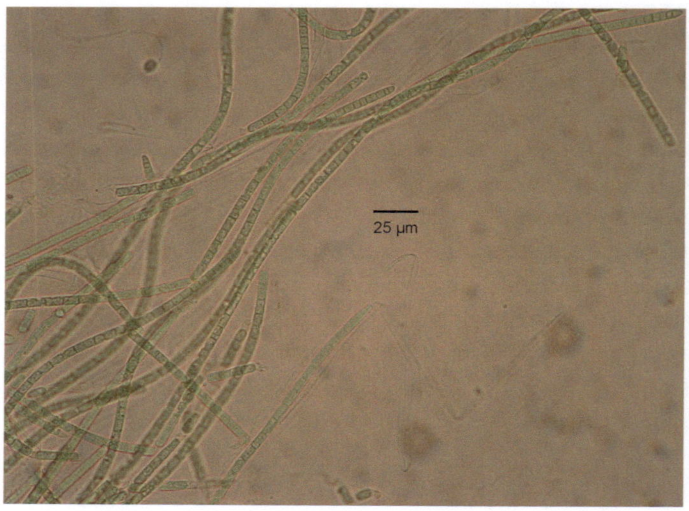

Fig. 3.12 Filaments of a young culture of *Phormidium autumnale*

respectively longer than broad). Trichomes and layers showed a blue-green to blackish-green color. In experiments, a relation between intensity of mucilage formation and the pressure of predating organisms was demonstrated (Pajdak-Stós et al. 2001). Apical cells of trichomes were slightly rounded, capitate, with rounded or truncate calyptra (thickened cap on the outer cell wall of an apical cell) (Hindák 2008). The variation within the taxon is enormous (e.g., Häder and Nultsch 1971; Teneva et al. 2005). Komárek and Anagnostidis (2005) proposed a retypification of the genus. The observed forms on building materials showed trichomes 4–5 μm broad and fitted to the morphological concept of *Phormidium autumnale* (Komarek and Anagnostidis 2005).

Komarek and Anagnostidis (2005) accepted this taxon as mainly limnic, while it was also documented in terrestrial and aerophytic habitats (e.g., Fehér 1936; Broady 1979, 1989a; Hoffmann 1986; Davey and Clarke 1991; Lukešová and Hoffmann 1996; Mataloni et al. 2000; Ohtani et al. 2000; Ciubuc 2005) as well as on surfaces of monuments and historical buildings (Ortega-Calvo et al. 1993; Ková̌cik 2000; Marcos Laso 2001; Darienko and Hoffmann 2003; Uher et al. 2005a), on Roman ruins (Ariño and Saiz-Jimenez 1996) and urban walls in Europe (Rindi and Guiry 2004). The aerophytic forms of *Phormidium autumnale* should be studied for their taxonomical and geographical diversity.

Phormidium autumnale comprises a group of taxa with preference for nitrogen-influenced habitats (Komárek and Anagnostidis 2005). The pH of the substrate varies between 4.06 and 5.43 according to Fehér (1936); the substrate pH lies between 6.2 and 8.2 after John (1942), measures 5.2 in January and 5.6 in May after Lukešová and Hoffmann (1996) and between 7.2 and 9.2 after Ohtani et al. (2000). *P. autumnale* is highly tolerant to desiccation and accumulates under water stress protective substances like Trehalose and Sucrose (Hershkovitz et al. 1991).

P. autumnale is motile on its substrate and shows a positive phototaxis (Nultsch 1961, 1962; Davey and Clarke 1991).

Phormidium schroeteri (Hansgirg) Anagnostidis (Syn.: *Oscillatoria brevis* Schröter)

Trichomes were about 8 μm broad without or with very delicate sheath. Cells were shorter than broad (except in apical area), not constricted at cross-walls with granular inclusions. Formerly known as *Oscillatoria brevis* Schröter (Komárek and Anagnostidis 2005), Hofbauer et al. (2003, Fig. 6) referred to this form as "*Oscillatoria* sp."

The taxon is already known from the surface of historical buildings and monuments (Ková̌cik 2000).

Genus *Symploca* Kützing

Trichomes are usually equipped with a distinct sheath and form upright shocks or bundles. Older trichomes may exhibit false branching. The genus *Symploca* is also described as form-genus XV *Symploca* Kützing (Castenholz et al. 2001c) of the subsection III (Castenholz et al. 2001a) to the cyanobacteria (Castenholz 2001c). The cosmopolitan genus is very close to *Phormidium*, its species are known to occur aerophytic, soil inhabiting and aquatic (Komárek and Anagnostidis 2005).

Symploca elegans Kützing (Fig. 3.13)

False branching occurred mainly in older cultures, and here, often also hernia-like excrescences were observed at the trichomes. Trichome cells were isodiametric or up to two (three) times longer than wide with a diameter of 1.5–2 μm.

A considerable geographical amplitude has been shown from temperate Europe (e.g., Hansgirg 1892) to tropical New Guinea (Johnson 1962). Mostly known from soil probes (e.g., Hansgirg 1882; Johnson 1962), the species was also collected from an aerophytic crust of a moderately old masonry and from air in Germany (Hofbauer 2007).

Family **Oscillatoriaceae** (S. F. Gray) Harvey, subfamily **Oscillatorioideae** Komárek et Anagnostidis

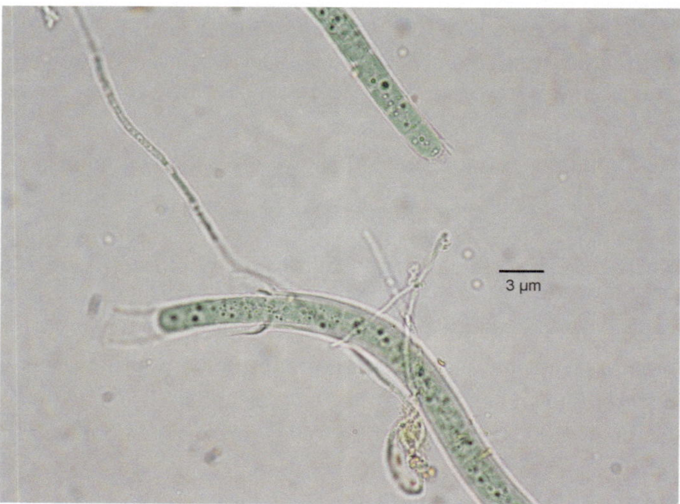

Fig. 3.13 Detached filament of a fascicle of *Symploca elegans,* raw culture of a sedimentation dish collected from outdoor air at the outdoor station of the IBP in Holzkirchen. The distinctly thinner threads comprise fungal hyphae

Genus *Oscillatoria* Vaucher

According to Komárek and Anagnostidis (2005) and Hindák (2008), the genus includes taxa with relatively broad (wider than 6–8 μm) motile trichomes and cells which are distinctly shorter than wide. During reproduction by hormogonia, the formation of necridia (dying cells) is characteristic. The genus *Oscillatoria* is also designated as form-genus IX Vaucher 1883 sensu Rippka, Deruelles, Waterbury, Herdman and Stanier 1979 (Castenholz et al. 2001d) of subsection III (Castenholz et al. 2001a) of Cyanobacteria (Castenholz 2001c). This cosmopolitan and very diverse genus is known from many different biotopes, e.g., limnic, aerophytic and on soil (Komárek and Anagnostidis 2005).

In cultures of older succession stages or of moist surfaces of building material, the following phenomenon was observed: After several weeks of cultivation, suddenly blue greens of *Oscillatoria*-type appeared in the culture vessel and died after a short time. It was not possible to produce clone cultures or to identify on species level. Such ephemeral appearance could also be a germination of resting cells of limnic forms.

Order **Nostocales** sensu Komárek et Anagnostidis

Family **Scytonemataceae** Komárek et Anagnostidis

Genus *Scytonema* Agardh ex Bornet and Flahault

The genus is characterized by trichomes with heterocytes, firm sheaths and formation of false branching, often as double branches. Limnic and terrestrial-aerophytic members are known. On calcareous rocks, the formation of "Tintenstriche" is a worldwide known phenomenon, together with different genera of Cyanoprokaryotes, especially *Scytonema*. The genus *Scytonema* is also designated as form-genus IX Agardh (Castenholz et al. 2001e) of subsection IV.I (Rippka et al. 2001b) of cyanobacteria (Castenholz 2001c).

Many taxa seduce yellow brown to dark pigments into the sheath, mainly Scytonemin (Garcia-Pichel and Castenholz 1991; Proteau et al. 1993). Scytonemin, found as sheath-pigment in different Cyanoprokaryota, shows a high UV extinction and obviously has a protective function against high UV radiation ("sunscreen pigment") (Garcia-Pichel et al. 1992; Dillon and Castenholz 1999). The production of Scytonemin correlates with environmental factors, notably desiccation (Dillon et al. 2002; Fleming and Castenholz 2007). The cosmopolitan genus is widely distributed with an emphasis on tropical regions (Komárek 2013). *Scytonema* is also known as photobiont in lichens, e.g., genera *Dictyonema* C. Ag., *Heppia* Naeg. and *Stereocaulon* Hoffm. (Tschermak- Woess 1983, 1988; Castenholz et al. 2001e).

Scytonema stuposum Bornet ex Bornet and Flahault (Fig. 3.14)

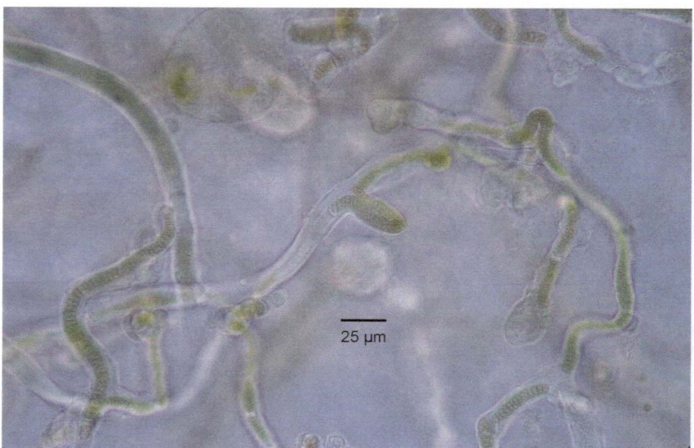

Fig. 3.14 Trichomes of *Scytonema* cf. *stuposum*, constrictions at cross-walls of cells are visible with false branching. (Hofbauer et al. 2003)

Filaments showed false branching and blue-green to yellow-green trichomes, about 15 µm wide. Sheaths were mainly colorless, only in old filaments slightly yellow brown. Single cells were shorter than wide, sometimes isodiametric, constricted at the cross-walls, heterocystes oval to rectangular and spread irregularly in the trichomes. In culture, the filaments grew upright. Dissolving and transgression of apical cells, as described by Chadha and Pandey (1979), were not observed. Identification was realized according to Geitler (1932) and Komárek (2013).

Family **Nostocaceae** Komárek et Anagnostidis

Genus ***Desmonostoc*** Hrouzek and Ventura

During most of its life cycle the rounded or amorphous colonies of *Desmonostoc* consist of long vegetative waved filaments forming intercalary and terminal heterocysts and akinetes. Single filaments are never surrounded by firm mucilage forming compact microcolonies. For further characters, see Komárek (2013). The presumably cosmopolitan genus contains both free-living and symbiotic (cycads, lichens) strains. *Desmonostoc* forms a sister group to *Nostoc*. So far only *D. muscorum* is validly described, but it is expected that in the future further species of *Nostoc* will be transferred to *Desmonostoc* (Komárek 2013).

Desmonostoc muscorum (Agardh) Hrouzek and Ventura (Fig. 3.16) (Syn.: *Nostoc muscorum* Agardh)

Vegetative cells were ca. 3.5–4 µm wide; the nearly spherical heteocytes grew up to 6 µm wide. Elliptic resting cells were often formed, up to 7.5 µm wide. Layers in culture demonstrated a verrucous surface; the mucilaginous sheaths of the trichomes were visible in the marginal area of the colony. Many daughter colonies were formed in culture.

During our study, the species was isolated from one crust of building material. *Desmonostoc muscorum* is widely distributed but has never been recorded on building surfaces so far.

Upper temperature limits were described as <48–60 ± 5 °C (Dodds et al. 1995; Al-Mousawi and Whitton 1983; Greenwood and Steenbergen 1976).

Genus *Nostoc* Vaucher ex Bornet and Flahault

Isodiametric or oval cells are distinctly constricted at cross-walls in long trichomes, with heterocystes and resting cells, embedded in mucilage. A special life cycle between phases of hormogonia and vegetative stages characterizes the genus (Komárek and Anagnostidis 1989; Hrouzek et al. 2005; Komárek 2013). The genus *Nostoc* is also designated as form-genus VIII Vaucher 1803 (Herdman et al. 2001c) of subsection IV.I (Rippka et al. 2001b) of cyanobacteria (Castenholz 2001c).

Species of *Nostoc* show a great morphological and ecological variability and live worldwide in different limnic as well as terrestrial and aerophytic habitats, some even under extreme conditions in hot and cold deserts. The mucilage of trichomes and colonies is often colored. Pigments like Scytonemin and mycosporin-like amino acids (MAA) with high UV absorption are produced (Scherer et al. 1988; Scherer 1991; Garcia-Pichel and Castenholz 1993). *Nostoc*-species are very tolerant against desiccation and can reactivate their physiology after long periods of drought quite fast (Scherer 1994), but they need liquid water (Lange et al. 1986). Some species of *Nostoc* are members of different symbioses, e.g., in mosses, ferns and lichens (like *Collema* Weber ex Wigg., *Stereocaulon* Hoffm. and *Peltigera* Willd.) (Tschermak-Woess 1988).

Nostoc commune Vaucher ex Bornet and Flahault (Fig. 3.15)

The sheath of the trichomes was without structure; only in the margin of the colonies, it was better expressed and slightly yellow brownish. The cells of trichomes were spherical, heterocytes spherical to slightly oval and only slightly bigger than vegetative cells. Contrary to free-living terrestrial samples of *N. commune* which show macroscopic layers, the material of the building surfaces presented its specific features by light microscopy and in culture. It was obvious that the thalli on inclined building surfaces did not

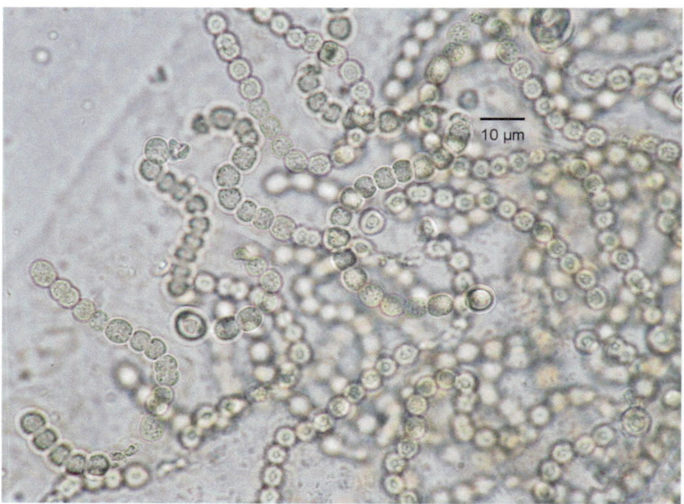

Fig. 3.15 Part of a colony of *Nostoc commune* with thick-walled heterocytes

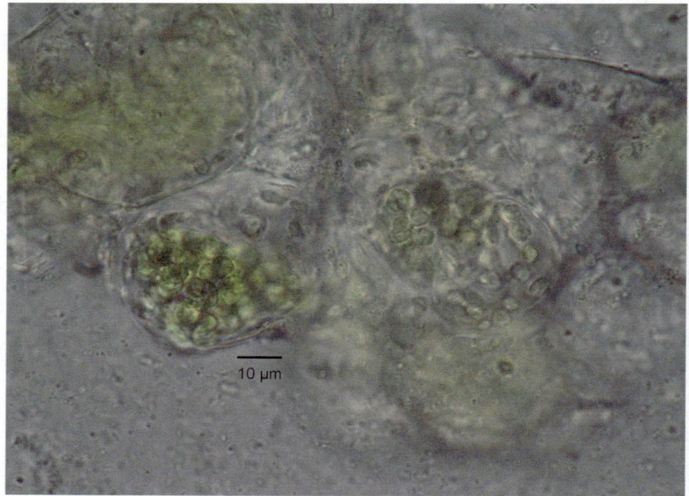

Fig. 3.16 Compact colonies of *Desmonostoc muscorum*

develop as in natural habitats, which can be attributed either on non-optimal environmental conditions or on a shorter time for development.

Nostoc commune was isolated from different older succession stages on buildings but could also be found among the initial colonization of modern building surfaces. Further, the taxon was observed in sedimentation samples. The species has already been documented on building surfaces in Europe; recently, it was reported on man-made structures in Svalbard (Raabová et al. 2016).

N. commune is an extreme example for tolerance against desiccation (Dodds et al. 1995). Even thalli stored for 63 years in a herbarium continued successfully to grow after remoistening (Shirkey et al. 2003). Scherer (1994) reported of material stored in a herbarium for 120 years which was brought to live again. *N. commune* also tolerates extremely low temperatures. According to Kappen (1993), a positive netto photosynthesis was measured until $-15\ °C$; the N-fixation in heterocytes was positive until $-7°C$. Dodds et al. (1995) demonstrated a lower temperature limit of $-1\ °C$. The pH of the substrate in terrestrial populations was reported from 8.0 to 8.2 (John 1942). Protective substances against UV radiation like Scytonemin and various MAA, which are transferred from the cells into the mucilage, were observed (Scherer et al. 1988; Scherer 1991). Their production depends on environmental factors like solar radiation and water stress (Hill et al. 1994). An effect of the radiation quality (wavelength) was detected as well (Ehling-Schulz et al. 1997; Sinha et al. 2003).

Nostoc ellipsosporum Rabenhorst

Trichomes were characterized by oval elliptic to cylindrical cells (ca. $4\ \mu m$ wide). Heterocytes were mainly elongate and slightly wider as vegetative cells. A gelatinous sheath was found around trichomes clearly visible in marginal areas of thalli.

This species was observed on building façades in various occasions together with other organisms.

In some of herein investigated samples, specific forms of *Nostoc* were isolated which showed different combinations of morphological features, not compatible to the known taxa. In most cases, these isolates showed aberrant growth, possibly because of non-axenic cultures as partial bacterial lysis of *Nostoc*-colonies, and morphological changes of trichomes were observed. Also, fungi provoked changes in thallus morphology of *Nostoc* (smaller, more compact thalli, development of many resting cells and reduced hormogonia). It was remarkable that some of the correctly identified isolates of *Nostoc* in cultures contained additional microorganisms (bacteria, fungi) which did not influence the characteristics of the thalli.

Family **Rivulariaceae** Komárek et Anagnostidis

Genus *Calothrix* Agardh ex Bornet and Flahault

Heteropolar trichomes with a basal heterocyte and distinctly attenuated at the ends are characteristics of this genus. The trichomes are covered by a distinctly stratified mucilaginous sheath, the filaments are not united into a gelatinous colony (like in *Rivularia* Agardh and *Gloeotrichia* Agardh) but free. The genus *Calothrix* is also designated as form-genus I Agardh (Rippka et al. 2001c) of subsection IV. II of cyanobacteria (Castenholz 2001c). The cosmopolitan heterogeneous genus *Calothrix* contains several morphotypes and ecotypes. It is distributed in marine, freshwater and aerophytic habitats and may also be the photosynthetic partner in lichens (Komárek 2013).

Calothrix parietina Thuret ex Bornet and Flahault

Cells near the basal heterocyte were about 6–8 μm wide, trichomes were attenuated toward their ends in hairs. Sheaths in free-living state were distinctly stratified; this feature developed in culture after a prolonged period of growth. Free-living material had a brownish-green color; in culture, it appeared blue green initially and grew olive green after 5–6 weeks. After transfer into fresh culture substrate, many hormogonia were observed.

Relatively common in nature, this form was only found once on the surface of a specimen. The species is described in limnic and also aerophytic habitats, e.g., calcareous rocks (Geitler 1932). *Calothrix parietina* is also known as symbiont in the lichen *Stereocaulon octomerellum* Müll. Arg. (Tschermak-Woess 1988).

During the initial colonization of building surfaces, especially under dew or high air moisture conditions, Cyanoprokaryota were rarely observed. Some aerophytic blue greens have a high tolerance against desiccation and are able to regenerate after a long time in dry conditions (Potts and Friedmann 1981; Scherer 1994). A positive netto photosynthesis runs in some taxa only from 96% (Lange et al. 1994) or 95% (Potts and Friedmann 1981; Palmer and Friedmann 1990) relative humidity onwards. Many of the documented Cyanoprokaryota produce protective substances against high solar radiation (especially UV radiation), and in addition, they have a high tolerance against high or low temperatures and are therefore well adapted to extreme habitats. This is supported by the production of gelatinous layers and sheaths which take up water from precipitation and support photosynthetic activity. Discussions over activities of Cyanoprokaryota in destructive erosion of stone material are ongoing. Some studies reported active intrusion of some species into the surface of calcareous stones, while others discussed the production of protective substances of endolithic forms against stone deterioration (Ehrlich 1998; Di Bonaventura et al. 1999; Saiz-Jimenez 1999; Pohl and Schneider 2005). On exposed calcareous rocks, natural erosion is a slow-going process, and on modern building façades, erosion processes caused by Cyanoprokaryota are not relevant over a period of 15–60 years.

Many investigations documented Cyanoprokaryota on building façades (Saiz-Jimenez et al. 1990; Ortega-Calvo et al. 1991, 1993, 1995; Palmer and Hirsch 1991; Ariño and Saiz-Jimenez 1996; Gaylarde and Gaylarde 2000; Lamenti et al. 2000; Ortega-Morales et al. 2000; Crispim and Gaylarde 2005; Hoppert et al. 2005), although mainly long existing successions on historical buildings and monuments were taken in account. Often the taxonomic results were not clear, and comparisons were difficult.

Some forms of Cyanoprokaryota exhibit a natural resistance against biocidal compounds. Resistances were also proven experimentally (Golden and Sherman 1984; Golden and Haselkorn 1985; Hirschberg et al. 1987; Jansson et al. 1987; Gingrich et al. 1988; Ajlani et al. 1989a, b; Dalla-Chiesa et al. 1997; Sajjaphan et al. 2002).

Regnum Eubacteria–Non-Oxygen Phosynthetic Active Groups

Gram-positive and gram-negative groups of bacteria are distinguished according to the structure of the cell walls (Garrity and Holt 2001; Sitte et al. 2002). In our study, other prokaryotic organisms (with the exception of Cyanoprokaryota) were recorded primarily in totals; only some strains were determined in detail. Bacteria were found together with microorganisms on nearly all investigated surfaces. Bacteria did not form a basis for further development of other organisms; they occurred predominantly independent. Previous studies into the flora of bacteria on building materials and buildings do exist. These studies used specially formatted stone and concrete blocks which were also exposed at the open area of the Fraunhofer Institute for Building Physics in Holzkirchen (Braams 1992; Mansch and Bock 1998). Parts of such samples have still been exposed there and were used for further investigations.

Surfaces of stones and buildings exposed to natural weather conditions are colonized by bacteria within the first months (Braams 1992; Mansch and Bock 1998; Papen et al. 2001). Beside Actinomycetes (s.l.) and chemoorganotrophic bacteria, also chemolithotrophic forms are important in the succession (May 2003). Chemoorganotrophic forms normally dominate in absolute numbers and their nutrification is supported by air pollution or other resources (e.g., excrements) (Saiz-Jimenez 1993, 1995, Sabbioni 1995; Papen et al. 2001). In Brazil, (Porto Allegre) Kiel and Gaylarde (2006) documented spore-forming bacteria of *Bacillus* as main part of the microflora with reference to biodeterioration. Gorbushina et al. (2002) reported bacteria on marble monuments in Russia. In relation to biodeterioration, particularly acidogene forms are important which can destroy the surface of buildings. Such gram-positive forms of bacteria (without taxonomic differentiation) were documented on sandstone monuments (Palmer et al. 1991).

The most important members of the chemolithotrophic microflora on stone buildings are endolithic, nitrificationous bacteria (Kauffmann 1953; Wagner and Schwarz 1965;

Krumbein 1968; Bock and Sand 1993; Mansch and Bock 1998). Their activity leads to the secretion of nitric acid and the formation of nitrates, as was demonstrated in laboratory experiments (e.g., Mansch and Bock 1994, 1996). In the context of nitrification (ammonium oxidation) in stones used in buildings, the taxa *Nitrosovibrio*, *Nitrosomonas*, *Nitrosospira* and *Nitrobacter* were isolated (Meincke et al. 1989; Bock et al. 1990; Koops and Möller 1992; Spieck et al. 1992; Krause-Kupsch 1993; Mansch and Bock 1998). Chemolithoautotrophic *Thiobacillus* spp. uses sulfur compounds and excretes sulfuric acid, but was so far seldom recorded on historical buildings (Bock and Sand 1993). Visible deterioration of bricks, concrete constructions, walls, etc., by chemolithotrophic and/or chemoorganotrophic bacteria were detected in wet situations with high nutrification and comfortable temperature conditions as, e.g., on brick constructions of wastewater buildings (Parker 1945; Jozsa et al. 1994).

It is not generally possible to separate chemoorganothrophic and chemolithoautotrophic bacteria, because some of the taxa are only facultative autotrophic and are using organic substances, which emphasizes the importance of organic contamination (Warscheid et al. 1991; Mansch 1994).

In our investigations, different bacteria were observed in the cultures of organisms of initial colonization. Beside non-pigmented forms also, colored taxa were regularly detected (protection against radiation) which tolerate oligotrophic conditions (compare also Wainwright et al. 1993). In spite of the observed presence of bacteria on the surface of building materials, no bacterial biofilm was visible in the first stage of colonization. Therefore, a differentiation between bacteria really growing on the substrate or just adsorbed forms (transported by wind, etc.) was difficult (see also e.g., Bock and Sand 1993; Eckhardt 1996).

From another point of view, it is remarkable that many bacteria precipitate calcareous and silicious material and the newly formed matrixes are used in bioremediation to enable restauration of historical buildings and monuments (May 2003).

Some algae and Cyanoprokaryota show better growth in culture or improve their physiological condition if cultivated together with special bacteria. The colonization of soft vinyl surfaces by *Leptolyngbya* and *Pleurochloris pyrenoidosa* Pascher was shown to improve through (undetermined) bacteria (Holmes 1986).

Regnum Archaea

Some authors demonstrated a possible colonization of archaea on surfaces of building materials (e.g., Rölleke et al. 1998; Piñar et al. 2001a, b, c) during investigations on biodeterioration of old, salt-incrusted surfaces of medieval frescoes indoors. On modern façades, such organisms are not regularly expected due to the very different ecological situations (high radiation, changing moisture, usually no salt incrustations). The cited studies are not documented with cultures. In our work, no archaebacteria were observed.

Life Form Eukaryota

Cells of the Eukaryota differ in some substantial characters from prokaryotic cells. A nucleus contains most of the DNA and is connected with a membrane system (endoplasmic reticulum, ER). In addition, different cell organelles exist. Mitochondria and plastids (only in plants and plant-derived organisms) are enclosed by membranes, contain their own DNA and are expected to be derived from prokaryotic symbionts. Typical structures of eukaryotes are further Golgi bodies, ribosomes and flagella with common ultrastructure.

Phycophyta (Algae) Division Sensu Lato

Division Rhodophyta
Class Rhodophyceae
Order Porphyridiales Kylin

Family **Porphyridiaceae** Kylin

Genus *Porphyridium* Nägeli

This genus appears with cells embedded in non-stratified mucilage. Cells are spherical or subglobose with a star-like chloroplast, which possesses a distinct pyrenoid in contrast to most other red algae. The nucleus is embedded in a sinus of the chloroplast like in the green algal genus *Trebouxia* Puymaly (Trebouxiophyceae). *Porphyridium*-taxa are cosmopolitan and can be found in marine, limnic and aerophytic habitats.

Porphyridium purpureum (Bory) K. M. Drew and R. Ross (Figs. 3.17 and 3.18) (Syn.: *P. cruentum* [Gray] Nägeli 1849)

The colonies appeared as blood-red or wine-red layers in natural conditions. Very few algae respectively Cyanoprokaryota (e.g., *Gloeocapsa sanguinea*) form similar mucilaginous layers, a clear distinction therefore needs microscopic investigation. An isolate of a habitat at the fortress of Kufstein/Tyrol formed dense gelatinous layers on basal areas of a wall of travertine with spherical to subspherical cells, which showed (6.3) 7–13 (13.8) μm diameter in free-living material, (7.4) 8.5–14 (18) μm in culture. Dimensions

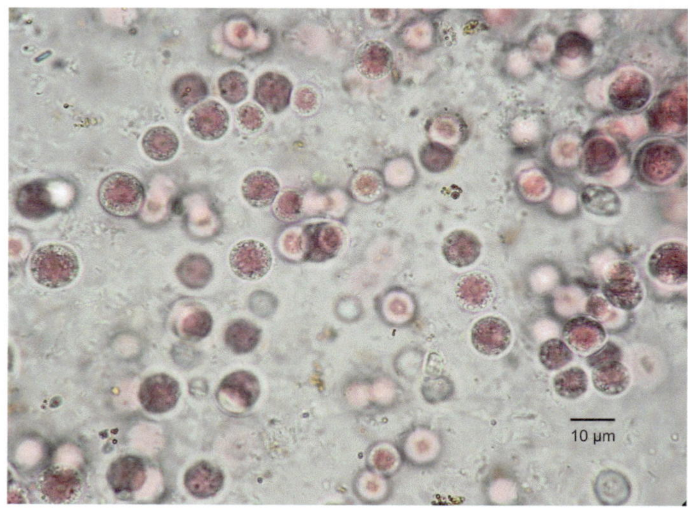

Fig. 3.17 *Porphyridium purpureum*, microscopic image of cells of fresh material grown in the natural habitat

coincided with data from literature (e.g., Ettl and Gärtner 1995, 2014). Asteroid chloroplasts had a central pyrenoid (slightly orange color) and an excentric nucleus between lobes of chloroplast. *Porphyridium purpureum* was found in fine layers of soil on the rough surface of the travertine blocks, together with other aerophytic algae and Cyanoprokaryota like *Phormidium autumnale*, diatoms (*Navicula* sp.), *Chlorella ellipsoidea* Gerneck agg., *Chlorella vulgaris* Beijerinck and others. Water and nutrient supply of the organisms was only indirect by capillary transport or splash water. At the zoological garden of Munich (Hellabrunn), another strain of *P. purpureum* grew indoors (subtropical house) and was also isolated from a travertine wall with wine-red mucilaginous covers. The alga was associated with similar organisms as in Kufstein and also only irrigated indirectly. The cell dimensions differed from the literature, varied in fresh material between (12) 14–16 (18) µm, in culture between (8.5) 12–21 (24.4) µm. A distinct vacuolization was visible in the cells (Fig. 3.18). Chloroplasts, pyrenoids and nuclei were similar to the previous isolate of Kufstein.

In contrary to observations in the past when *Porphyridium* was documeted regularly from urinated surfaces (e.g., Hansgirg 1892; Dalla Torre and Sarnthein 1901; Gams 1950), it is supposed that *P. purpureum* is not abundant any more in the present cultured landscape. The reasons may be raised hygienic demands, and changes in city planning which have terminated many former habitats. As colonizer of modern building parts, for instance façades, the species therefore is confined to situations that fit to its special physiological needs.

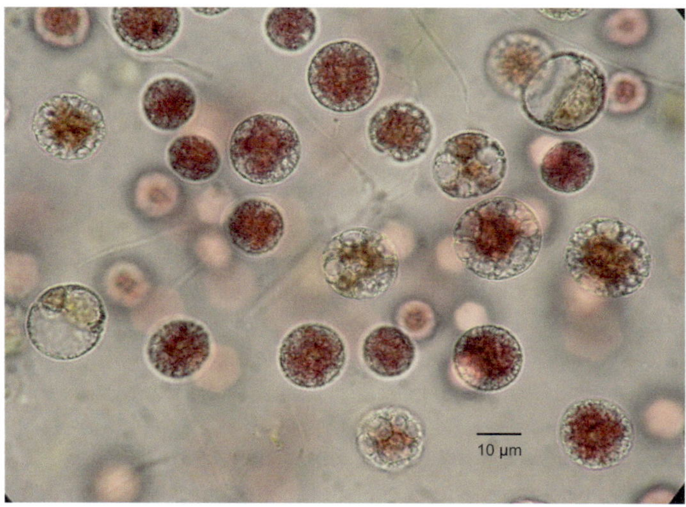

Fig. 3.18 *Porphyridium purpureum*, vacuoles in the cell periphery. Very big vacuoles and brownish chloroplasts mark degenerated cells

Observations of *P. purpureum* were made in two different localities, as discussed above (see also Hofbauer et al. 2006), at the fortress in Kufstein on travertine and on an indoor wall at the zoological garden of Munich. Water and nutrient supply of *P. purpureum* were in both locations only indirect by capillary transport or splash water. Geitler (1944b, p. 325) noted that *P. purpureum* prefers habitats not directly irrigated by rainwater and characterized the species as "ombrotrophe Alge" and "trockenresistente Atmophyten."

P. purpureum is regarded as very nitrophilic and salt tolerant (Gams 1927; Ettl and Gärtner 2014). The optimum temperature for growth is 27 °C (Rieth 1966). The strains of our study grew well on cultures with BBM-Agar (Bold's Basal Medium; Bischoff and Bold 1963). The majority of strains in culture collections (SAG, Göttingen, Germany; Schlösser 1994), UTEX (Austin, Texas, USA; Starr and Zeikus 1993) and CCAP (Cambridge, UK; UKNCC 2001) are cultivated on special *Porphyridium*-media (Andersen 2005).

When *P. purpureum* gets dominant in an association and is macroscopically visible, it belongs to the association *Porphyridietum cruenti*, the most nitrophilic algal association (Gams 1927). According to nomenclatural changes, it should now be named *Porphyridietum purpurei* (Hofbauer et al. 2006).

Division Stramenopiles (Heterokonta)
Class Bacillariophyceae

Bacillariophyceae or diatoms are distinguished from other photosynthetic stramenopiles by a gametic life cycle (Graham et al. 2009). Their cells are enclosed in a rigid box (frustule or theca) composed of amorphous silica. The top and bottom of the frustule (epi- and hypotheca) overlap like a Petri dish. Additionally silica bands are included. A thin organic membrane (Diatotepin) covers the frustule. Chloroplasts have a triple membrane and a naked pyrenoid. Assimilation pigments are chlorophyll a and c which are often masked by accessory pigments (e.g., fucoxanthin). Assimilation products are chrysolaminarin (does not stain with IPI Lugolian solution) and lipid. Reproduction is vegetative by dividing frustules with formation of two new walls internally which become the new hypothecae of the two daughter cells. Diatoms occur as single cells, chains of cells or colonies and are widely distributed in freshwater, marine and aerophytic environments (Bellinger and Sigee 2015). They are perhaps the class with the highest species diversity (about 285 recorded genera and 10,000–12,000 species); sometimes even millions of species are assumed (Norton et al. 1996; Reviers 2003). In freshwater environment, they form a major part of the algal biomass and are a substantial contributor to primary productivity (Bellinger and Sigee 2015).

Diatoms were recorded on historical buildings and monuments (Anagnostidis et al. 1983; Darienko and Hoffmann 2003, Uher et al. 2005b). These habitats have been exposed over centuries, or the algae were found in ditches filled with rainwater on the surfaces of such habitats. Diatoms were also documented on historical fountains (Peraza Zurita et al. 2005). The taxa recorded here were exclusively documented on vertical surfaces of building façades (Hofbauer et al. 2003).

Order Pennales (Schütt) West orth. mut. Karsten
Family **Fragilariaceae** Hustedt

Genus *Fragilaria* Lyngbye

The frustules of genus *Fragilaria* have no central raphe (channel in thecae for secretion of mucus which promotes movement on solid surfaces) and no internal septa. The cells contain two plate-like chloroplasts. Frustules are often joined to ribbon-like chains (Krammer and Lange-Bertalot 1988b; Bellinger and Sigee 2015). The genus is cosmopolitan with marine, freshwater, soil and aerophytic species.

Fragilaria sp. (Fig. 3.19)

In several surface growth situations diatoms were observed, among them a *Fragilaria* sp. (Fig. 3.19) with small frustules of thin and reduced wall structure. In fresh material, no clear identification was possible. Cultures were established, but further observations are needed to clarify the taxonomic status of the strain.

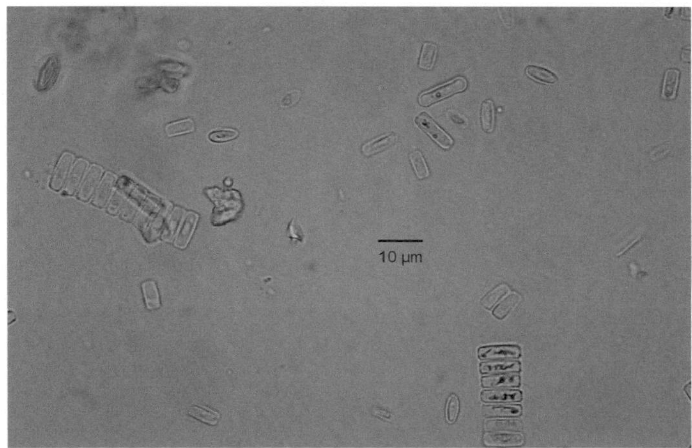

Fig. 3.19 Chains and single cells of *Fragilaria* sp., of fresh living material. Top right: cells of *Pinnularia* sp

Uher et al. (2005b) recorded *Fragilaria construens* (Ehrenberg) Grunov on monuments in Bratislava.

Family **Bacillariaceae** Ehrenberg

Genus *Hantzschia* Grunow

Hantzschia belongs to slightly bended elongate forms with raphes on margins of both parts of frustules (channel-raphe). During cell division, no successive diminishing of daughter cells takes place (Geitler 1980). The cosmopolitan genus is found in diverse habitats, also aerophytic.

Hantzschia amphioxys (Ehrenberg) Grunow (Fig. 3.20)

Elongate, slightly bended raphe on one margin with ribs (fibulae) which were in the middle more distant than others. The species was in form and size very variable. Cells were up to 210 (300) μm long, 5–10 (15) μm broad (Krammer and Lange-Bertalot 1988).

The species is cosmopolitan and aerophytic, common between mosses, on moist soil, also terrestrial.

In one surface sample (AV19), species of diatoms, mainly living cells of *Hantzschia amphioxys*, were found (Fig. 3.20). This seems to be the first documentation of *Hantzschia amphioxys* on new, only a few years old building parts surfaces. *H. amphioxys* is known from the surface of cultural monuments, like the Parthenon in Athens

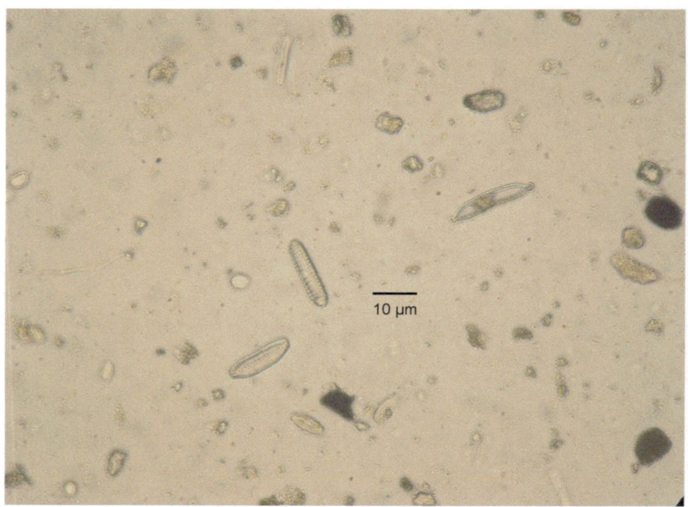

Fig. 3.20 *Hantzschia amphioxys* and *Pinnularia* sp., living material

(Anagnostidis et al. 1983), from calcareous surfaces and mortar of historical buildings in Ukraine (Darienko and Hoffmann 2003) and from monuments in Bratislava (Uher et al. 2005b).

For populations on soil the pH of the substrate lies between 8.0 and 8.2 (John 1942).

Family **Naviculaceae** Kützing

Genus *Navicula* Bory

Valves are elliptical to lanceolate with central raphe; cell apices are rounded or subcapitate. Cells usually have two plate-like chloroplasts, sometimes a single H-shaped chloroplast. It is a large genus in limnic and terrestrial habitats. Some smaller genera have been divided, e.g., *Craticula, Diadesmis, Luticola* (Bellinger and Sigee 2015).

Navicula sp.

Found in habitats of *Porphyridium purpureum* (Bory) K. M. Drew and R. Ross together with other algae. No further differentiation was possible.

Genus *Pinnularia* Ehrenberg

A large genus which has linear, elliptic or lanceolate cells with rounded or capitate poles and a central raphe. Striae (lines on the valves) are rib-like and mostly coarse; usually, there are two plate-like chloroplasts. The genus is cosmopolitan and widespread in many split off habitats but also aerophytic on damp surfaces, between mosses, etc.

Pinnularia sp. (Fig. 3.19)

In several samples, many empty frustules of *Pinnularia* sp. were found, but less living cells. No cultures were obtained.

According to the literature *Pinnularia borealis,* Ehrenberg was documented on marble of the Parthenon in Athens (Anagnostidis et al. 1983) and on monuments in Bratislava (Uher et al. 2005b), *Pinnularia microstauron* (Ehrenberg) Cleve on surfaces of historical buildings in Bratislava (Uher et al. 2005b). Hofbauer et al. (2003) reported a finding of *Melosira roeseana* Rabenhorst (*Orthoseira roeseana* [Rabenhorst] O' Meara, Centrales, Melosiraceae; Krammer and Lange-Bertalot 1991) on building surfaces, although characterization of the habitat needs further observations.

Class Xanthophyceae

This class exhibits organisms at all levels of organization from monadoid to siphonous (Hoek van den et al. 1995). Their characteristic features are discoid yellow-green or green chloroplasts with chlorophyll a and chlorophyll c (chlorophyll b is absent), with accessory pigments like β-carotene and xanthines. The main storage polysaccharide is chrysolaminarin (no starch!) which cannot be stained with IPI. Flagella are heterokont and inserted close to the apex of the zoids, an eyespot lies within a chloroplast. Some species form endogenous silica impregnated cysts of two unequal halves (like box and lid), the cell wall may also consist of two overlapping halves or one piece.

Xanthophyceae are found mainly in freshwater, but also in terrestrial and some in marine habitats. Some aerophytic forms were found on building surfaces and are listed below.

Order Mischococcales Fott
Family **Botryochloridaceae** Pascher

Genus ***Chlorellidium*** Vischer et Pascher

Members of these genus occur as single cells, diads or tetrahedral cell groups, which can be united in colonies which are strongly sticking together, even if single autosporangial or zoosporangial cells are already empty. Two or more discoid parietal chloroplasts are present. Reproduction is mainly by autospores, which form tetrads after the release from the sporangium. Also, single autospores develop into juvenile cells. Zoospores are with or without stigma. The genus is known as a soil alga and was previously recorded in Europe and New Zealand.

Chlorellidium astigmatum Schwarz (Fig. 3.21)

Cells measure 12–18 (28) μm with very thick walls up to 4 μm, zoospores are without eyespot. The taxon was originally described on soil (Schwarz 1979; Ettl and Gärtner 2014).

The observed form (Fig. 3.21) presented many free single cells which were quite big, but also compact tetrads and colonies were formed. Juvenile cells often showed one chloroplast, whereas adult cells showed numerous discoid chloroplasts of different size. Zoospores were without eyespot. Adult cells and colonies had thick cell walls. In our study, the species was found only once on one of the additional investigated habitats, and it was not observed on other building surfaces.

Family **Botrydiopsidaceae** Ettl

Genus ***Botrydiopsis*** Borzi

Members of the genus form multinucleate coenoblasts of enormous size which are mainly globous to ellipsoidal or irregular (Deason and Bold 1960). The cell wall is usually thin but in some taxa thicker or with local thickenings. In adult cells, central vacuoles are formed and many discoid or spindle-shaped chloroplasts lie in parietal position. In some forms, staples of plastids were observed (Tschermak-Woess 1979). Two taxa were described with pyrenoids (Trenkwalder 1975). Asexual reproduction is with auto- and zoospores, and sexual reproduction is unknown. A main diagnostic feature is the

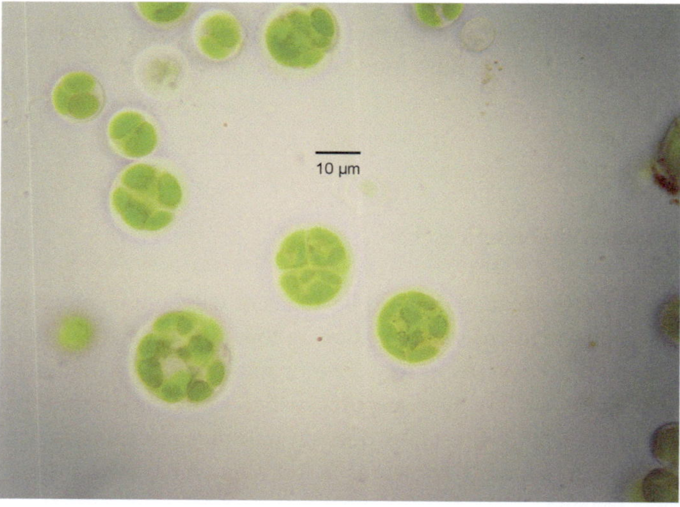

Fig. 3.21 *Chlorellidium* cf. *astigmatum* with thick cell walls, forming groups and small colonies

eyespot, which is lacking in some species. Within the genus, soil algae prevail, but some species also occur in aerophytic or freshwater biotopes.

Botrydiopsis intercedens Pascher (Fig. 3.22)

The observed material coincided with the description of the type in Ettl and Gärtner (2014). Coenoblasts were globous, plastids discoid to polygonal, parietal or turned into the cell lumen or into the plasma. Plastids were sometimes in staples (Tschermak-Woess 1979). Vacuoles and orange oil droplets were sometimes present. The cell wall was mostly thin, in empty sporangia often thicker (also observed by Vischer [1945]). A cell wall consisting of two parts, as described by Johnson (2002), was not observed. Zoosporangia open with an irregular crack; the heterokont zoospores had typically two chloroplasts and a distinct eyespot. Eyespots were already visible in the sporangia before the release of zoospores (Fig. 3.22; see also Lokhorst and Segaar 1989).

Botrydiopsis intercedens was found on the surface of a specimen sample in Holzkirchen after 3 years' weathering exposure.

Genus ***Excentrochloris*** Pascher

Adult coenoblasts are ellipsoidal to pear-shaped to irregular. The cell wall is irregular on one side or on both sides thickened, stratified, sometimes with appendix. Chloroplasts

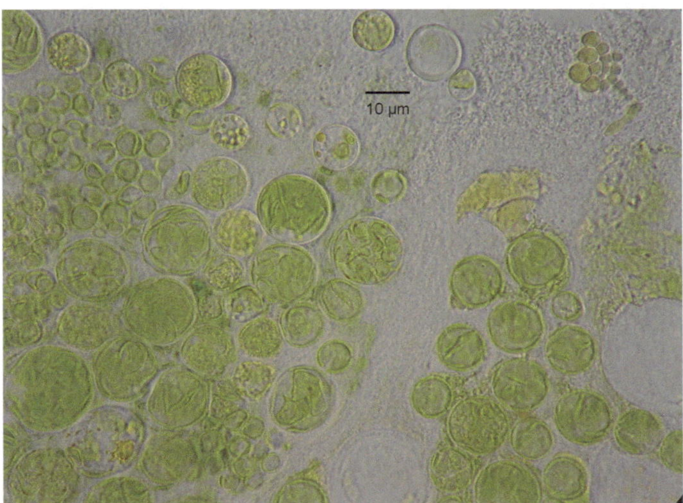

Fig. 3.22 Cells of *Botrydiopsis intercedens* with rounded polygonal chloroplasts. Top left: autospores of different size. Center: a zoosporangium with visible eyespots in the zoospores before release. In some cells, also orange oil droplets are visible

are numerous, lenticular, without pyrenoid. Reproduction is by zoospores which have one chloroplast and eyespot. Zoospores are arranged peripheral in the sporangia before release, as it is with autospores (Ettl 1978; Ettl and Gärtner 2014).

Excentrochloris fraunhoferiana Hofbauer, Gärtner, Rennebarth, Sedlbauer, Mayer et Breuer (Figs. 3.23 and 3.24)

Coenoblasts were usually single pear-shaped, ellipsoidal, lageniform or fusiform, rarely spherical (Fig. 3.23). In adult coenoblasts, a division in two parts was sometimes visible (Fig. 3.24). Cells were up to 96 μm long and up to 87 μm broad, zoospores were 4.4–6 μm long and 2.3–3.8 μm broad. Chloroplasts had a lens or oval disk shape, sometimes in staples. For further details of morphology, reproduction and discussion of similarities with other taxa, see Hofbauer et al. (2011). *Exentrochloris fraunhoferiana* was isolated from the surface of a specimen which was exposed 3 years in open field conditions at the IBP in Holzkirchen, Bavaria.

Hofbauer et al. (2003) additionally reported a finding of *Chloridella neglecta* (Pascher et Geitler) Pascher, Pleurochloridaceae Pascher, on outer building parts in Germany.

Order Tribonematales G. S. West
Family **Xanthonemataceae** Kostikov (Tribonemataceae Pascher p.p.)

Genus *Xanthonema* Silva

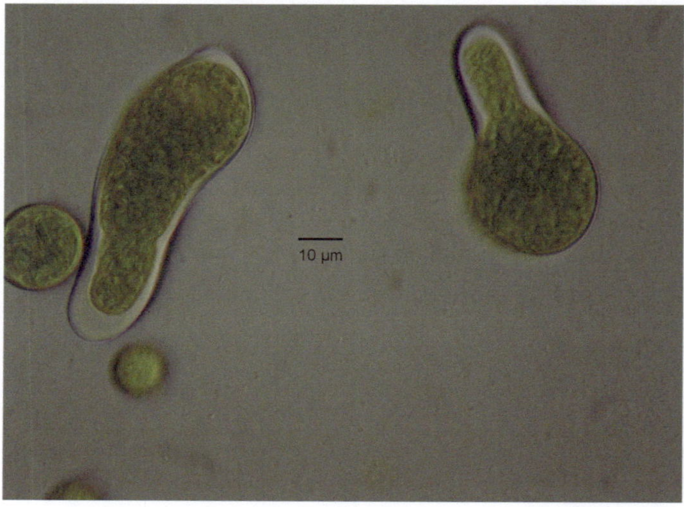

Fig. 3.23 Pear-shaped coenoblasts of *Excentrochloris fraunhoferiana* with irregular thickenings and stratification of cell wall. Left: a juvenile coenoblast with staples of plastids

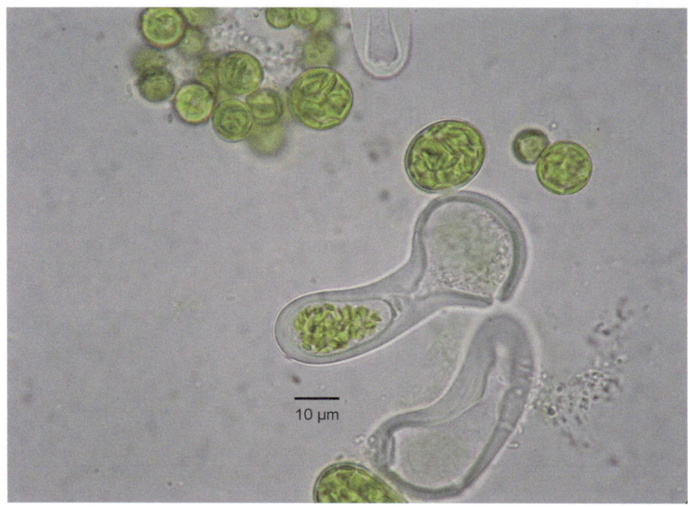

Fig. 3.24 Coenoblasts of different age, one sporangium divided into two parts, one part already empty, the other still vegetative. Bottom: an empty sporangium shows thick and stratified sporangial wall

The genus comprises unbranched filamentous algae without H-pieces in the cell wall (difference to genus *Tribonema* Derbes and Solier). Cylindrical cells have thin cell wall and different numbers of chloroplasts. A cosmopolitan genus known from freshwater, soil and aerophytic habitats.

Xanthonema solidum (Vischer) Silva (Figs. 3.25 and 3.26) (Syn: *Heterothrix solida* Vischer)

Filaments disintegrated in culture into single cells and short pieces (Figs. 3.25 and 3.26). Cells were 5 µm broad and 5–12 µm long, with many chloroplasts. *Xanthonema solidum* is known as a common soil alga (Ettl and Gärtner 2014). In our study, the species was only observed in driving rain samples, but we expect it to be also found on building surfaces.

Family **Heterococcaceae** Silva

Genus ***Heterococcus*** Chodat

The genus comprises mainly terrestrial algae with many described taxa. Thalli consist of branched filaments growing in two or three dimensions. Old thalli easily dissolve into barrel formed or globular single cells with several chloroplasts without pyrenoids.

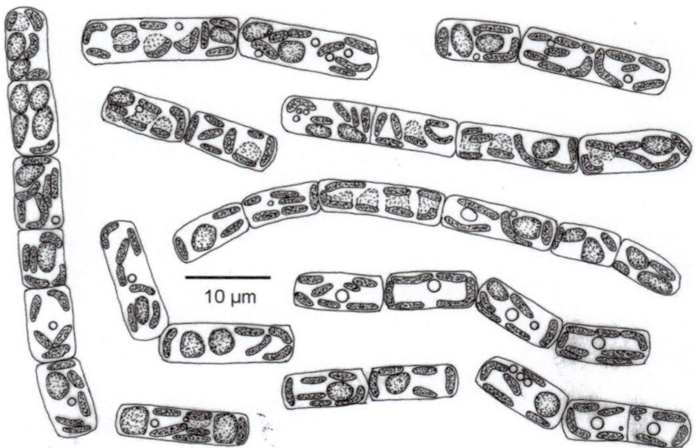

Fig. 3.25 *Xanthonema solidum* (isolate from driving rain) with filaments of different cell numbers

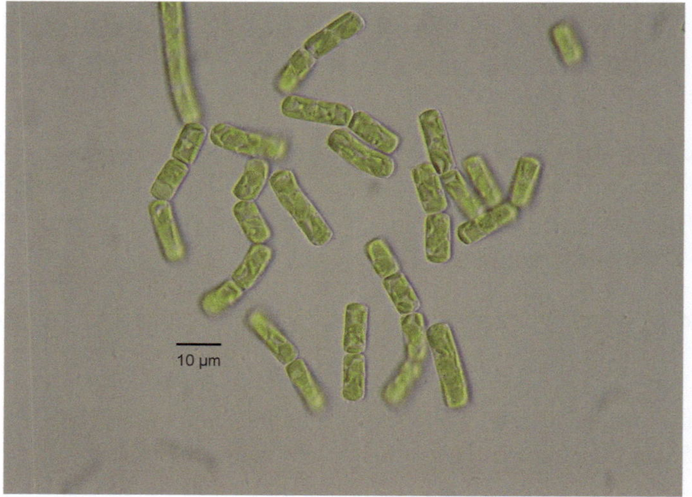

Fig. 3.26 *Xanthonema solidum*, longer cells are undergoing division

Reproduction is by zoo- and autospores. Identification succeeds only with cultures and observation of complete life cycle (Lokhorst 1992; Ettl and Gärtner 2014). A classical cosmopolitan soil alga, aerophytic and endolithic occurrences are also known.

Heterococcus spp.

From several samples, algae of *Heterococcus*-type with erected filaments were isolated, but could not be identified to species level.

Class Eustigmatophyceae

A small class of green to yellow-green coccoid algae resembling Xanthophyceae; however, distinct cytological features separate it as a single taxonomic group (Hibberd and Leedale 1970, 1971, 1972; Andersen et al. 1998; Hofbauer 2015). Main features are: The lack of chlorophyll b and c, the dominant xanthophyll is violaxanthin (Whittle and Casselton 1975a, b; Karlson et al. 1996), the outer membrane of plastids typically has no connection to the nuclear envelope, and plastids typically possess stalked polyhedral pyrenoids (secondary missing in some taxa). An orange to reddish lipid body can often be recognized in vegetative cells of several taxa. Reproduction is only asexual with auto-spores and zoospores which have one or two very unequal subapical flagella and extra-plastidial eyespot. About 30 taxa were described in freshwater and terrestrial habitats; taxonomy is still in exploration (Hofbauer 2015; Eliáš et al. 2017).

Order Eustigmatales Hibberd
Family **Eustigmataceae** Hibberd

Genus *Eustigmatos* Hibberd

Cells are globular or slightly irregular with smooth cell wall. In older cells the polyhedral pyrenoid (which looks like a crystalloid structure) is distinct. Reproduction is by auto- and zoospores. The cosmopolitan alga was previously isolated from soil and aerophytic habitats.

Eustigmatos magnus (J. B. Petersen) Hibberd (Fig. 3.27) (Syn.: *Pleurochloris magna* J. B. Petersen)

Adult cells measured >10 µm (up to 17 µm) in diameter, with a parietal incised chloroplast. Pyrenoid was polyhedral of crystalloid character, and orange-red oil droplets were in the plasma. Zoospores were small (up to 12 µm long).

Found in a sample of an additional location at the base of a wall influenced by splash water and soil particles. This was the first documentation on building surfaces (Hofbauer 2003); the alga was also reported on gravestones in Ukraine (Darienko and Hoffmann 2003).

Division Chlorophyta
Class Chlorophyceae sensu strictu (sensu Reviers)

According to current literature (e.g., Reviers 2002; Graham et al. 2009), the chlorophyta are regarded as more distantly related to the ancestry of land plants. This monophyletic group has diversified into a variety of organization types which has been the main

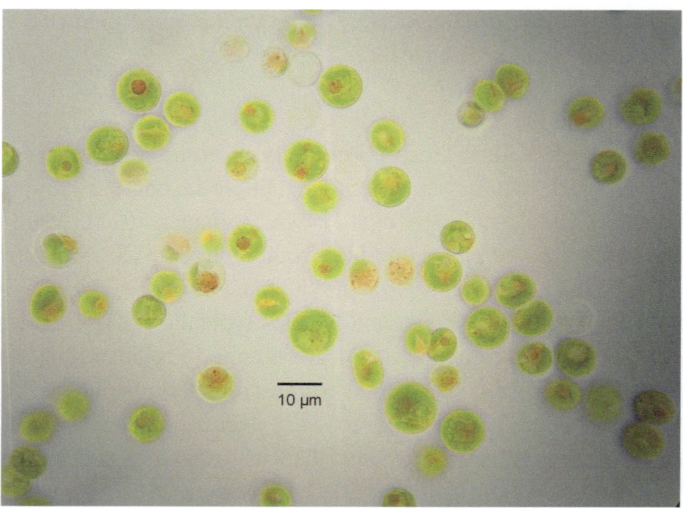

Fig. 3.27 *Eustigmatos magnus* in culture. Cells with incised chloroplast, reddish oil droplets and polyhedral pyrenoid

base for classification until the last century. Today, notably molecular features provide a more natural classification because the groups (lineages) as shown by Graham et al. (2009) have undergone parallel evolution of organization types. The class chlorophyceae includes all types of organization among green algae. Several orders are assigned to the class on basis of molecular and ultrastructural traits, but taxonomy of genera in many cases is still incomplete and provisional and needs further investigation.

Order Chlamydomonadales Fritsch
Family **Chlamydomonadaceae** G. M. Smith

Genus *Chlamydomonas* Ehrenberg

Biflagellate unicells have a rigid cell wall, a parietal usually cup-shaped chloroplast with a pyrenoid (one or several) and typically with an eyespot. Additionally, asexual stages without formation of flagella (palmella-stages, akinetes) are known. Asexual reproduction is by successive mitotic divisions resulting in 2, 4, 8 or 16 daughter cells (zoospores), sexual reproduction runs in all different types (isogamy, anisogamy, oogamy). More than 500 species were described and artificially grouped by Ettl (1976). A new revision transferred some species to new genera (Pröschold et al. 2001; Lewin 2003; Pröschold 2005). Of more than 500 known species, about 70 live in the soil or are aerophytic (Ettl and Gärtner 2014).

Chlamydomonas sp. (Fig. 3.28)

This taxon was detected in one sample of a flat roof together with *Haematococcus pluvialis* Flotow em. Wille forming a palmella-stage. A similar alga was also isolated from an air sedimentation sample. The isolate belonged to the group "*Euchlamydomonas*" sensu Ettl (1976) with a cup-shaped chloroplast and basal pyrenoid (Fig. 3.40). *Chlamydomonas* sp. is not common on building surfaces and obviously prefers habitats with at least small amounts of liquid water.

Family **Haematococcaceae** (Trevisan) Marchand

Genus ***Haematococcus*** C. A. Agardh em. Flotow

Cells are biflagellate, ellipsoidal to globular with protoplast attached to the cell wall by cytoplasmic strands, the chloroplast is cup-shaped with one to four pyrenoids. Reproduction is similar to *Chlamydomonas*. In drying or starving (insufficient N supply) conditions, akinetes are formed with accumulation of astaxanthin. They survive complete desiccation and germinate under favorable conditions. The cosmopolitan genus *Haematococcus* occurs in small and often periodic water accumulations, birdbaths, stone and concrete pools, etc. About five species are described (Pocock 1937, 1960; Ettl 1983).

Haematococcus pluvialis Flotow em. Wille (Fig. 3.29) (Syn: *Sphaerella lacustris* [Girod] Wittrock, *S. pluvialis* [Flotow] Wittrock, *Haematococcus lacustris* [Girod] Rostafinski, incl. *H. longissima* Ettl)

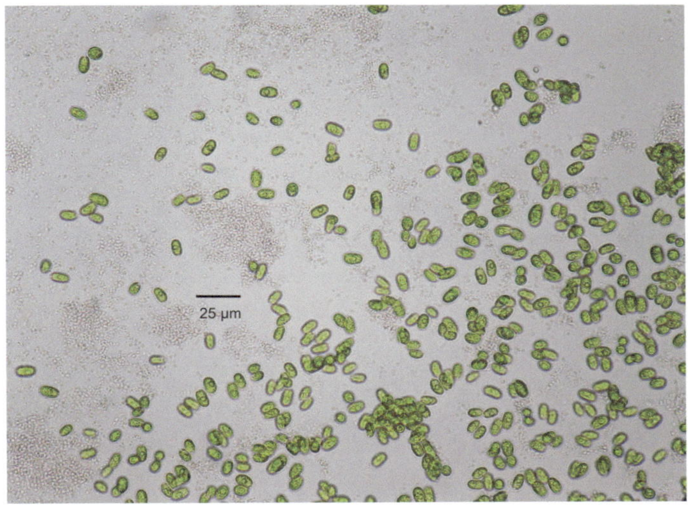

Fig. 3.28 *Chlamydomonas* sp., air sedimentation sample of the outdoor area at the IBP

The most common species of the genus is distributed worldwide. Life cycle starts with green flagellates (Fig. 3.29), followed by nonflagellate orange-red akinetes (which are still able to divide) containing Astaxanthin. Akinetes transform again into biflagellate swarmers under favorable wet conditions (Hazen 1899; Elliot 1934; Kakizono et al. 1992; Fan et al. 1994). Astaxanthin formation is favored by lack of Nitrogen (Cygan 1970).

The species was identified and isolated from different habitats, a flat roof in Berlin, a periodic water ditch at a street in Innsbruck, a window sill at a building in Holzkirchen and building parts made of aerated concrete. There is no doubt that this alga prefers mainly horizontal but limnic habitats, and as an exception, it also occurs on vertical building surfaces with a thin water film (Ariño and Saiz-Jimenez 1996). Its occurrence can be hindered by adequate construction, meaning protection against rainfall and/or surfaces where the water can easily run off.

The accumulation of astaxanthin in *Haematococcus* has been studied biotechnologically for many years, and continuous experimental research is performed worldwide. Astaxanthin is commercially used as food colorant and food supplement, in cosmetics and aquaculture (salmon) (e.g., Boussiba and Vonshak 1991; Grung et al. 1992; Gurin et al. 2003; Cifuentes et al. 2003). At the Fraunhofer Institute for Interfacial Engineering and Biotechnology (IGB), a photobioreactor for cultivation of *Haematococcus pluvialis* (and other algae) was developed (Degen et al. 2001; Degen 2002).

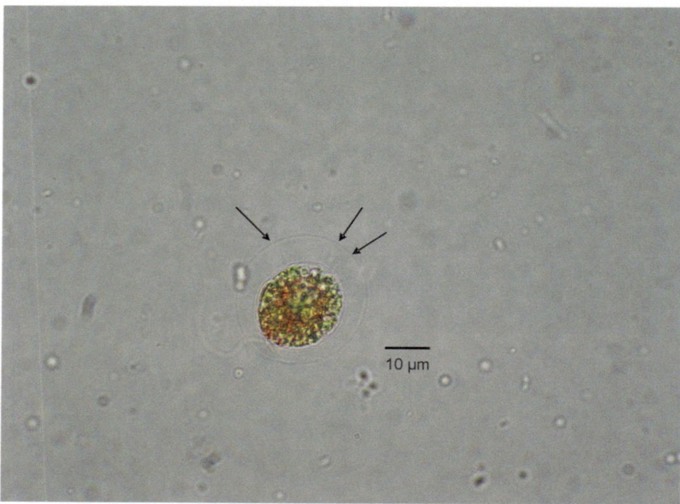

Fig. 3.29 *Haematococcus pluvialis,* isolated from a building surface at IBP in Holzkirchen. Protoplast connects to the cell wall by fine protoplasmic extensions (arrows). One flagella and orange-red granules (astaxanthin) are visible in the protoplast

Order Chlorococcales Marchand (mut. charact.)

A classical order based on organismal morphology, but now replaced by a new taxonomy reflecting the relations among morphologically variable organisms by molecular phylogenetic analyses. Therefore, some nonflagellate unicellular genera (formerly placed in Chlorococcales and other orders) are now placed within Chlamydomonadales and Sphaeropleales (see Leliaert et al. 2012, with extensive literature).

Family **Chlorococcaceae** Blackman et Tansley

Genus *Chlorococcum* Meneghini

Single cells are uninucleate, sometimes found in irregular masses. The chloroplast is parietal cup shaped with at least one pyrenoid. Two contractile vacuoles are often present. Asexual reproduction is by zoo- and aplanospores, zoospores are biflagellate, walled, resembling *Chlamydomonas*. Also, isogamous sexual reproduction occurs resulting in zygotes which have a thick and structured cell wall (Brown and McLean 1969; Gärtner and Ettl 1988; Gärtner and Ingolic 1995, 1997). The very diverse and cosmopolitan genus occurs in many different habitats, marine, freshwater, soil and aerophytic.

Chlorococcum infusionum (Schrank) Meneghini (Syn.: *Lepraria infusionum* Schrank, *Chlorococcum humicola* [Nägeli] Rabenhorst p.p. sensu auct., *Ch. chlorococcoides* [Korschikoff] Philipose, *Hypnomonas chlorococcoides* Korschikoff, *Tetracystis chlorococcoides* [Korschikoff] S. Watanabe; incl. *Chlorococcum infusionum* var. *macrostigmaticum* L. Moewus)

Cells were globular, measured 10–40 μm and had two contractile vacuoles. Chloroplast was cup shaped with few lobes and one pyrenoid with a starch sheath consisting of several irregular starch granules. Aplanospores had thick but nonsculptured walls.

Chlorococcum infusionum was identified from different samples of Bavaria (Hofbauer et al. 2003). As the most common species of the genus, it is distributed worldwide in different habitats, mostly terrestrial but was isolated from limnic habitats too (e.g., Cowell 1960; Girgin et al. 2004; Ariyadej et al. 2004). It was also found under low light conditions with other algae and Cyanoprokaryota in Baradla cave near Aggtelek in Hungary (Claus 1955, 1958). For further literature, see Bold (1931) and Ettl and Gärtner (1988a).

Chlorococcum acidum Archibald et Bold (Fig. 3.30)

Cells measured about 18–20 μm. The chloroplast was cup shaped with many incision, the pyrenoid had a continous starch sheath, sometimes more than one pyrenoid was found. Additional starch granules were sometimes accumulated in the chloroplast. Zoospores and aplanospores were isolated from an air sediment sample and from probes

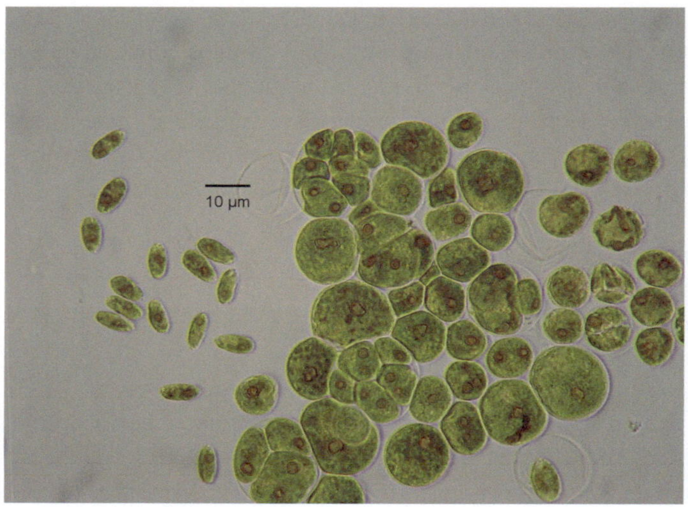

Fig. 3.30 Vegetative cells, sporangia and zoospores of *Chlorococcum acidum* of driving rain sample. Pyrenoids stained with IPI are multiplicated in some cells

of driving rain (Fig. 3.30). This species was not known in Germany and was first documented in soil samples by Hofbauer (2007).

Chlorococcum sp.

Vegetative cells of *Chlorococcum* sp. were found in some samples of the "old" and "new" series. The species could not be determined because of lack of zoospores and short living cultures.

During cooperation with the Swiss Federal Laboratories for Materials Science and Technology (EMPA) in St. Gallen, a further species of *Chlorococcum,* ***Chlorococcum lobatum*** (Korschikov) Fritsch and John, was identified from building surface material (alga isolated by Dr. M. Nay; Hofbauer et al. 2003).

Order Protosiphonales Ettl et Komárek
Family **Deasoniaceae** Kostikov (Actinochloridaceae Korschikoff p.p.)

Genus *Deasonia* Ettl et Komárek

Cells are of *Chlorococcum*-type, but are always multinucleate (coenoblasts) and have spongiomorphous chloroplasts with many fissures and stripes and one (or several) pyrenoid(s). Contractile vacuoles occur only in zoospores. Reproduction is asexual by walled zoo- and autospores, sexual reproduction is not observed so far. The cosmopolitan genus is found in freshwater, soil and aerophytic habitats.

Deasonia multinucleata (Deason et Bold) Ettl et Komárek (Fig. 3.31)

Cells in juvenile cultures had a diameter between 6–30 μm, which was even bigger in adult cultures. The chloroplast was spongiomorphous with an excentric pyrenoid and a discontinuous starch sheath and big compact starch grains (Fig. 3.31). Zoospores were not observed, no permanent culture was achieved.

Deasonia multinucleata was isolated from an old sample which is the first record on building surfaces.

Order Scenedesmales Kostikov (Sphaeropleales Luerssen em. Deason, Silva, Watanabe et Floyd p.p.)
Family **Scenedesmataceae** Oltmanns

Genus *Scenedesmus* Meyen

Characteristic coenobia of 2–16 (32) cells are arranged in one or two lines. Cells are elongate to spindle shaped with smooth or sculptured wall and protuberances, bristles or warts according to the species. Chloroplast is parietal sometimes lobed or irregularly, one pyrenoid with starch envelope. Reproduction is by 2–16 (32) autospores which form new coenobia within the mother cell. The formation of coenobia (or single cells) is obviously also depending on predator pressure (Lürling and Van Donk 1997, 1999; Lürling and Beekmann 1999). In addition, Vischer (1927) showed that cultivation temperature influences the morphology of *S. basiliensis* Chod.: characteristic chains were formed between 7 and 10 °C; at temperatures between 17–20 °C, mainly single cells were found. About 100 taxa of the cosmopolitan genus were described, mostly occurring in

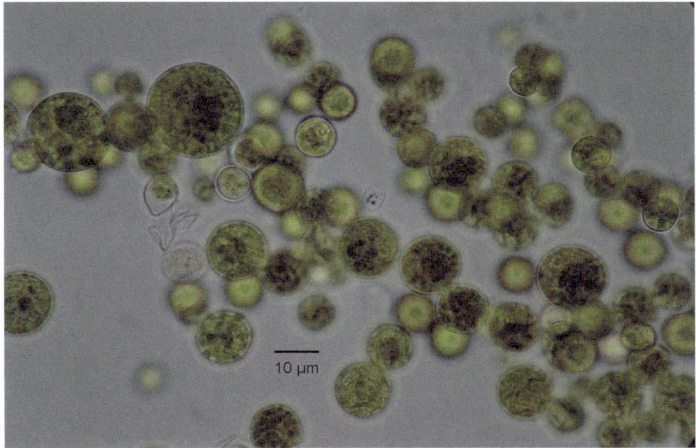

Fig. 3.31 *Deasonia multinucleata*, isolated from a surface sample of a specimen, pyrenoid and starch grains stained with IPI. (Hofbauer et al. 2003)

freshwater; only some were documented in aerophytic localities. For the taxonomy of the genus, see Hegewald and Silva (1988), Hanagata (1998), Hegewald and Hanagata (2000, 2002).

Scenedesmus acutus Meyen (Fig. 3.32) (Syn.: *Scenedesmus obliquus* [Turpin] Brébisson; non *Scenedesmus obliquus* [Turpin] Kützing; *Scenedesmus tetradesmiformis* [Woloszinska] Chodat sensu Chodat, *S. crassus* Chodat, *S. scenedesmoides* Chodat)

Coenobia were without spines, formed with 4–8 spindle-shaped cells and connected with other cells in about half of their length. Outer walls of the external cells were in juvenile stages concave, later convex. On agarized medium (BBM), the isolate of *S. acutus* developed after inoculation irregular large single cells and after 3 weeks untypical coenobia of two and four cells. Also much orange-brown oil was accumulated which covered the inner architecture of the cell. Only in liquid BBM medium, typical coenobia developed with visible features of chloroplast, etc., and allowed determination (Fig. 3.32).

In our study, this alga was found once at one additional investigation locality in a periodically wet habitat (flat roof). *Scenedesmus acutus* is known as an aerophytic and soil alga (Ettl and Gärtner 2014), but also found in limnic habitats (e.g., Ariyadej et al. 2004). The species was seldom recorded on building habitats: Rifón-Lastra and Noguerol-Seoane (2001) documented the alga on granite stones of historical buildings in Spain.

Genera ***Scotiellopsis*** Vinatzer em. Kalina et Punčochářová, ***Coelastrella*** Chodat and ***Graesiella*** Kalina et Punčochářová

These widely distributed (cosmopolitan) genera are characterized by single cells with a peculiar wall structure visible only partially by light microscope better by scanning electron microscope (SEM). Hegewald and Hanagata (2000, 2002) revised the taxa from all

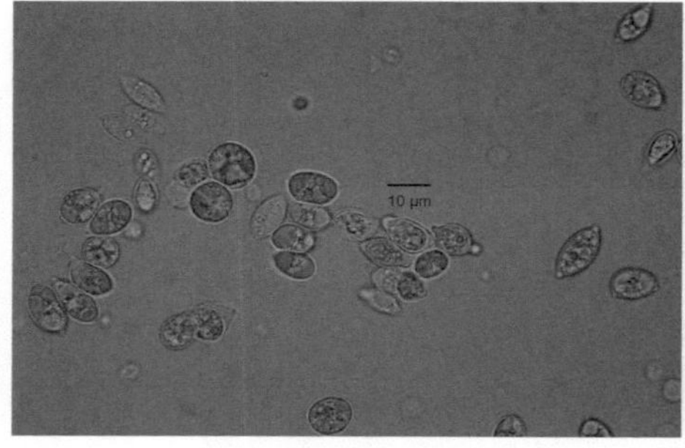
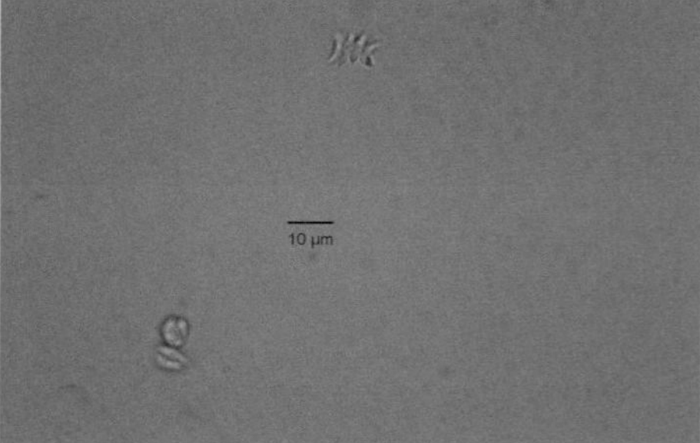

Fig. 3.32 *Scenedesmus acutus*. Left: cultivation on agarized BBM-medium, single cells and untypical coenobia of two and four cells, accumulation of orange-brown oil. Right: culture in liquid medium with characteristic coenobia

three closely related genera and proposed the single uniting genus *Coelastrella* Chodat sensu Hegewald and Hanagata (within Scenedesmaceae-Sphaeropleales). Based on clear and stable morphological criteria and also on a practical point of view, we propose to keep the three genera (or subgenera of *Coelastrella*), which were also suggested by Tschaikner et al. (2007).

Genus ***Graesiella*** Kalina et Punčochářová (*Coelastrella* sensu Hegewald et Hanagata p.p.)

This genus was created by Kalina et Punčochářová (1987) for a taxon of *Chlorella fusca*-group (*Chlorella fusca* Shihira et Krauss var. *vacuolata* Shihira et Krauss) with oval to globular cells showing a smooth wall in light microscopy but fine ribs or a network of ribs in SEM and having no polar cell wall thickenings (which exist in *Scotiellopsis*-taxa). The chloroplast has one pyrenoid (as in *Scotiellopsis* and *Coelastrella*) which features a starch sheath. The taxonomic status and extend of the genus is not yet clear; therefore, no general information on its distribution can be given.

Graesiella emersonii (Shihira et Krauss) Nozaki, Katagiri, Nakagawa, Aizawa et Watanabe (Figs. 3.33, 3.34, 3.35 and 3.36) (Syn.: *Chlorella emersonii* Shihira et Krauss)

Cells were found singular or in small groups (mainly four cells kept together by remnants of the mother cell wall) and were slightly asymmetric, ovoid to globular. Cell dimensions measured between 3 μm (autospores) up to 12 μm (adult cells); juvenile cells were slightly acute on one pole (Figs. 3.33 and 3.36). Using light microscopy, no

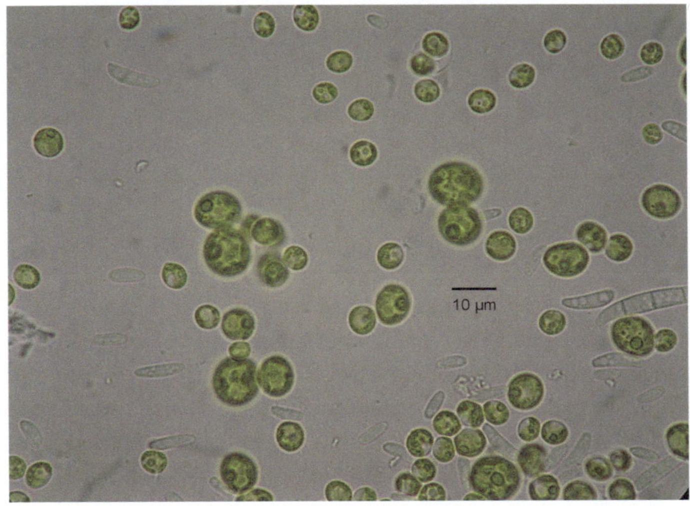

Fig. 3.33 *Graesiella* cf. *emersonii*, cells of different age, chloroplast with prominent pyrenoid

Fig. 3.34 *Graesiella* cf. *emersonii*, even staining with IPI lugolian solution and Lactophenol showed no cell wall sculpture using light microscopy. Some cells with thick wall and accumulation of orange oil

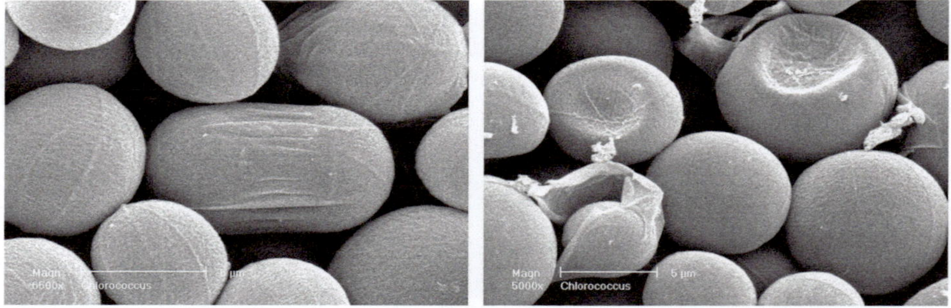

Fig. 3.35 *Graesiella* cf. *emersonii*, using electron microscopy (SCAN) the fine meridional and net-like ribs on the cell wall were visible. Left center: juvenile autosporangium with stretched wall. (Scan pictures: W. Kofler, University of Innsbruck)

cell wall sculptures were visible even with staining procedures. In older cells, the wall could be thickened. Using SEM (prepared with thanks by W. Kofler, Institute of Botany, University of Innsbruck) meridional and other fine cell wall ribs from pole to pole were visible, sometimes forming a net-like sculpture. The cells were uninucleate; the chloroplast was parietal, forming polygonal portions in adult cells. One pyrenoid had a starch sheath mainly in two portions (Fig. 3.36). Reproduction was by 2–4–8 autospores. Colonies in culture displayed a green color, and after 5 weeks, color changed to orange by oil accumulation in the cells (formation of vacuoles).

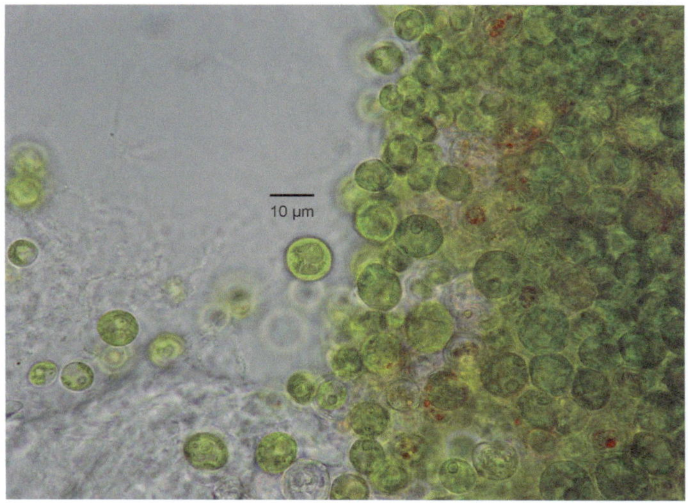

Fig. 3.36 *Graesiella* cf. *emersonii*, chloroplast with prominent pyrenoid and starch sheath (light microscopy 1000-fold, oil immersion)

The species was found as a crust on concrete surface.

Family **Bracteacoccaceae** Kostikov (Neochloridaceae Ettl et Komárek em. Deason, Silva, Watanabe et Floyd p.p.; Deason et al. 1991; Kostikov et al. 2001)

Genus ***Bracteacoccus*** Tereg

Cells are globular and multinucleate (coenoblasts), arranged single or in groups with some parietal, lenticular chloroplasts without pyrenoid. Reproduction is by autospores or zoospores which have heterokont flagella; sexual reproduction is unknown. Phylogenetic investigations on some isolates exist (Wolf et al. 2002; Friedl and O'Kelly 2002; Fučíková and Lewis 2012), but taxonomy of the whole genus remains unclear. Kostikov et al. (2001) proposed the new family Bracteacoccaceae. Different taxa of *Bracteacoccus* were previously documented in building habitats (e.g., Darienko and Hoffmann 2003; Hofbauer 2003). Recently *Bracteacoccus minor* (Chodat) Petrová and *B. minutus* Schwarz were documented on man-made structures in Svalbard (Raabová et al. 2016).

Bracteacoccus sp. (Fig. 3.37)

The species was isolated from different investigated habitats and surfaces of several specimens, which were exposed 3 years in Holzkirchen. The coenoblasts were 9–12 (20) μm in diameter, chloroplasts were polygonal, and adult cells accumulated orange-red oil (Fig. 3.54, degeneration?). The isolate possibly resembled *Bracteacoccus engadinensis*

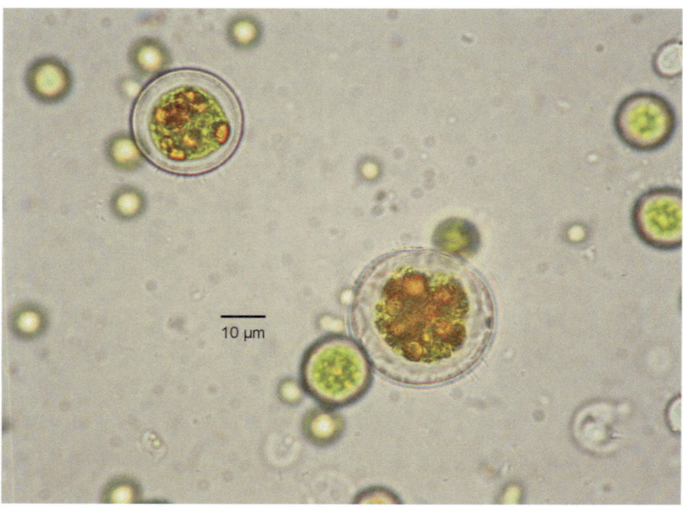

Fig. 3.37 Old cells of *Bracteacoccus* sp. with thickened walls and oil accumulation

(Kol and F. Chodat) Starr, but no zoospores were observed. Hofbauer et al. (2003) previously reported *Bracteacoccus engadinensis* on established building surfaces in Germany.

Class Ulvophyceae Mattox et Stewart
Order Trentepohliales Bourrelly

Family **Trentepohliaceae** De Toni

The single family has only aerophytic forms with branched trichal organization of the thallus (López-Bautista et al. 1998, 2006a, b; López-Bautista and Chapman 2003). The thallus is mainly heterotrichal, sometimes with basal pseudoparenchymatic layer. Beside free-living aerophytic species, many lichenized taxa as well as parasitic forms exist (López-Bautista et al. 2002; Chapman and Waters 2004). Five genera have been described (Rindi et al. 2006). In our investigations, only the genus Trentepohlia was observed on building surfaces, and in Great Britain also *Printzina* Thompson et Wujek was reported on man-made structures (Rindi and Guiry 2003; John et al. 2011).

Genus ***Trentepohlia*** Martius (Syn: *Chroolepus* Agardh; *Physolinum* Printz)

Trentepohlia is characterized by prostrate and erected systems of repeatedly branched filaments; apical cells bear occasionally a cellulose cap. Cells are cylindrical to barrel shaped or even globular with disk-like or band-shaped chloroplast without pyrenoid. Orange or red carotenoid pigments accumulate within the cells, covering the chloroplast morphology. These pigments have been proposed to be a protection against high radiation, as plants in sunny habitats are more intensely orange colored than in shade. Cells are connected by plasmodesmata. Cell division is via phragmoplast (Chapman et al. 2001; López-Bautista et al. 2003). The specific reproduction system includes cycles of autospores and zoospores. Form and position of zoosporangia (which can be actively hurled off) and their basal cells within the thallus are characteristic features for identification. The genus *Trentepohlia* is widespread on different substrates in all parts of the world with a main distribution in tropical regions of the Old World. Some species were identified as symbionts in lichens (Tschermak-Woess 1988; Chapman and Waters 2004). The cosmopolitan genus includes more than 40 species (López-Bautista et al. 2002; Rindi et al. 2006). Bachmann (1992) and Bachmann and Gärtner (1992) studied free-living and lichenized species in Central Europe (Northern Italy/Trentino, North Tyrol, Bavaria). Ettl and Gärtner (1995, 2014) provided a survey of taxonomy and geography of European taxa. *Trentepohlia* species have previously been documented in building habitats in different regions of the world (e.g., Jaag 1945; Wee and Lee 1980; Lee and Wee 1982; Loh and Lee 1987; Wee 1988; Wakefield et al. 1996; Loh 2002; Rindi and Guiry 2002, 2003; Rindi et al. 2003, 2005, 2006; Gaylarde et al. 2006). For Germany, no systematic recording of the alga on buildings was available.

Trentepohlia aurea (Linnaeus) Martius (Syn.: *T. velutina* Kützing 1843; *T. maxima* Karsten; *T. germanica* Glück)

The alga appeared in felt-like layers of orange-red to golden-yellow color, with prostrate and erect branched filaments. The cell wall was stratified; end cells of filaments had cellulose caps, sometimes shifted aside. Zoosporangia were mainly single, seldom two together, ovoid, with bended basal cell. Cells were about 15–25 μm broad and 30–40 (50) μm long.

In our study, *Trentepohlia aurea* was identified once together with Cyanoprokaryota on a rough concrete wall as a dominant cover. The species is common and distributed in cold, temperate and tropical regions. It is the most common but very variable species in Central Europe with many described forms (Printz 1939; Ettl and Gärtner 1995, 2014). Other records of the species on old concrete surfaces exist from Austria (Bachmann 1992, Bachmann and Gärtner 2002), Ireland (Rindi and Guiry 2002) and France (Rindi et al. 2003).

Actual data on ecophysiology of *T. aurea* in natural habitats are scarce. Howland (1929) presented observations on the water balance which should be checked. According to Renner (1932), the lowest relative air humidity for growth of *T. aurea* is 95%. Different cultivation experiments for extraction of carotenoids were conducted by Abe et al. (1998, 1999).

Trentepohlia iolithus (Linnaeus) Wallroth (Fig. 3.38) (Syn.: *T. koerberi* De Toni 1889; *Chroolepus iolithus* Agardh)

The species forms orange brown to grayish red, up to 2 mm thick coverages on the substrate, without a difference between prostrate and erect filaments. Filaments are irregularly branched; branches are arranged mainly parallel to main filaments. Cells have thick stratified walls (Ettl and Gärtner 1995, 2014). Our investigated isolate formed short-branched filaments which disintegrated in short pieces or single cells. Only once a suspending cell with zoosporangium was observed, which differentiated our material against related taxa as *T. odorata* (Wiggers) Wittrock and *T. umbrina* (Kützing) Bornet. Obviously, the reproduction runs by hypnoblasts (akinetes), on the substrate characteristic "rain stripes" were visible (Fig. 3.38). Growth in culture was very slow, and so far, no reproductive stages were observed.

There have been records of the species on building surfaces in Ireland, Great Britain and France (Rindi and Guiry 2002; Barberousse et al. 2006; Rindi 2007) which were on old substrates without specification of the age of the substrate. Hofbauer observed the taxon additionally on façades in Scotland, Madeira, Slovenia, Italy and Switzerland.

In nature, this taxon is mainly recorded on siliceous rocks but also on calcareous and dolomite substrates (Aboal et al. 2002). It is still unknown if the occurrence on different substrates represents two separate taxa or the wide ecological amplitude of one taxon.

The alga was found on three different surfaces with mineral coatings of "older" specimens and on a variety of "new" specimens exposed 3 years to outdoor conditions (Fig. 3.38). Hofbauer (2007) documented *T. iolithus* for the first time as a member of the initial colonization of new modern façades.

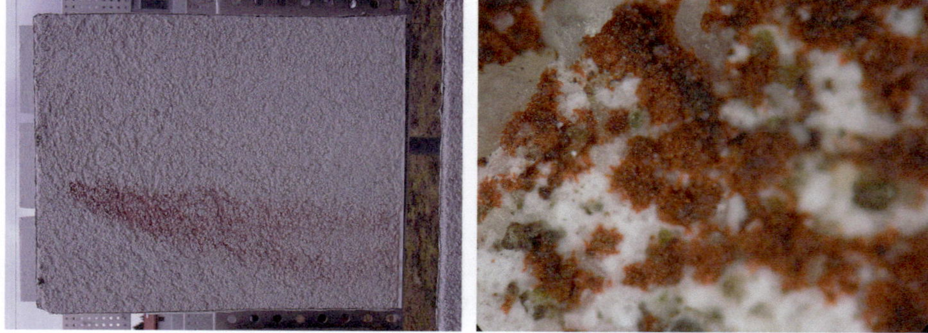

Fig. 3.38 *Trentepohlia iolithus,* growing on a new specimen (about 30 cm broad). Left: typical water stripes. Right: surface of the layer (ca.1 × 1 cm²). (Hofbauer et al. 2006; Hofbauer 2007)

Data on the ecophysiology of the species is still poor. A peculiar feature of the alga on building substrates is the enormous thickening of the cell wall which is less expressed in culture. The cell walls of the cells of main filaments show an external funnel-shaped stratified layer (Printz 1939: Ettl and Gärtner 1995) which could be helpful in the uptake of water into the thallus.

A characteristic feature of *T. iolithus* is the strong red color of the thallus produced by special types of carotenoids like β,ε-Carotin-2-ol (Buchecker et al. 1973). In moist conditions or after destruction of cells, the plant smells like violets (β-Ionone). According to literature, the species *Trentepohlia odorata* (Wiggers) Wittrock is widely distributed on buildings in South Asia (e.g., Wee and Lee 1980, 1982; Wee 1988; Loh 2002); however, this species can easily be misidentified with *T. iolithus* (which is frequently growing on building substrates in Europe) when reproductive stages are missing. Cultivation experiments have been conducted with the Asian taxon (e.g., Ho et al. 1983; Ong et al. 1992; Tan et al. 1993) though comparative taxonomical and physiological investigations still lack.

Order Gloeotilales Ettl et Gärtner nom. prov. (Protosiphonales sensu Kostikov p.p.)

Family **Gloeotilaceae** Ettl et Gärtner nom. prov. (Cylindrocapsaceae sensu Kostikov; incl. Cylindrocapsales sensu Reviers)

Genus *Geminella* Turpin

Cells are arranged in chains in a common homogenous mucilage (pseudofilaments) which easily dissolves, and the cells become single. Cells are ellipsoidal to cylindrical with one parietal chloroplast which fills half of the lumen, one pyrenoid. Reproduction is by fragmentation of filaments. Zoospores are unknown (Ettl and Gärtner 1995, 2014). The cosmopolitan genus mainly occurs in freshwater habitats; only the following species is also known in aerophytic habitats (Ettl and Gärtner 2014).

Geminella terricola J. B. Petersen (Fig. 3.39) (regarded as a synonym of: *Interfilum terricola* [J. B. Petersen] Mikhailyuk, Sluiman, Massalski, Mudimu, Demchenko, Friedl and Kondratyuk; Mikhailyuk et al. 2008)

Cells were embedded in thin mucilage, ellipsoidal to short cylindrical with rounded poles, about 5–10 μm long, 5–7 μm broad, arranged in singles or doublets (Fig. 3.39). The chloroplast was parietal, with lobed margin and a distinct pyrenoid.

Geminella terricola is known from soil, also from Antarctic areas (Broady 1982; Ettl and Gärtner 1995, 2014) and as a bark alga (Khaybullina et al. 2010). The species was also

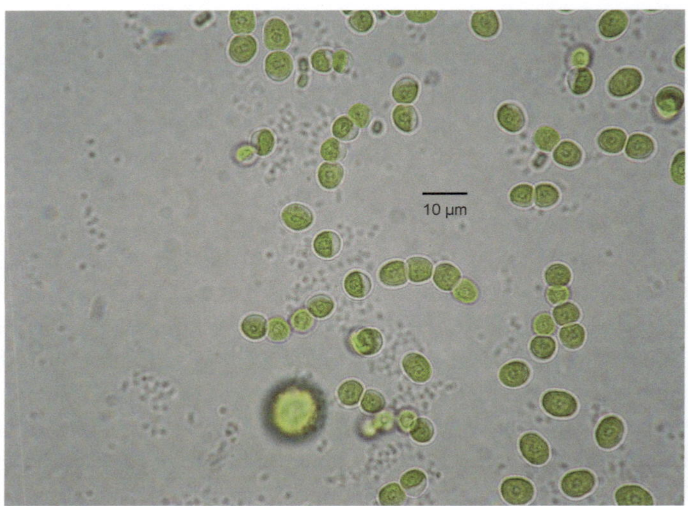

Fig. 3.39 *Geminella terricola,* in culture of a sedimentation sample, different pseudofilaments and cell doublets

found on historical buildings and monuments in Ukraine (Darienko and Hoffmann 2003) and was first documented in sedimentation and driving rain samples in Bavaria/Germany by Hofbauer (2007).

Class Trebouxiophyceae

Order Chlorellales Komárek et Fott sensu Kostikov (Chlorellales Komárek and Fott p.p.)

Family **Chlorellaceae** Brunnthaler sensu Kostikov (Chlorellaceae Brunnthaler p.p.)

A family which exclusively shows vegetative reproduction by autospores, without flagellated stages and without sexuality.

Genus *Chlorella* Beijerinck

The genus is cosmopolitan with many described species. It is one of the most common planktic and aeroterrestric algae, also occuring as photobiont in lichens and protozoa and some other organisms of microfauna (Tschermak-Woess 1948, 1988b; Douglas and Huss 1986; Kessler et al. 1988; Lewis and Muller-Parker 2004; Ettl and Gärtner 2014; Shubert and Gärtner 2015). Cells are uninucleate, globular, ellipsoid or ovoid with cup-shaped or plate-like parietal chloroplast with or without pyrenoid. Reproduction is asexual only by autospores. Since the first monographs, the heterogeneity of the genus was recognized (e.g., Shihira and Krauss 1965; Fott and Nováková 1969; Komárek and Fott 1983; see Ettl and Gärtner 1995). The genus was split by Kalina and Punčochářová (1987) with some new genera (see also Ettl and Gärtner 1995, 2014). Biochemical and molecular results (e.g., Friedl 1997; Krienitz and et al. 2004; Bock et al. 2011) showed that members of *Chlorella* belong to different classes (Chlorophyceae and Trebouxiophyceae), and further investigations on existing cultures and new isolates led again to new combinations and new genera (e.g., Darienko et al. 2010; further literature in Ettl and Gärtner 2014; Darienko and Pröschold 2019). Note that due to the difficulties in identification of *Chlorella*-isolates by light microscopy, we followed the taxonomy of Ettl and Gärtner (2014) and quoted the newest taxonomical and nomenclatural changes in brackets. Publications of the new names can be found in the list of references.

Members of the genus play an important role in the initial succession on façades, especially taxa of the group of *C. ellipsoidea* Gerneck s.l.

Chlorella ellipsoidea Gerneck s.l. (Syn.: *Chlorella saccharophila* [Krüger] Migula var. *ellipsoidea* [Gerneck] Fott et Nováková, *Chloroidium ellipsoideum* [Gerneck] Darienko, Gustafs, Mudimu, Rad Menendez, Schumann, Karsten, Friedl et Pröschold 2010)

Cells were ovoid to ellipsoidal, and the chloroplast was cup shaped with lobed margin, one pyrenoid. Reproduction was by 4–64 autospores of various sizes.

The species has been recorded in different forms on soil and bark in Europe and Japan and on historical buildings and monuments in Ukraine (Darienko and Hoffmann 2003). It is one of the common members of the first pioneer organisms on building surfaces.

Chlorella ellipsoidea s.l. shows a high tolerance against acidity and still grows at pH 2–3, with an optimum at 5–7 (Kessler and Kramer 1960; Kessler 1965, 1972). Salt resistance lies at about 4–6% NaCl concentration (maximum), which is very high compared with other taxa of the genus (Kessler 1974, 1992).

Different authors described forms of *Chlorella ellipsoidea* s.l. as separate taxa (e.g., *C. trebouxioides* Punčochářová, *C. angusto-ellipsoidea* Hanagata et Chihara). We tried to follow this concept in our study but sometimes overlapping morphological features occurred. In such situations, we used the term "*Chlorella ellipsoidea* Gerneck s.l.", resembling e.g., isolates from old specimens and from additional habitats. Kalina (1996) proposed for the taxonomic group around *C. ellipsoidea* (including also the genus *Viridiella* Albertano, Pollio et Taddei; Albertano et al. 1991) the new genus *Glaphyrella* (with the subfamily Glaphyrelloideae within Chlorellaceae), but this was so far not finalized according to the Code (nomen nudum).

Chlorella angusto-ellipsoidea Hanagata et Chihara, a morphospecies of **Chlorella ellipsoidea** s.l. (Figs. 3.40 and 3.41) (Syn.: *Chloroidium angusto-ellipsoideum* [Hanagata et Chihara] Darienko, Gustavs, Mudimu, Rad Menendez, Schumann, Karsten, Friedl et Pröschold 2010)

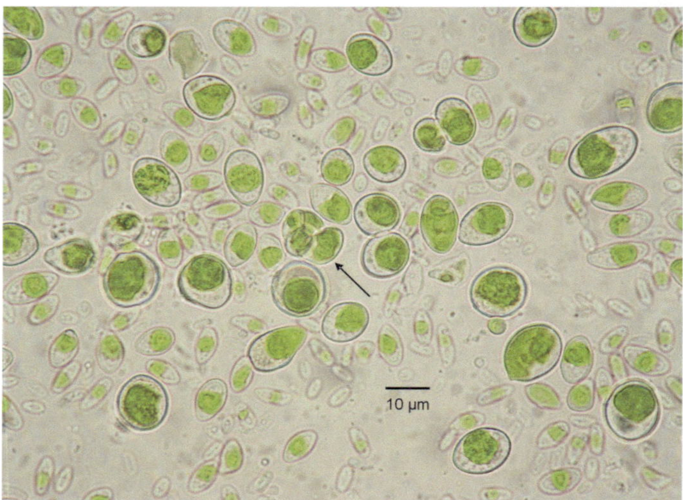

Fig. 3.40 *Chlorella angusto-ellipsoidea*, adult cells with central chloroplast, autospores of different size (arrow)

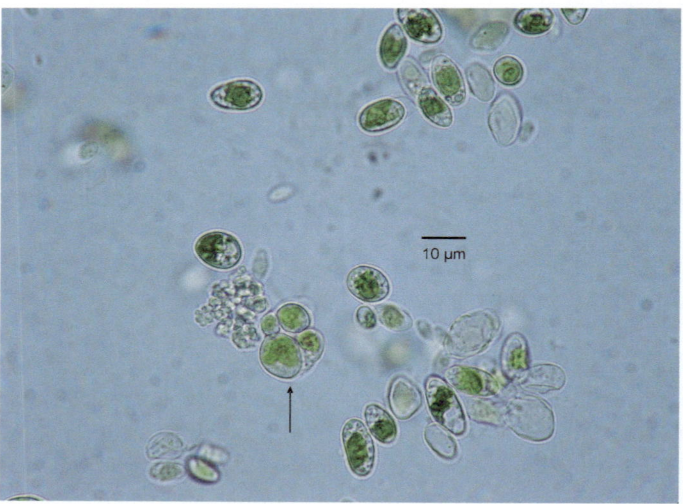

Fig. 3.41 *Chlorella angusto-ellipsoidea*, elongate cells and autosporangium with autospores of very different sizes (arrow)

Adult cells were ovoid, seldom globular and measured 15–18 μm, the chloroplast was mainly central and sometimes with incisions. Autospores (4–8–16) were relatively slim, all of the same size or one much bigger than the others (Figs. 3.40, 3.41). Colonies on agar were dark green and smooth. According to Hanagata et al. (1997), the taxon has autospores of nearly the same size, but contrary to their description, their figures showed autospores of obvious different size.

C. angusto-ellipsoidea was isolated from different specimens after 2 and 3 years' exposure in Finnentrop and Holzkirchen. No ecophysiological data are known. Until recently, this taxon was only described in aerophytic habitats. It is well adapted to changing temperatures and desiccation (Mikhailyuk et al. 2003; Darienko et al. 2010).

Chlorella trebouxioides Punčochářová, morphospecies of ***Chlorella ellipsoidea*** Gerneck s.l. (Fig. 3.42) (Syn.: *Chloroidium ellipsoideum* [Gerneck] Darienko, Gustavs, Mudimu, Rad Menendez, Schumann, Karsten, Friedl et Pröschold 2010pp.)

Cells were ellipsoidal without distinct vacuoles, up to 15 (17) μm long. The chloroplast in juvenile cells was parietal belt-like, median or cup-shaped, polar, with a distinct

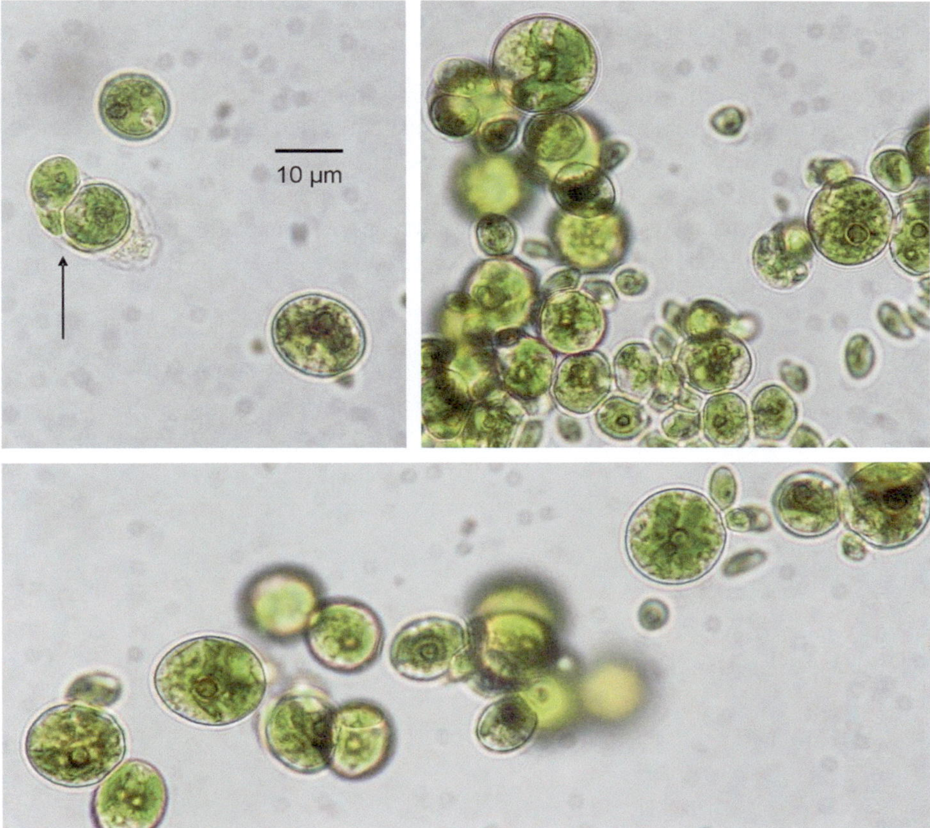

Fig. 3.42 *Chlorella trebouxioides*. Adult cells with central lobed to stellate chloroplast with pyrenoid (stained with IPI). Top left: autosporangium with autospores of different sizes (arrow)

pyrenoid. In adult cells, the chloroplast was more central, incised and irregularly lobed or stellate (Fig. 3.42), the pyrenoid had a thick starch layer with many starch grains. Reproduction was by 4–8–16 autospores of different size (Fig. 3.42). Autospores were ellipsoidal, slim to ovoid. Colonies on agar showed a dark-green color and were dry with a rough surface.

The taxon was observed on different specimens after 2 and 3 years' exposure in Bavaria and was also isolated from sedimentation samples. It is one of the common and early appearing pioneer algae on building surfaces.

As with *C. angusto-ellipsoidea,* no data about the ecophysiology and distribution exist (Mikhailyuk et al. 2003). A part of investigations on *Chlorella ellipsoidea* s.l. could possibly deal with *C. trebouxioides* though this has to be checked.

Chlorella luteoviridis Chodat (Fig. 3.43) (Syn.: *Chlorella aureoviridis* Meyer, *Ch. mutabilis* Shihira et Krauss, *Heterochlorella luteoviridis* Neustupa, Nemcová Eliás et Skaloud [see Neustupa et al. 2009], *Jaagichlorella luteoviridis* [Chodat] Darienko and Pröschold [see Darienko and Pröschold 2019])

Juvenile cells were ellipsoidal, and adult cells were globular with a size up to 10–12 (14) μm with vacuoles. The chloroplast was parietal cup like, often filling only half of the cell lumen, pyrenoid with starch granules. Autospores had different sizes.

In our study, the species was isolated from different specimens after 3 years' exposure in Germany. The alga is known on soil and in aerophytic habitats in Europe and Japan

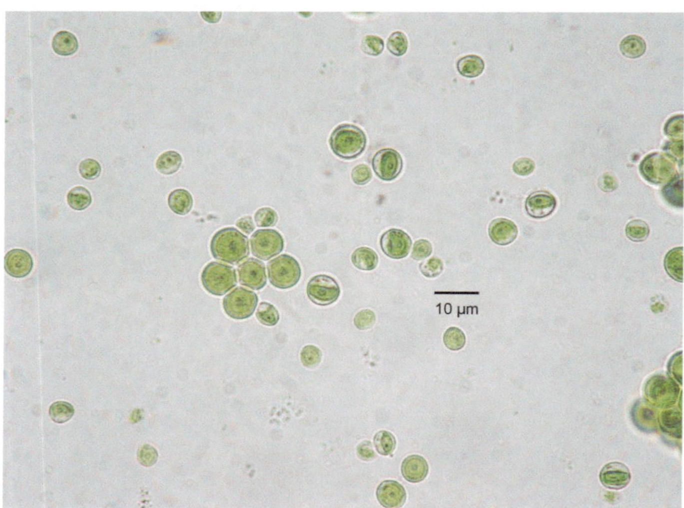

Fig. 3.43 *Chlorella luteoviridis*, chloroplast distant from the cell wall, with pyrenoid

(Ettl and Gärtner 1995; Hanagata et al. 1996) but has not been documented on building façades.

A high salt resistance of a maximum of 5% NaCl concentration (Kessler 1974, 1992) like in *Chlorella ellipsoidea* s.l. was found and a high resistance against acid (down to pH 3) (Kessler 1965). The upper temperature limit is about 28 °C (Kessler 1985). According to Häubner et al. (2006), the alga can grow in a temperature range from <1 to <35 °C.

Chlorella minutissima Fott et Nováková (Fig. 3.44) (Syn.: *Chlorella homosphaera* Skuja var. *minor* Kufferath, *Mychonastes homosphaera* [Skuja] Kalina et Punčochářová)

Cells of the observed form were very small and had a diameter of 3–4 μm, globular. The chloroplast was cup shaped, filling 2/3 of the cell, without a pyrenoid (Fig. 3.44). (Oil?) droplets accumulated in the cells. Two to four autospores of equal size were kept together by the gelatinous dissolving mother cell wall for a time. The genus *Mychonastes* was erected because of the fine network of ribs on the cell wall (not visible by light microscopy) by Simpson and van Valkenburg (1978). Krienitz et al. (2011).

In our study, the alga was isolated from specimens after 2 years' exposure in Holzkirchen and in an epilithic association with the zoologic garden Hellabrunn in Munich. The taxon is mainly known on soil (Ettl and Gärtner 1995) but was also found on historical buildings in Ukraine (Darienko and Hoffmann 2003). It has not been documented yet as a member of the flora on building façades in Middle Europe.

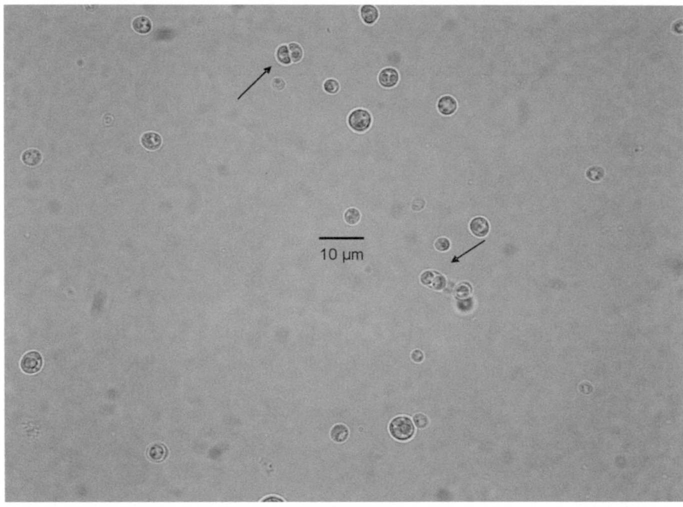

Fig. 3.44 Chlorella *minutissima*, vegetative cells and autosporangia with two and four daughter cells (arrows)

Isolates investigated by Kessler (1992) showed a lowest pH of 5.5, the highest tolerated salt concentration was 1% NaCl, and the highest temperature in vegetative stages was 32 °C.

Chlorella saccharophila (Krüger) Migula (Fig. 3.45) (Syn.: *Chloroidium saccharophilum* [Krüger] Darienko, Gustavs, Mudimu, Rad Menendez, Schumann, Karsten, Friedl et Pröschold 2010)

Cells were ellipsoidal with parietal chloroplast without incisions and a naked pyrenoid. Autospores were 4–8 (32) ellipsoidal to ovoid, all of the same size (Fig. 3.45). It was sometimes difficult to distinguish the species from *Chlorella ellipsoidea* s.l. (when pyrenoid not stained with IPI).

The alga was isolated from a specimen after 3 years' exposure in Holzkirchen and after 2 and 3 years' exposure in Ernsthofen/Oberramstadt. Additionally, it was observed in different habitats and in a sample of driving rain. Isolates are also known from marine water (Beuf et al. 2000).

The uppermost temperature limit tolerated by vegetative cells of *Chlorella saccharophila* was reported at 28 °C by Kessler (1985). The taxon is tolerant against acid and grows at pH 2–3 (Kessler 1965, 1967).

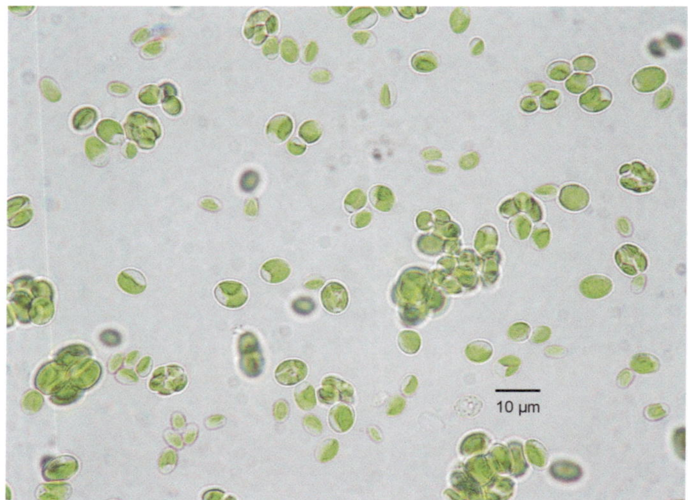

Fig. 3.45 *Chlorella saccharophila* isolated from a specimen after 3 years' exposure in Holzkirchen. Vegetative cells with cup-shaped chloroplast and autosporangia showing equal-sized autospores

Chlorella vulgaris Beijerinck (Fig. 3.46) (Syn.: *Chlorella terricola* Gollerbach)

The cells of our isolate were globular, up to 10 μm in diameter with thin cell wall and cup-shaped chloroplasts which had a big pyrenoid with starch grains (Fig. 3.46). Old cells developed a big central vacuole. Reproduction was usually by 4 (up to 16) autospores, slightly ellipsoidal and of the same size.

In our study, the alga was isolated from the surface of old specimens as well as new specimens after 2 and 3 years' exposure in Holzkirchen and Ernsthofen/Oberramstadt and also from many other habitats. In addition, *Chlorella vulgaris* was regularly observed in sedimentation and driving rain samples. It is a common aeroterrestric alga which often occurs on building substrates. It was also documented on historical buildings and monuments in Ukraine (Darienko and Hoffmann 2003).

Isolates of *C. vulgaris* are acid resistant between pH 3.5 and 4.5 and tolerate salt concentrations up to 3–4% NaCl. Temperature maximum lies between 28 and 32 °C (Kessler 1992), pH of substrates in nature range between 4.6 and 8.2 (John 1942). Apparently, the ecological amplitude of *Chlorella* vulgaris is very wide. The alga was also regularly isolated from limnic habitats (e.g., Ariyadej et al. 2004).

Family **Stichococcaceae** Kostikov

Genus ***Stichococcus*** Nägeli

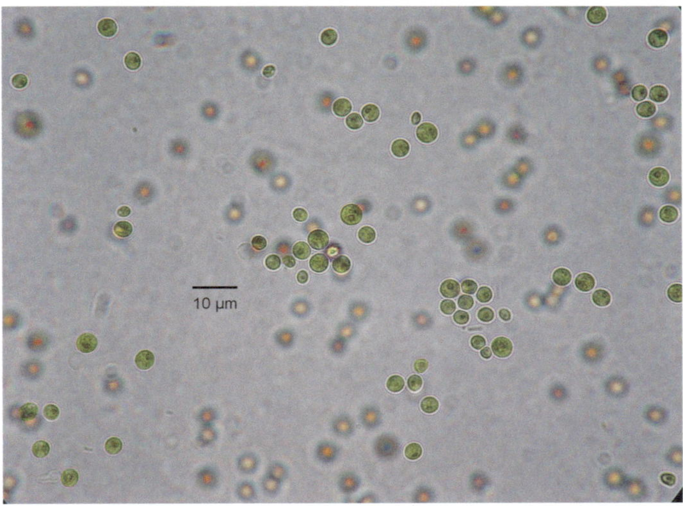

Fig. 3.46 *Chlorella vulgaris* isolated from a driving rain sample with relatively small cells in a young culture, with cup-shaped chloroplast and pyrenoid (starch sheath colored with IPI)

This genus occurs with single cells or pairs of cells or short but easily disintegrating filaments (short chains). Cells are more or less cylindrical to almost globular, sometimes slightly bended, with rounded free ends. A single parietal chloroplast which fills the cell only partially has a small (naked) pyrenoid or no pyrenoid. Vacuoles lie at the cell poles. Reproduction is by cell division; no flagellate stages are known. Diagnostic features like length of filaments and dissolving of filaments are not very valuable, because they depend on the temperature of the culture (e.g., Vischer 1927). Molecular investigations could help to clarify the taxonomy which is so far mainly based on morphological criteria (Henley et al. 2004). Recently, different edaphic strains were compared on a morphological basis (Fahrthofer 2004). According to molecular studies, the genus is included in the *Prasiola*-clade of Trebouxiophyceae (Leliaert et al. 2012). *Stichococcus* is rather diverse and contains some cosmopolitan forms that occur in very different habitats, including marine, freshwater and aerophytic (also with symbiotic relations). A new taxonomic concept is provided by Proeschold & Darienko (2020).

Stichococcus bacillaris (Nägeli) (Fig. 3.47) (Syn.: *S. minor* Nägeli; *S. membranaefaciens* Chodat; *S. dubius* Chodat; *S. chloranthus* Raths; *S. pallescens* Chodat; *S. coniocybes* Letellier; *S. nivalis* Chodat; *S. viridis* Nakano)

Cells were in average 2–3 times longer than broad (2–2.5 μm broad, 5–6 [8]) μm long). Old cells had vacuoles in the cell poles, and additionally, fat droplets occurred (Fig. 3.47). The pyrenoid was with IPI visible in the center of chloroplast.

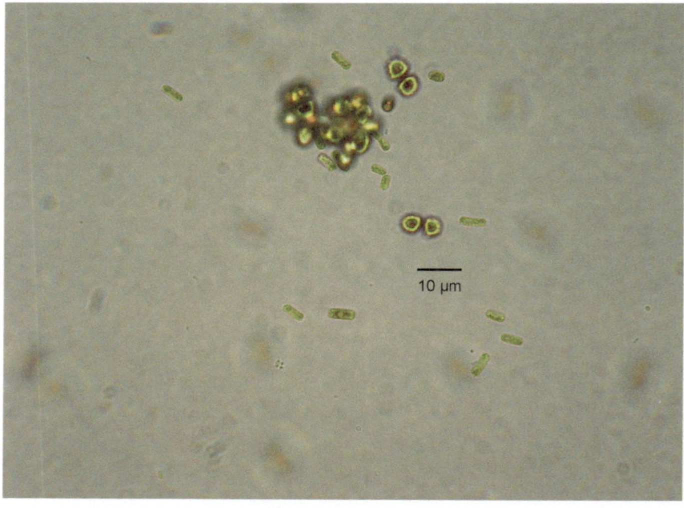

Fig. 3.47 *Stichococcus bacillaris* (cylindrical cells) and *Geminella terricola* (globular) in culture. Pyrenoids stained with IPI

In our study, *Stichococcus bacillaris* was regularly observed on surfaces of old and new specimens (after 2 and 3 years' exposure at all locations) and was further found on surface growth of buildings. The species was also detected in sedimentation samples and driving rain samples. The species has been documented on historical buildings and monuments (Ariño and Saiz- Jimenez 1996; Gorbushina et al. 2002; Kováčik 2000; Darienko and Hoffmann 2003) and was regularly observed in initial colonization of building façades but less often than *Stichococcus minutus* Grintzesco and Péterfi. It is one of the most common aeroterrestric alga in nutrient-rich habitats (Ettl and Gärtner 1995).

According to Edlich (1936), the minimum relative air humidity for the development of the alga is 90–100%, which coincides with investigations performed on *Stichococcus* sp. by Häubner et al. (2006) and Karsten et al. (2007). Substrate pH for soils varies from 4.6 to 8 as determined by John (1942) and for isolates from acid waters from pH 2.6– 3.1 (Gimmler 2001). *Stichococcus bacillaris* was also found in low light habitats as the Baradla cave near Aggtelek in Hungary together with Cyanoprokaryota and other algae (Claus 1955, 1958). Recently, the species was documented on man-made structures in Svalbard (Raabová et al. 2016).

Stichococcus chlorelloides Grintzesco et Péterfi

Cells were ellipsoidal to globular or slightly cylindrical with rounded cell poles. The chloroplast was parietal, open cylindrical, and a pyrenoid was not observed. Dimensions: 3.5–4 µm broad, 4–6 (10) µm long. The taxon is not very common.

S. chlorelloides was found only twice, once on the surface of a specimen exposed in Heggen/Finnentrop for 2 years and a second time on a specimen exposed in Holzkirchen for 3 years. The species was documented recently on man-made structures in Svalbard (Raabová et al. 2016). So far, ecophysiological data are missing.

Stichococcus exiguus Gerneck (Figs. 3.48 and 3.49 (Syn.: *Pseudostichococcus monallantoides* var. *exiguus* (Gerneck) Pröschold & Darienko)

Cells were ca. 2–3 µm broad and had different lengths up to 20–25 µm. Cells of juvenile cultures were slightly bended (Fig. 3.49). Sometimes, filaments of two cells of different length were observed. The chloroplast was parietal with an incision where the nucleus is situated. The pyrenoid was visible with IPI, though sometimes visible without staining (Fig. 3.48). Adult and old cells had vacuoles at the cell poles and oil droplets. In older cultures, cells developed more straight and were shorter which made discrimination to *Stichococcus bacillaris* difficult.

Mainly described on soil, the species was first recorded as a member of the colonization of façades and the aerophytic flora in Germany by Hofbauer (2007). *Stichococcus*

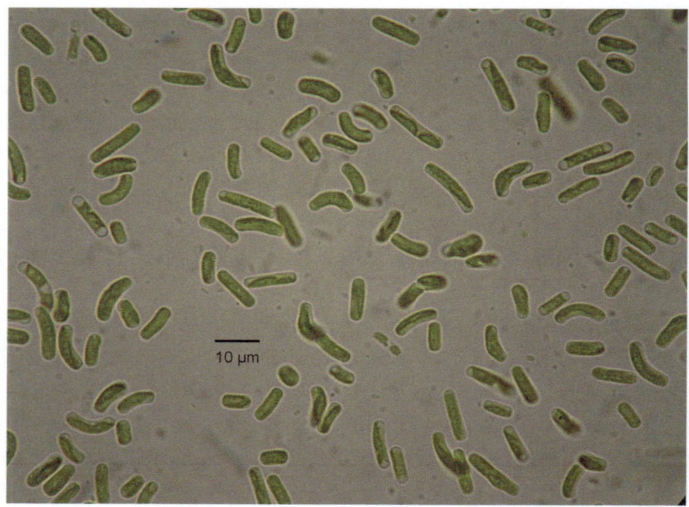

Fig. 3.48 *Stichococcus exiguus* with bended cells and parietal chloroplast

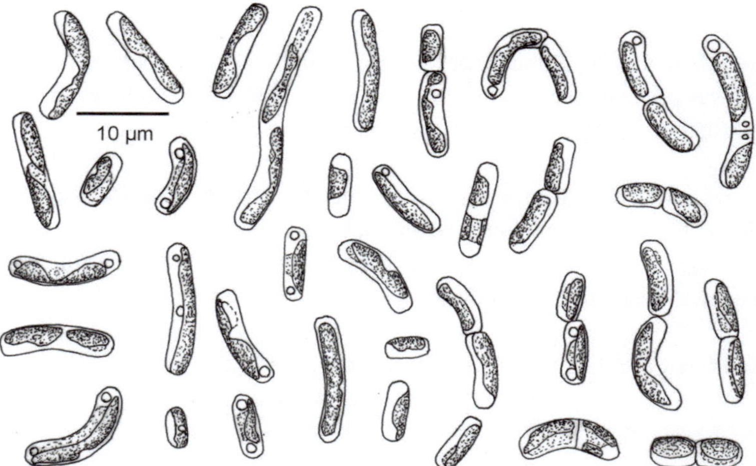

Fig. 3.49 Single cells and two-celled filaments of *Stichococcus exiguus*. Right bottom: pyrenoids depicted

exiguus seems to be uncommon and was isolated from the surface of a specimen exposed in Holzkirchen for 2 years, as well as from an additional sample and from driving rain.

Stichococcus minutus Grintzesco et Péterfi (Figs. 3.50 and 3.51)

Cells were short and barrel shaped with rounded poles (2–3 μm broad and 4–6–8 μm long), the chloroplast was small cup shaped and often at one cell pole. In the center of the chloroplast, a bright structure was visible which obviously represents the naked pyrenoid (Figs. 3.50 and 3.51). Additionally, vacuoles and oil droplets were detected in the cells.

In our study, *Stichococcus minutus* was observed on the surfaces of new specimens in all locations after 2 and also 3 years' exposure. In addition, the alga was found in many of the additional habitats and in a sample of driving rain. The taxon is one of the common pioneers (like *Diplosphaera* sp.) and is also found in older surface communities. The alga is known from different aerial habitats and distributed widely (Ettl and Gärtner 1995, 2014). The species was reported on building surfaces in Europe and recently on man-made structures in Svalbard (Raabová et al. 2016).

Order Choricystidales Kostikov
Family **Choricystidaceae** Kostikov

Genus *Choricystis* (Skuja) Fott 1976

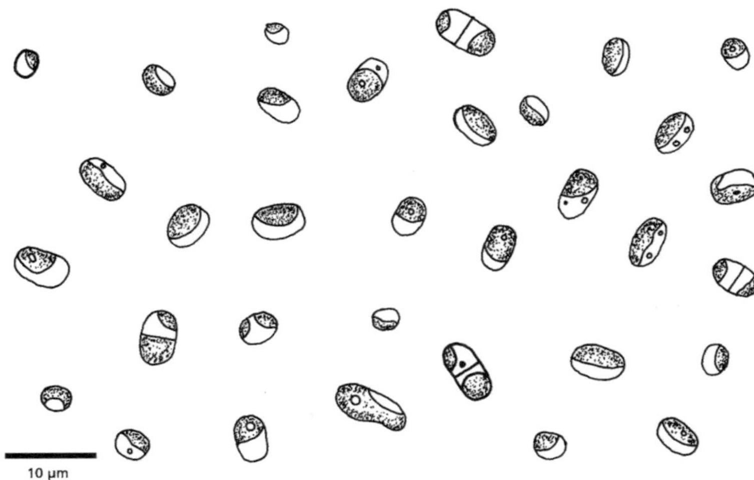

Fig. 3.50 Cells of *Stichococcus minutus* with typical short rounded cylindrical to globular cells and division stages. Pyrenoids are small and not visible in all cells. The taxon belongs to the smallest eukaryotic aerophytic algae

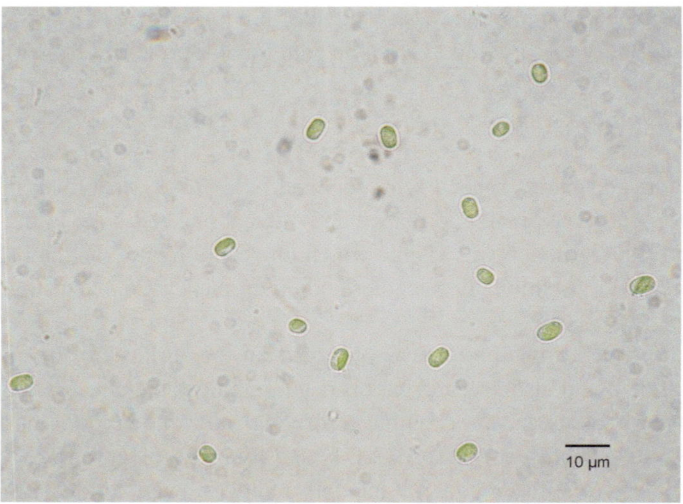

Fig. 3.51 *Stichococcus minutus* with typically formed vegetative cells

Cells are single or in pairs and have an ellipsoidal, straight or curved shape. The cell wall is thin, and no colloid is produced. The chloroplast is parietal, without pyrenoid. Reproduction is with two or four autospores, and cell division is oblique. The genus is widely distributed, occurring mainly in freshwater. Two species are also known from aerophytic habitats (Ettl and Gärtner 2014).

Choricystis minor (Skuja) Fott var. ***gallica*** (Bourelly) Komárek (Figs. 3.52 and 3.53)

Cells were small ellipsoid-ovoid, mainly asymmetrical shaped, often with one cell pole thicker than the other. Dimensions: 1–2–5 μm broad, 5–8–10 μm long. The chloroplast was parietal, covering half of the cell, containing starch granules (IPI staining) but no pyrenoid. Reproduction was by characteristic autosporulation (Figs. 3.52 and 3.53) with oblique protoplast division. This resembled the genus *Pseudococcomyxa* Korshikov (with mainly four autospores and a gelatinous cap on one cell pole). In culture, *C. minor* var. *gallica* was much more sensitive to drought than, e.g., *Chlorella* spp.

So far, *C. minor* var. *gallica* was mainly observed on bark and soil in Europe and Japan (Ettl and Gärtner 1995). Barberrousse et al. (2006) reported the alga on surfaces of buildings without referring to the age of the substrate. For the first time, it was documented as part of initial succession on modern building surfaces and as component of airborne algal flora and in rainwater by Hofbauer (2007). *Choricystis minor* var. *gallica* was regularly found on surfaces of various specimens in Heggen/Finnentrop after 2 and 3 years' exposure, although infrequently in Ernsthofen/Oberramstadt and rarely in Holzkirchen (only

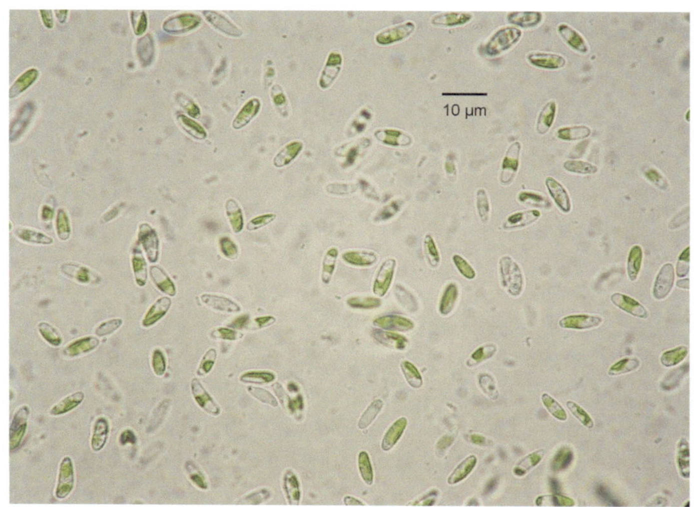

Fig. 3.52 **Choricystis** *minor* var. *gallica* isolated from the surface of a specimen, exposed in Heggen (Sauerland) for 3 years

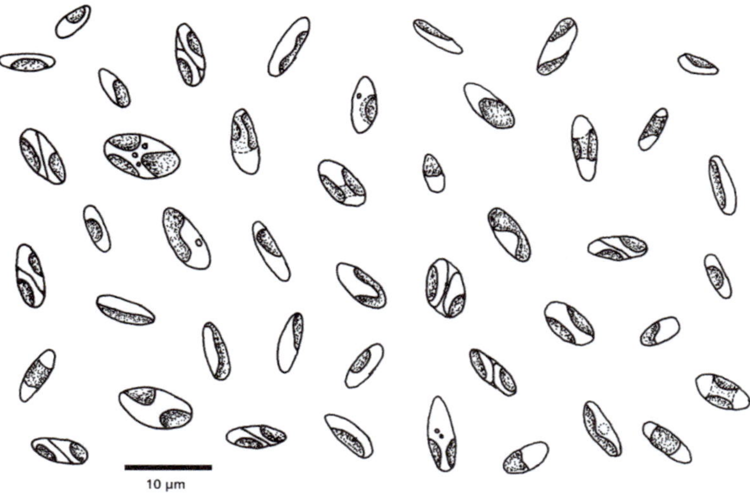

Fig. 3.53 *Choricystis minor* var. *gallica*, some cells with division stages

once after 3 years' exposure). In sedimentation probes and samples of driving rain in Holzkirchen, the taxon was regularly part of background germ load.

Family **Radiococcaceae** Fott ex Komárek em. Kostikov, Darienko, Lukešová et Hoffmann (Kostikov et al. 2002)

The taxonomy of the family follows Kostikov et al. (2001). A new concept of the group with further genera included was given by the same authors (Kostikov et al. 2002). According to molecular studies (Wolf et al. 2003), some taxa are related with other families, so the clear status of Radiococcaceae needs further investigations.

Genus *Coenochloris* Koršikov

This genus forms colonies with stratified or homogenous mucilage; cells are ellipsoidal to ellipsoidal-elongate with one parietal chloroplast with pyrenoid. Reproduction is with 4 to 8 (16) autospores. The autosporangial wall is ruptured in two to several pieces, embedded in the mucilage (Kostikov et al. 2002). It is a cosmopolitan genus occurring in freshwater with only few (ca. 3) forms also known from soil and aerophytic habitats mainly from the southern hemisphere (Ettl and Gärtner 2014).

Coenochloris sp.

A form of *Coenochloris* was found on the surface of a special type of building stone specimen at the outdoor area in Holzkirchen (a so-called Asterix), the clonal culture could not be kept and therefore no further identification was possible.

Genus *Gloeocystis* Nägeli

Cells have mucilaginous sheaths and are arranged single or in groups in a common, more or less colorless, sometimes stratified mucilage. Cells are globular to ellipsoidal and have a parietal chloroplast with one pyrenoid. Reproduction is by 2–8 (16) autospores. The genus is cosmopolitan with at least three aerophytic species, also phycobiont in lichen symbioses (Ettl and Gärtner 2014).

Gloeocystis cf. *polydermatica* (Kützing) Hindák (Fig. 3.54) (Syn. *Sporotetras polydermatica* [Butcher] Kostikov, Darienko, Lukešová et Hoffmann)

This species formed convex colonies, with single cells or groups of four cells embedded in a more or less concentric stratified mucilage (Fig. 3.54). Cells were ellipsoidal, one side more convex. One parietal chloroplast filled half of the cell and had one pyrenoid, sometimes difficult to see, surrounded by small starch grains. Reproduction led typically to four autospores which were embedded in a common mucilage but soon developed their own mucilaginous layers.

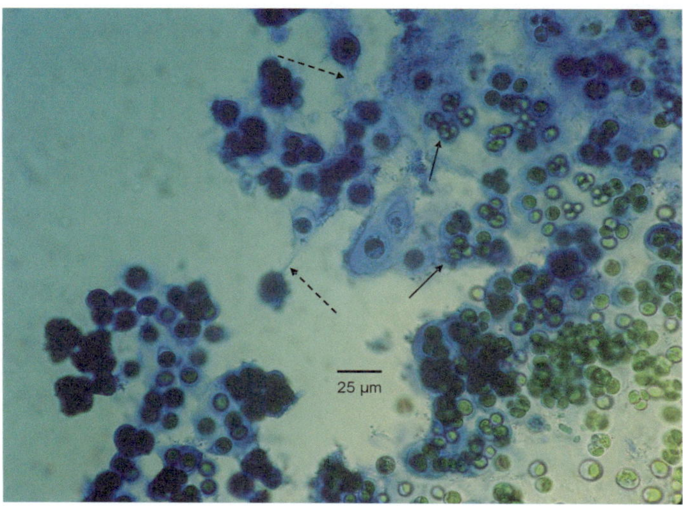

Fig. 3.54 *Gloeocystis polydermatica,* mucilage stained with methylene-blue

Gloeocystis polydermatica is a cosmopolitan taxon (Hindák 1978; Komárek and Fott 1983), living in aerophytic habitats but also on soil and as phycobiont (Ettl and Gärtner 1995). On building surfaces, it was first reported by Hofbauer (2007).

Genus *Radiococcus* Schmidle

The genus forms gelatinous colonies like *Gloeocystis*, but the autosporangial walls remain in pieces within the mucilage. Adult cells are globular, with one chloroplast and one (sometimes several) pyrenoid(s). Reproduction is by 4 to 16 autospores (Kostikov et al. 2002). The cosmopolitan genus is mainly known from freshwater habitats with the exception of the following species (Ettl and Gärtner 2014).

Radiococcus papuanus (S. Watanabe) Kostikov, Darienko, Lukešová et Hoffmann (Fig. 3.55) (Syn.: *Palmogloea papuana* S.Watanabe; *Gloeocystis papuana* [S.Watanabe] Ettl et Gärtner)

Single cells sat in a gelatinous matrix, and cell groups were also surrounded by unstratified mucilage. Juvenile cells were ellipsoidal. The chloroplast was parietal, cup shaped with a small pyrenoid. Also, two pyrenoids were observed (Fig. 3.55). Adult cells measured about 10 µm in diameter (Ettl and Gärtner 2014). So far, reproduction by four autospores was observed, and remnants of autosporangial walls were found in the mucilage.

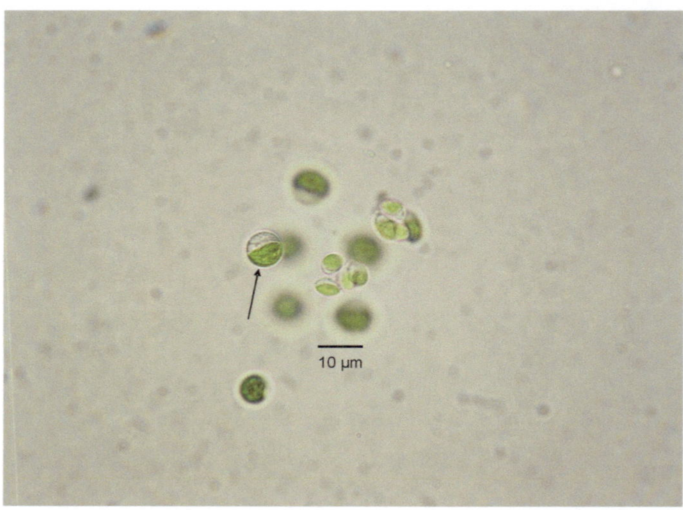

Fig. 3.55 *Radiococcus papuana*, single cells and autospores (4). Adult cell with 2 pyrenoids (arrow)

Radiococcus papuana was described on soils in Papua-New Guinea (Watanabe 1979). The taxon was recorded for the first time in a driving rain sample in Germany by Hofbauer (2007) and an occurrence on building surfaces is anticipated.

Order Microthamniales sensu Kostikov
Family **Leptosiraceae** Kostikov

Genus *Leptosira* Borzi

The genus forms cushion-like thalli consisting of prostrate and erect uniseriate filaments, which are irregularly branched, with short side branches. Cells are irregularly shaped with a parietal chloroplast with one or several pyrenoid(s). Reproduction is by biflagellate zoospores, aplanospores and isogametes. The difficult cosmopolitan genus occurs mainly terrestrial, aerophytic and phycobiontic (Ettl and Gärtner 2014).

Leptosira terricola (Bristol) Printz (Fig. 3.56) (Syn.: *Gongrosira terricola* Bristol, *G. australis* Phillipson, *Pleurastrum terricola* [Bristol] D. M. John)

Thalli (cushions) were formed by prostrate and erect filaments, the cells of the former being more globular, of the latter short cylindrical; the end cells were rounded. Sometimes, almost sarcinoid cell groups were found. Cells were 6–10 µm broad and up to 18 µm long and had one parietal chloroplast with one to two pyrenoid(s). Zoosporangia were big, up to 18 µm broad, globular to ovoid or irregular. Zoospores were 7–10 µm long (Fig. 3.56).

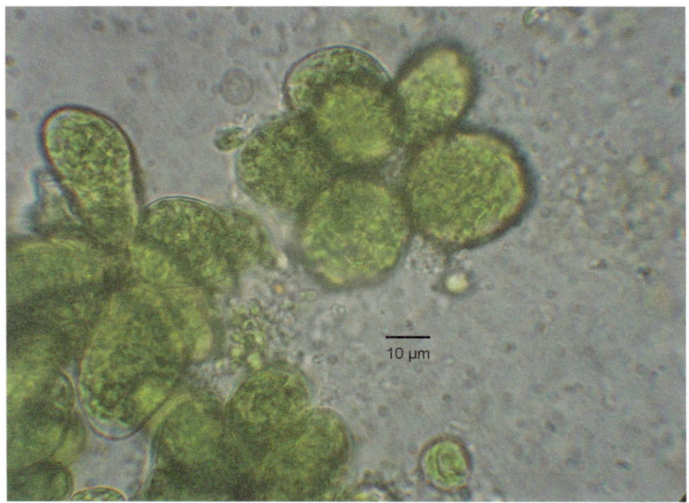

Fig. 3.56 *Leptosira terricola*, sporangia

In our study, *Leptosira terricola* was observed on the surface of specimens exposed in Heggen/Finnentrop for 2 and 3 years and in different samples of the territory of Germany. *L. terricola* is known on soil (Bristol 1920; Phillipson 1935; James 1935; Lund 1947; Printz 1964; Starmach 1972; Broady 1979; Ettl and Gärtner 1995). An occurrence on the surface of building materials was first reported by Hofbauer et al. (2003).

Order Prasiolales Chadefaud

A more or less isolated order within the chlorophyceae, which after moleculargenetic studies is now positioned within the class Trebouxiophyceae (Friedl and O´Kelly 2002; Rindi et al. 2007; Leliaert et al. 2012).

Family **Prasiolaceae** Vischer

Genus *Prasiolopsis* Vischer

The monotypic aerophytic genus forms branched filaments which are uniseriate in juvenile condition, later pluriseriate with thick cell packets. Cells have a central asteroid chloroplast with pyrenoid. End cells of the filaments are elongate, and basal cells are more or less isodiametric. The form is known from Europe, Asia and South America (Brazil).

Prasiolopsis ramosa Vischer (Fig. 3.57) (Syn.: *Pleurococcus vulgaris* Meneghini sensu Chodat, *P. vulgaris* Meneghini emend. Brand)

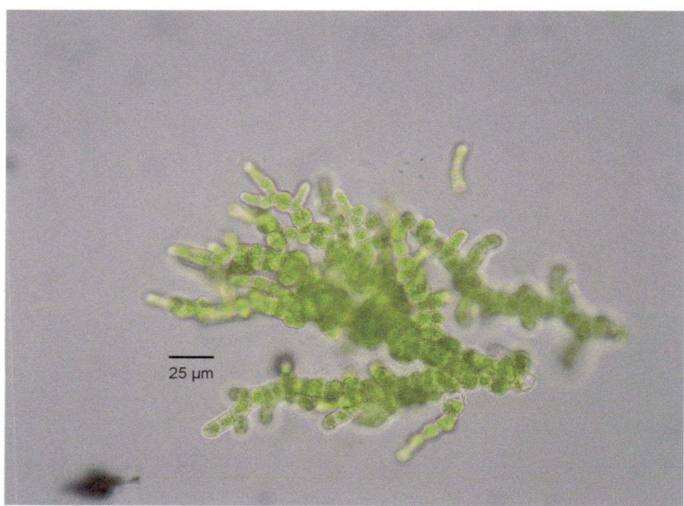

Fig. 3.57 *Prasiolopsis ramosa*, thallus with elongated end cells and isodiametric basal cells with asteroid chloroplast. (Hofbauer 2007)

The genus is monotypic; therefore, the description of the genus matches also entirely the species. Fig. 3.57 shows the typical morphology.

According to literature, (e.g., Kornmann and Sahling 1974) *Prasiolopsis ramosa* is widely distributed in Europe though only few cultures exist. We found the species on external walls of a church in Innsbruck, a habitat which was obviously exposed to environmental factors for a long time. Although it is expected to be a member of the pioneer flora of modern surfaces on façades, the taxon has not yet been documented as such.

Genus *Prasiola* Agardh

The juvenile algae form uniseriate filaments, soon developing into multiseriate band- to fan-like thalli. Cells have an axial asteroid chloroplast with central pyrenoid. Vegetative reproduction is by disintegration of thalli and aplanospores. Sexual reproduction is oogamous, but unknown in terrestrial and aerophytic forms. The first monograph by Knebel (1935) still presents a fundamental work on this genus. Molecular studies of different *Prasiola*-clones were conducted by Rindi et al. (2007) although the differentiation of species within *Prasiola* is somewhat unclear. The cosmopolitan genus includes marine, freshwater, terrestric and aerophytic forms.

Prasiola crispa (Lightfoot) Kützing (Fig. 3.58) (Syn.: *Schizogonium crispum* Gay, *Gayella polyrhiza* Rosenvinge)

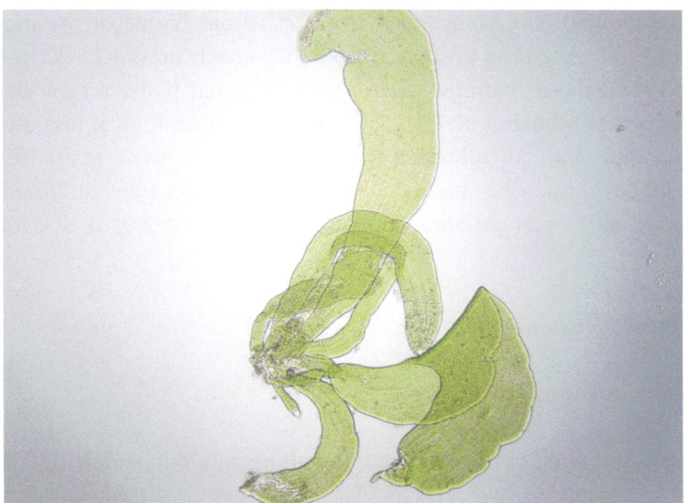

Fig. 3.58 *Prasiola crispa*, several thalli arise from one basal part, there the cell walls are thicker (stabilization). Natural size is ca. 2 mm

Prasiola crispa produced band- to blade-like thalli with one thick cell layer (Fig. 3.58), which grew up to several mm length. Single cells were rectangular to polygonal with a central asteroid chloroplast containing a big pyrenoid.

P. crispa was collected from wall corners in Kufstein/Austria. As the organism prefers moist situations with high input of organic nitrogen, it is common on places. e.g., impregnated with dog urine like the base of tree trunks. Thus, the alga is not a regular constituent of the flora of modern building surfaces but develops quite well and more regularly on older surfaces, if conditions are favorable. According to literature, habitats of the alga include walls and other anthropogenic substrates (Schmidle 1895; Grabherr 1941; Neustupa 1998; Rindi et al. 1999, 2004). The recorded habitats in literature are mostly characterized by high organic nitrogen and regular water supply, often in shaded situations. Recently, the species was reported from man-made structures in Svalbard (Raabová et al. 2016).

P. crispa is distributed worldwide and is free-living and known as phycobiont (Broady 1989b; Izaguirre and Pizarro 1998; Kovačik and Pereira 2001). Apart from tree bases and building surfaces, it also likes riverbanks and coastal regions. Rindi (2005) reported *P. crispa* in the intertidal zone of the Mediterranean coast. According to Barkman (1969), the association *Prasioletum crispae* is typical for nitrogen rich habitats as may be found at bird rocks.

According to Schofield and Ahmadjian (1972) *P. crispa* is one of the most nitrophilic algae. Becker (1982) measured positive net photosynthesis until $-15\ °C$ in *P. crispa* (see also Vincent 1988); these findings should be reevaluated if the recent changes in taxonomy are accepted (Moniz et al. 2012). With its broad ecological amplitude, *P. crispa* tolerates desiccation and salt influence (Jacob et al. 1992; Bock et al. 1996). The alga accumulates mycosporine-like amino acids (MAA) which play an important role in resistance against ultraviolet radiation (Lud et al. 2001; Karsten et al. 2005a).

Order Trebouxiales Friedl
Family **Desmococcaceae** Kostikov

Genus *Apatococcus* Brand em. Geitler

The genus forms sarcinoid cell packets and in culture sometimes short filaments. In fresh material, also aggregates of two cells (diads) between cubical packets (tetrads) are common. The cell wall is smooth ore slightly rough, sometimes thickened in age. Juvenile cells have one parietal chloroplast, and in adult cells, the chloroplast has a more or less lobed margin or is divided in two parts without pyrenoid. Reproduction is by aplanospores or by dissolving of cell packets. Zoospores are also known (Gärtner and Ingolić 1989). Recently, the genus was confirmed as a phycobiont in the lichen *Fuscidea* (Zahradníková et al. 2017). The genus is known as cosmopolitan and strictly aerophytic.

Apatococcus lobatus (Chodat) J. P. Petersen (Fig. 3.59) (incl. *A. lobatus* f. *minor* Brand, *A. fuscideae* A. Beck et Zaharadníková; Syn.: *A. vulgaris* Brand; *Pleurococcus lobatus* Chodat; *Pleurastrum lobatum* Printz)

This alga formed sarcinoid aggregates of cells in diads or tetrads with sometimes thick cell walls (Fig. 3.59), and in culture, it occasionally formed filaments (Vinatzer 1975; Gärtner and Ingolić 1989). In one of our isolates, short-celled filaments were observed. The chloroplast was parietal, with lobed margin and often in two portions. Zoospores without stigma were previously observed (Ettl and Gärtner 1995), however not found in our isolates. The size of the cells was 7–14 (20) µm.

The results by Zahradníková et al. (2017) showed that several cryptic taxa may occur within *A. lobatus*. In the same work, also a new species, *A. fuscideae* A. Beck and Zahradníková, was described. The exclusive diagnostic morphologic character, a reticulate chloroplast, mentioned by Zahradníková et al. (2017) was already depicted by Vischer (1960) for *A. lobatus*. Therefore, genetic differences of the new taxon remain to be characterized. As the status of the new form is not clarified, we placed it within the synonymy of *A. lobatus*.

Fig. 3.59 *Apatococcus lobatus,* sarcinoid cell packages, cells with irregularly lobed chloroplast (old culture)

In our study, *Apatococcus lobatus* was observed on surfaces of old specimens, on different surfaces of new specimens after 3 years' outdoor exposure (in Heggen/Finnentrop and Ernsthofen/Oberramstadt), in many additional surface growth situations on buildings and in a sedimentation sample. Regarding outdoor exposure *A. lobatus* seldom belongs to the very first "pioneer algae," although on older surfaces it can be dominant. On surface coatings of buildings, a preference for mineral types is observed.

A. lobatus is one of the most common aerophytic algae and is cosmopolitan (Gärtner 1994; Sitte et al. 2002). The alga lives in high abundance on bark, wood, rock and soil (Gärtner 1974; Schwarz 1975; Vinatzer 1975; Shubert and Starks 1978; Starks and Shubert 1978; Cambra and Hernandez-Mariné 1998; Michailjuk 1999; Kostikov et al. 2001; John et al. 2011) and even on leafs of plants like spruce needles (Neustupa and Albrechtova 2003). It appears regularly on artificial substrates like walls and plaster (Ettl and Gärtner 1995; Darienko and Hoffmann 2003; Gärtner and Stoyneva 2003; Hofbauer et al. 2003), however not as prevalent as on natural surfaces.

A. lobatus is well adapted to aerophytic habitats. Its net photosynthesis runs from 68% relative humidity, has an optimum at 97–98% (Bertsch 1966) and does not need the influence of liquid water. Recently, it was shown that the alga also profits from mixotrophic conditions (Gustavs et al. 2016), which the authors consider a further adaption to aerophytic habitats.

A. lobatus crusts in nature are hydrophobic (Geitler 1942; Gärtner and Ingolić 1989). Among bark algae, *A. lobatus* is the dominant species in the association *Apatococcetum lobati* Gärtner and Ingolić (Gärtner and Ingolić 1989) which represents a vegetation community often named "*Pleurococcus Protococcus* community" or "*Pleurococcetum vulgare*" (e.g., Ellenberg 1986).

Genus ***Desmococcus*** Brand em. Vischer

Cells form cubic sarcinoid packages, in juvenile stages also with short uniseriate and branched filaments. The chloroplast is parietal, small, cup shaped with a small pyrenoid which is hardly visible in light microscopy even after staining with IPI (Gärtner and Ingolić 1989). Reproduction is by vegetative cell division. Aplanospores or zoospores have four flagella. Aplanosporangia have a *Trochiscia*-like verrucous surface. Zoosporangia are also ornamented or smooth (Broady and Ingerfeld 1993). Three species of the cosmopolitan genus are described to occur in aerophytic habitats, on bark, soil and rock (Ettl and Gärtner 2014).

Desmococcus olivaceus (Pers.) Laundon (Fig. 3.60) (Syn: *Pleurococcus vulgaris* Meneghini sensu Nägeli; *P. naegeli* Chodat; *Desmococcus vulgaris* Brand emend. Geitler; non *Pleurococcus vulgaris* Meneghini sensu Brand; *Protococcus viridis* Agardh)

The morphology of our isolates represented the type species (Gärtner and Ingolić 2000). In fresh juvenile cultures, short-branched filaments were distinct (Fig. 3.60), and in adult cultures, mainly cell packets or single cells were visible. Only small aplanosporangia

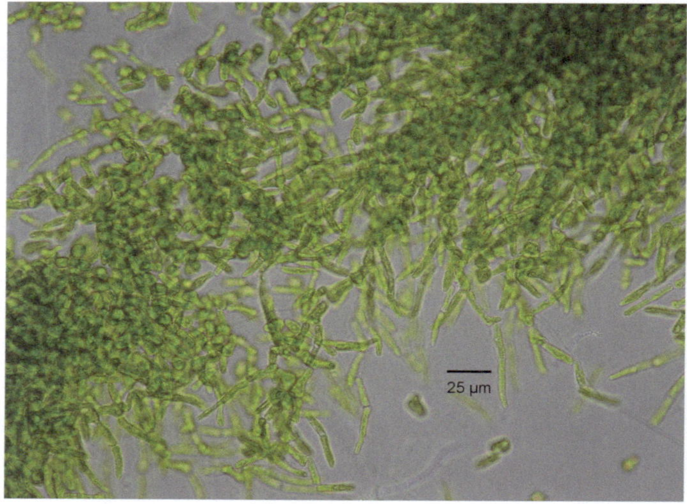

Fig. 3.60 *Desmococcus olivaceus*, juvenile culture with branched uniseriate filaments

with fine surface ornamentation were found, as described, e.g., by Vischer (1960). The small pyrenoid was well visible in the parietal, cup-shaped chloroplast, especially in cells of filamentous stages.

Desmococcus olivaceus is a common aerophytic alga which prefers habitats with relatively moist conditions. It has been described on soil, wet rock and bark (Geitler 1942; Vischer 1953, 1960; Printz 1964; Bourrelly 1966; Gärtner 1994). Søchting (1997) reported it from needles of *Picea abies* in Denmark and Poland. The species has previously been documented on historical buildings and monuments (Kováčik 2000; Darienko and Hoffmann 2003). In our study, *D. olivaceus* was observed from several samples.

Genus *Diplosphaera* Bialosuknia em. Vischer

Cells are mainly irregular globular or slightly elongated, and typical diads are formed after division, developing into sarcinoid aggregates which disintegrate easily into single cells. The chloroplast is parietal (similar to *Desmococcus*) with a pyrenoid not well visible in light microscopy. Reproduction is by disintegration of thalli and diads and by aplanospores. Aplanosporangia have a smooth wall, aplanospores are slightly asymmetrical-rounded to elongated. Zoospores are unknown (Ingolić and Gärtner 1989; Gärtner and Ingolić 1990; Ettl and Gärtner 1995). The presumably cosmopolitan genus is only desribed in aerophytic habitats and lichen symbioses.

Diplosphaera chodatii Bialosuknia em. Vischer (Fig. 3.61) (Syn.: *Stichococcus diplosphaera* Chodat, *S. chodatii* [Bialosuknia] Heering, *Protococcus chodatii* [Bialosuknia] R. et F. Chodat)

Colonies in culture consisted of many small celled packets which easily disintegrated. Masses of aplanospores and typical cell-diads were found in the marginal parts of colonies (Fig. 3.61). Juvenile aplanospores were slightly irregular-subsphaerical, 4 µm in diameter, adult cells measured up to 7 µm. Pyrenoid in adult cells was visible after staining with IPI.

Diplosphaera chodatii was first documented as an important part of the initial succession on modern façades by Hofbauer (2007). *D. chodatii* was investigated from the surface of different new specimens exposed in Holzkirchen and Heggen/Finnentrop over 2 and 3 years. Additionally, the species was found in different surface growth situations, in some sedimentation samples and in a sample of driving rain.

The taxon was originally described as a lichen phycobiont but was also found free living as an aerophytic organism (Ettl and Gärtner 1995). It is widespread and a typical partner in lichen symbiosis, especially in Verrucariaceae. A study by Fontaine et al. (2013) reported considerable genetic variation in symbiotic populations between different continents, Europe and North America.

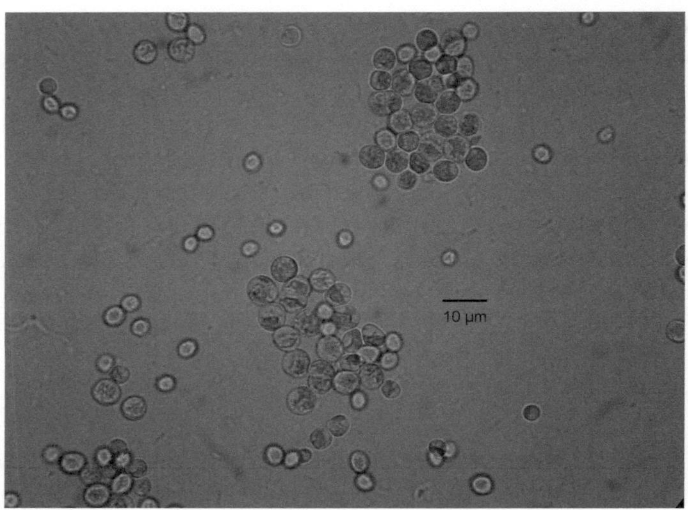

Fig. 3.61 *Diplosphaera chodatii*, diads, aplanospores and single cells of disintegrating sarcinoid aggregates

Mattox and Bold (1962) und Watanabe (1979) defined a bacilliform species as *Stichococcus chodatii* (Bial.) heering, which did not agree with the type. Kostikov et al. (2001) and Darienko and Hoffmann (2003) separated *D. chodatii* Bialosuknia emend. Vischer and *Stichococcus chodatii* (Bial.) Heering as independent taxa, but without a clear definition of these taxa. Both algae are recorded on soil and historical buildings and monuments in Ukraine (Kostikov et al. 2001; Darienko and Hoffmann 2003). The species was further documented on man-made structures in Svalbard (Raabová et al. 2016).

***Diplosphaera* sp.** (Fig. 3.62)

This form is one of the most common algae in our investigation on old and new samples of all outdoor locations after 2 and 3 years' exposure and also found in analyses of many additional habitats (including a sample of driving rain from Holzkirchen). This organism is not only one of the first pioneer algae but also regularly found on older surface situations. In the pioneer phase, it appears among *Chlorella* ssp. and *Stichococcus* ssp. The morphology resembles somewhat *D. chodatii,* but with incomplete features. The colonies were smooth with less mucilage, and mainly, globular to slightly irregular aplanospores (ca. 3 µm) were formed. These aplanospores were similar to *Stichococcus minutus* Grintzesco et Péterfí, however were more bulgy and never had the same regular cylindrical shape. In culture also, single cells and diads developed, and after some months of cultivation, additional packets of cells like in *D. chodatii* were observed (Fig. 3.62). This taxon showed quite stable morphological characteristics although further investigations are needed to clarify its position.

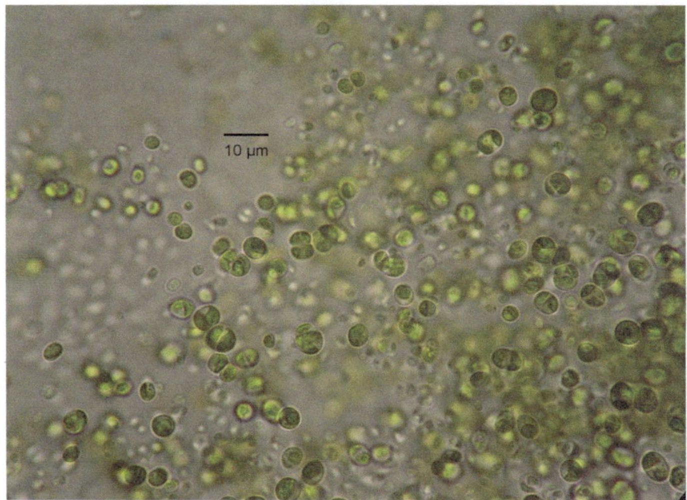

Fig. 3.62 *Diplosphaera* sp., with many aplanospores and cell packets of 3 months old culture

Family **Dictyochloropsidaceae** Kostikov (Myrmeciaceae Ettl et Gärtner p.p.)

Genus *Dictyochloropsis* Geitler em. Tschermak-Woess

This genus has relatively big uninucleate cells (up to 45–50 µm). The chloroplast is more or less parietal but divided in many articulate ridges, ribs and lobes, sometimes net like, without a pyrenoid. Reproduction is by autospores (only several per cell) and zoospores (many per cell) without stigma. Four species have been described, aerophytic and phycobiontic (Geitler 1966; Tschermak-Woess 1980a, 1984; Nakano and Isagi 1987; Škaloud et al. 2005). Formerly positioned among Chlorophyceae, the genus belongs according to molecular data to Trebouxiophyceae (Friedl 1995). All known species of the genus live aerophytic, terrestrial or phycobiontic (Ettl and Gärtner 2014).

Dictyochloropsis cf. *splendida* Geitler (Fig. 3.63)

Cells measured up to 35 µm. Reproduction was with autospores (ca. 6 µm). Found in a sedimentation sample (Fig. 3.63), it is not unlikely that this form also grows on building surfaces.

Family **Myrmeciaceae** Ettl et Gärtner em. Kostikov

Genus *Elliptochloris* Tschermak-Woess

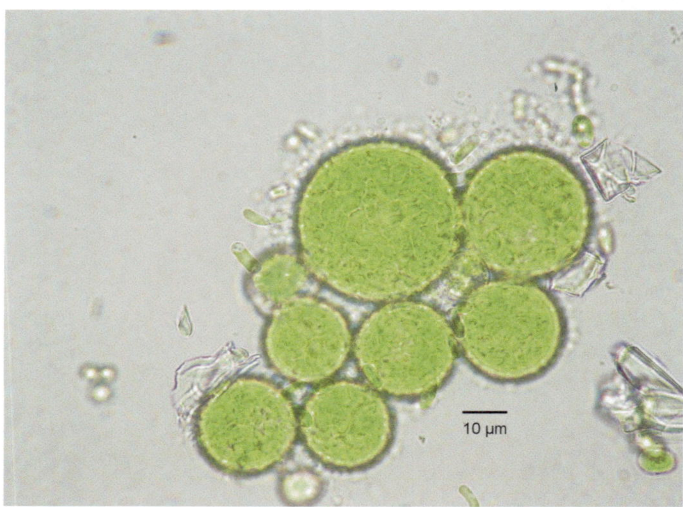

Fig. 3.63 Vegetative globular cells of *Dictyochloropsis* cf. *splendida* with reticulate chloroplast and singular cells of *Stichococcus bacillaris* (short cylindrical) from culture of a driving rain sample 5

Cells are singular, globular to elliptic, uninucleate. The chloroplast is often lobed and incised with or without pyrenoid. The genus is characterized by two different types of autospores (several [2–4] globular-ellipsoidal and many [up to 32] elongated, bacilliform). Five species have been described, but not all with cultures (Watanabe 1977; Tschermak-Woess 1980b; Hoffmann et al. 2007). All known species of the cosmopolitan genus were isolated from soil or aerophytic habitats.

Elliptochloris reniformis (S. Watanabe) Ettl et Gärtner (Figs. 3.64 and 3.65) (Basionym: *Chlorella reniformis* S. Watanabe; Syn: *Palmellococcus reniformis* [S. Watanabe] S. Watanabe)

Cells were globular (up to 15 μm in diameter) to ellipsoidal (6–10 μm broad, 11–14 μm long) with one parietal chloroplast. Juvenile chloroplasts were cup shaped; adult chloroplasts incised and lobed without pyrenoid. Starch grains were visible (with IPI). Reproduction was by two different types of autospores (as mentioned in the description of the genus) (Figs. 3.64 and 3.65).

Known as a member of the soil algal flora, this alga was also reported on gravestones in Ukraine (Darienko and Hoffmann 2003). It was first documented as part of initial flora successions on modern building façades by Hofbauer (2007). The alga was isolated from

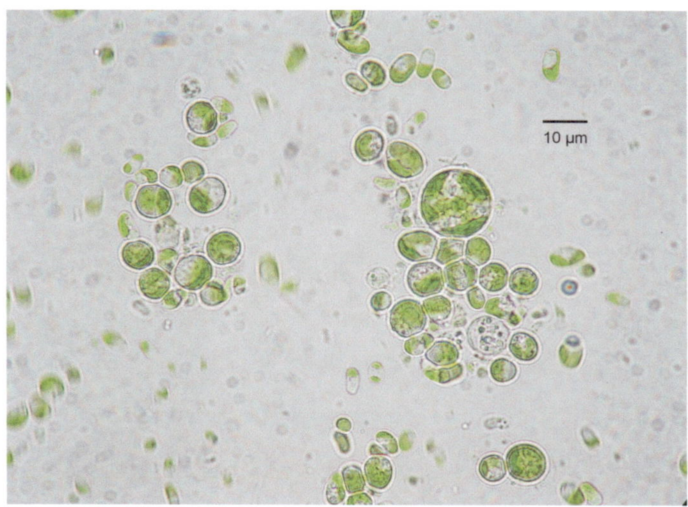

Fig. 3.64 *Elliptochloris reniformis* with different cell types. The biggest cell with one parietal, lobed chloroplast

specimens exposed in Heggen/Finnentrop for 3 years and also from driving rain experiments and sedimentation samples in Holzkirchen.

Elliptochloris subsphaerica (Reisigl) Ettl et Gärtner (Figs. 3.66 and 3.67) (Basionym: *Pseudochlorella subsphaerica* Reisigl; Syn: *Chlorella reisiglii* S. Watanabe)

In our isolates, cells were globular (up to 22 μm) or ellipsoidal (7–16 μm × 9–12 μm). The chloroplast was parietal and cup shaped, in adult cells often lobed with one pyrenoid (seldom two) which had a sheath of starch grains. Reproduction was by two different types of autospores (globular type and ellipsoidal-elongate to cylindrical or kidney-shaped type). The observed morphological characters coincide with *E. subsphaerica* (Figs. 3.66 and 3.67).

The species was observed on different soils and also isolated from surfaces of historical buildings and monuments in Ukraine (Darienko and Hoffmann 2003). Until documented by Hofbauer (2007), the alga was not known as pioneer organism on surfaces of modern buildings. The alga was isolated from specimen surfaces exposed in Holzkirchen for 3 years and from a driving rain sample.

Family **Trebouxiaceae** Friedl

Genus ***Trebouxia*** Puymaly

Fig. 3.65 *Elliptochloris reniformis*, different cell and autospore types (drawings of living material)

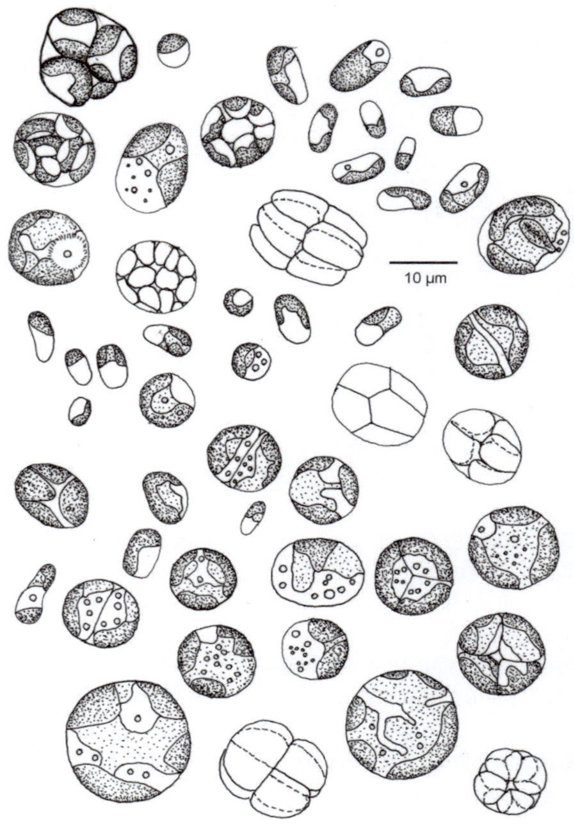

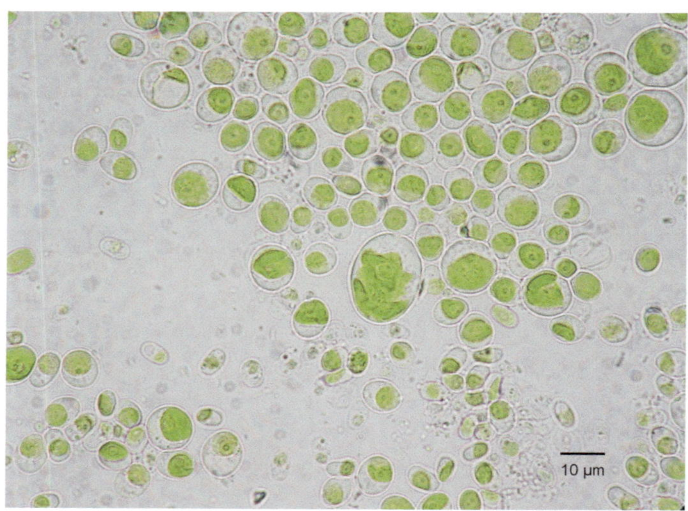

Fig. 3.66 *Elliptochloris subsphaerica*, chloroplasts with pyrenoid

Fig. 3.67 *Elliptochloris subsphaerica*, vegetative cells of different age and division stages with two types of autospores

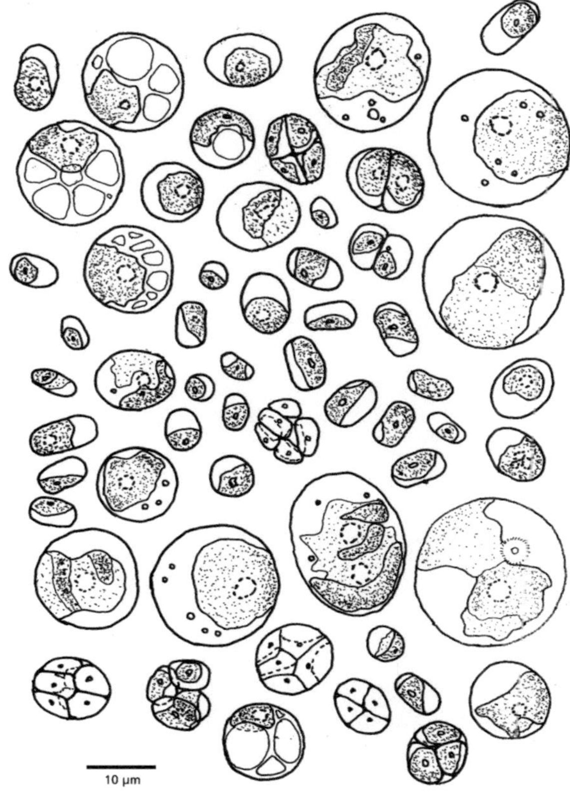

Cells are single or in groups, globular to slightly ellipsoidal or irregular, with a more or less central star-shaped and irregularly lobed chloroplast and one to several naked pyrenoid(s). Starch in granules is found within the chloroplast. The nucleus lies in a very typical sinus of the chloroplast, eccentric in the cell. Reproduction is by autospores and naked biflagellate zoospores with or without eyespot. Based on molecular genetic investigations (Škaloud and Peksa 2010) the genus *Asterochloris* (Tschermak-Woess 1980) was resurrected, and some new combinations were created. For further literature about the genus *Trebouxia,* see Ettl and Gärtner (2014). The cosmopolitan genus is predominantly known from lichen symbioses; only few taxa have been recorded as free-living in aerophytic habitats (on bark).

Trebouxia arboricola Puymaly (Fig. 3.68) (Syn.: *Cystococcus humicola* Nägeli sensu Treboux)

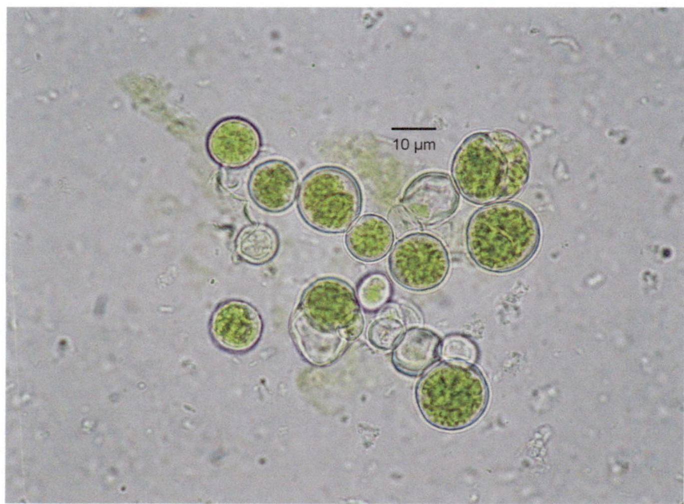

Fig. 3.68 *Trebouxia arboricola* of a mature culture. Cells with a massive chloroplast, its surface minor incised. The culture is contaminated by bacteria

Juvenile cells were kept together by the sporangial wall; adult cells measured ca. 12–14 µm. The chloroplast was compact and coarsly lobed, and the pyrenoid was angular. Zoospores were not observed in our culture.

Trebouxia arboricola was found once by investigation of an additional sample location (Fig. 3.68). The substrate was an ETICS surface several years old which was strongly stained green to orange green. *T. arboricola* dominated the association, lichens were not recorded, and the alga formed a "bloom." It was obviously free living in the investigated algal crust on ETICS; typical cells were found in the sample of nature without contact to fungal hyphae or lichen thalli. *Trebouxia arboricola* is one of the several free-living species but is also known from lichen symbiosis (Ettl and Gärtner 1995, 2014).

***Trebouxia* sp. (Fig. 3.69)**

In different surface samples, algae of *Trebouxia*-type were found but not determined at species level. An isolated alga of an ETICS surface from Nuremberg (Fig. 3.69) showed typical *Trebouxia* character: The cells were up to 20 µm in diameter, and the chloroplast was structured and formed many slightly clavately lobes. One central naked pyrenoid was observed in each chloroplast. Reproduction was by 8–16 autospores. Zoospores in culture were not observed. Only several isolates of this type grew in the cultures, and there was never a dominant appearance in the investigated crusts (preparations of nature). It could be possible that the alga developed from fragments of lichen thalli or from soredia, but on the surface of the investigated façade no lichens were detected.

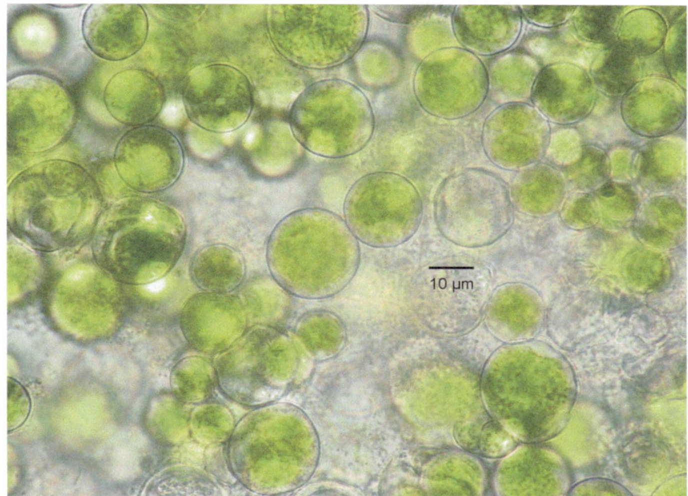

Fig. 3.69 *Trebouxia* sp., comparable big cells with lobed chloroplast

Streptophyta Divisions sensu lato

Division Streptophyta – Charophyta (Algae)

According to the actual scientific opinion, the division Streptophyta is separated from all other green algae (Chlorophyta) (Pickett-Heaps and Marchant 1972; Reviers 2003; Leliaert et al. 2012). The division Streptophyta includes algae (Charophyta) and land plants. Reviers (2003) used the term "Streptophycophyta" to characterize the algal forms included in the division.

Class Klebsormidiophyceae
Order Klebsormidiales Stewart et Mattox

The filamentous algae separate in several taxa easily into single cells. With exception of genus *Hormidiella* Iyengar and Kanthamma no special basal cells are formed, and in some cases, fixing mucilage is produced around the basal cells and at some points along the filaments (Lokhorst 1996). Asexual reproduction is by zoospores, vegetative cell division or disintegration of filaments. Zoospores are naked, biflagellate, and the flagella is inserted subapical and released through a wall pore of sporangium.

Family **Klebsormidiaceae**

Genus ***Klebsormidium*** Silva, Mattox et Blackwell

A worldwide distributed genus which is found in aeroterrestric but also freshwater habitats. Altough numerous studies were performed within the last 20 years, the taxonomy and nomenclature of about 15 described species is still in discussion, due to the morphological plasticity of isolates as well as the existence of cryptic species (Lokhorst 1996; Škaloud 2006: Rindi et al. 2008, 2011; Škaloud and Rindi 2013; Ryšánek et al. 2016). Even photosynthetic plasticity as studied on morphologically identic species of cultures by Karsten et al. (2017, and references therein) showed the difficulties in determination of field samples (ecotypes?). Reproduction runs in all species by fragmentation of filaments, autospores and zoospores (1 per cell, without eyespot). Sexual reproduction by isogamy is not well known. The genus nowadays becomes more and more important for biotechnology (see review by Stoyneva-Gärtner et al. 2019). The very diverse and cosmopolitan genus can be found in very different habitats, e.g., marine, freshwater, soil and aerophytic.

Klebsormidium flaccidum (Kützing) Silva, Mattox et Blackwell (Figs. 3.70 and 3.71) (Syn.: *Hormidium flaccidum* [Kützing] Braun; *Stichococcus flaccidus* [Kützing] Gay; *Chlorhormidium flaccidum* [Kützing] Fott; *Chlorhormidium flaccidum* var. *nitens* [Kützing] Farooqui)

The investigated strains formed long filaments on agar and in liquid cultures. On agar plates, the filaments lay parallel in curves. Cells were cylindrical with one parietal chloroplast which was an open cylinder and filled about the length of the cell and about 2/3 of the cell circumference (Figs. 3.70 and 3.71). Cells were about 6.5–7 μm broad and one to several times longer. The pyrenoid lay in the center of the chloroplast, surrounded by starch grains. Sometimes the pyrenoid was elongated, or even two pyrenoids were visible (before cell division?). The nucleus was located opposite the chloroplast in a plasmatic bridge. Reproduction was mainly by fragmentation of filaments or aplanospores. Even in liquid cultures zoospores were seldom developed. Sexual reproduction was not observed.

A very variable aerophytic alga which is cosmopolitan on different substrates (Ettl and Gärtner 1995; Kováčik 2000; Darienko and Hoffmann 2003) and was also found together with blue greens and other algae in the Baradla cave at Aggtelek/Hungary (Claus 1955, 1958).

Klebsormidium flaccidum was detected nearly on all old specimens, on additional samples and in analyses of driving rain samples although it is not among the very first colonizers of modern of façades. On surfaces which were at least 3 years old, the alga was regularly found as part of the microbial surface crust. Obviously, the alga likes nutrients and is common in places which are influenced by dripping water and bird excrements.

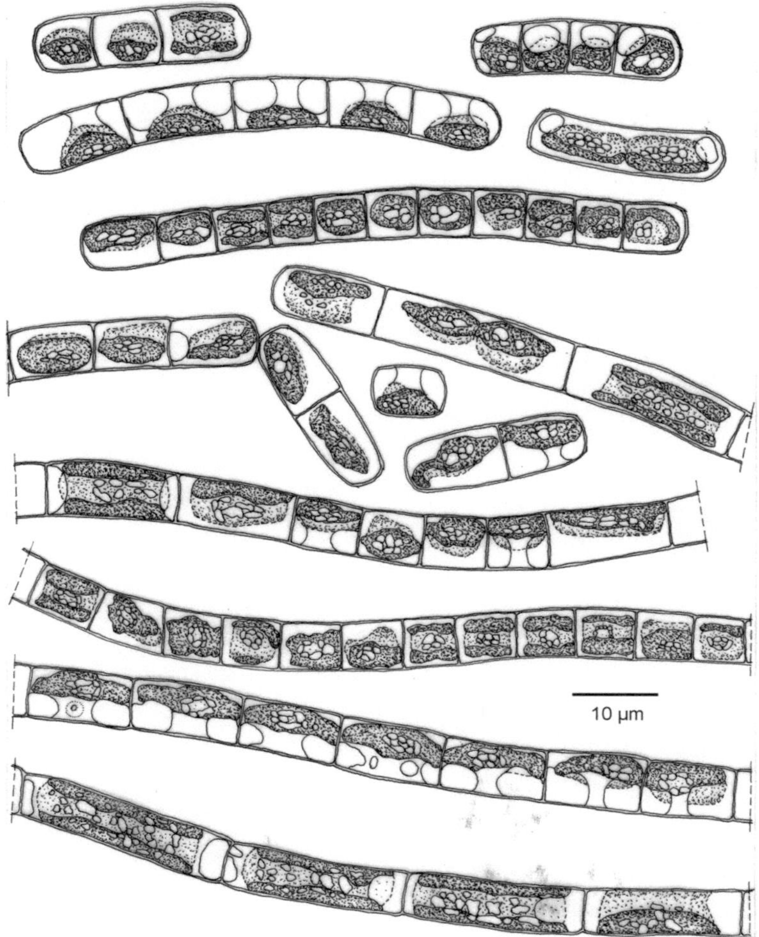

Fig. 3.70 Fragments and filaments of *Klebsormidium flaccidum*

During investigations of several strains, it was observed that the morphology of the cells and the tendency of the filaments to disintegrate were strongly dependent on culture conditions. Fragmentation was supported by a low N:P ratio (Dřímalová and Poulíčková 2003; Poulíčková et al. 2001). Kouřil et al. (2001) investigated a thermotolerant strain and found its expected maximum temperature for growth to be around 48.3 °C. Mattox (1971) showed by trials with several isolates that dried cultures after heating to 100 °C could be revived.

Klebsormidium sterile (Deason et Bold) Silva, Mattox et Blackwell (Figs. 3.72 and 3.73)

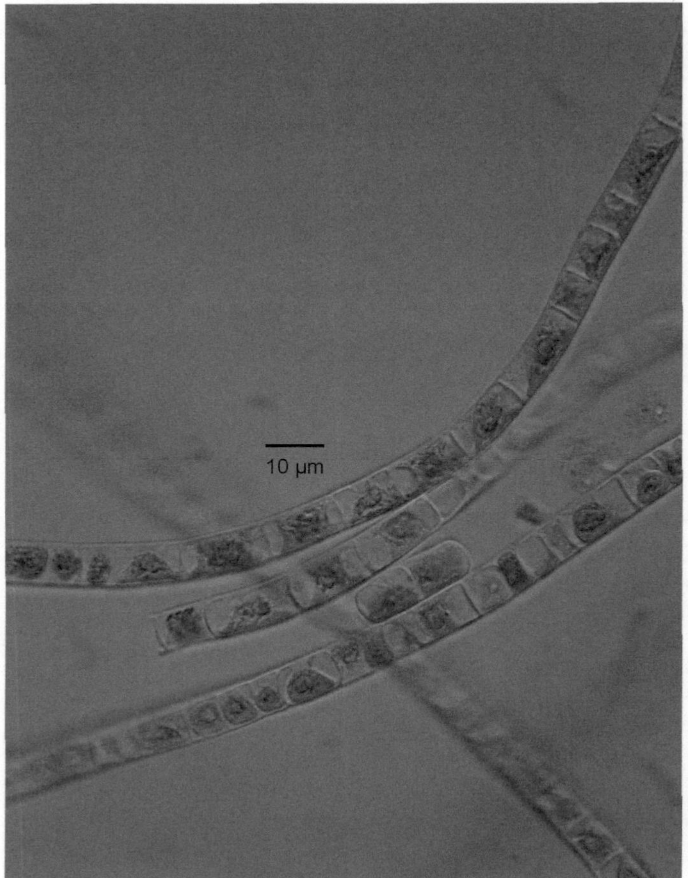

Fig. 3.71 Filaments of *Klebsormidium flaccidum*, pyrenoid stained with IPI

This alga developed long filaments in culture, which disintegrated in older cultures into short fragments and single cells. Cells were 1.5–3 × longer than broad (Mattox and Bold 1962). Cells were 4–5 μm broad and up to 15 μm long. The chloroplast was parietal like an open cylinder, covered in juvenile cells more than half of the cell length and was shorter in adult cells. The pyrenoid was distinct and surrounded with many small starch grains (Figs. 3.72 and 3.73).

Klebsormidium sterile is known from different soils (Deason and Bold 1960; Ettl and Gärtner 1995) and from historical buildings and monuments in Ukraine (Darienko and Hoffmann 2003). It was documented on surfaces of modern façades for the first time by Hofbauer (2007). The taxon was found in several samples of the additional surface situations.

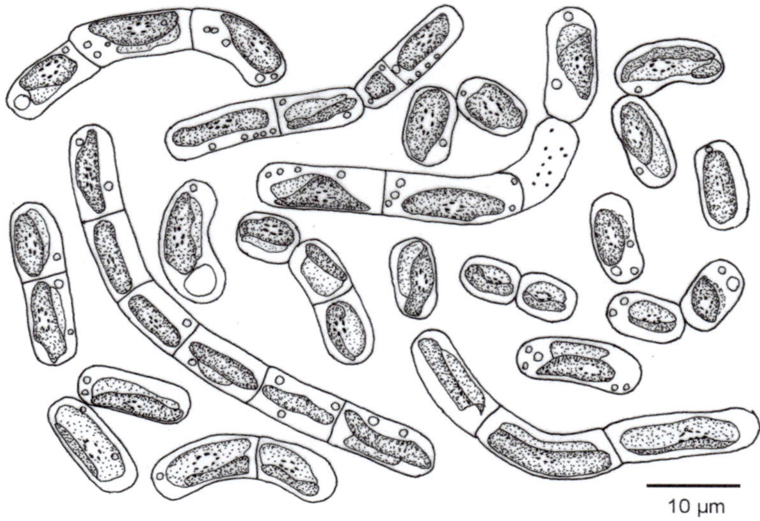

Fig. 3.72 *Klebsormidium sterile* with short filaments, cell doublets and single cells of an older culture

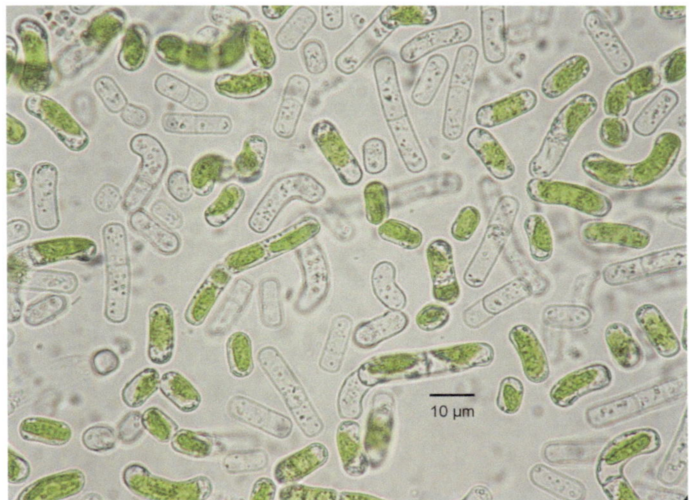

Fig. 3.73 *Klebsormidium sterile* of a 4 weeks old culture with many single cells, cell doublets and dead cells

Klebsormidium sp. (Fig. 3.74)

During investigation of the surfaces of a series of specimens which were exposed in Holzkirchen for 2 years and of several additional samples, algae of *Klebsormidium* type were observed (Fig. 3.74). However, these could not be assigned to the known aerophytic species. Further observations are necessary.

Hofbauer et al. (2003) reported *Klebsormidium montanum* (Skuja) S. Watanabe. on established building surfaces in Germany.

Class Zygnematophyceae
Order Zygnematales Borge et Pascher

Family **Mesotaeniaceae** Oltmanns

Genus ***Cylindrocystis*** Meneghini

Cells are cylindric or oval with rounded ends and no constriction on the outline. The cell wall is smooth. Every half-cell has one asteroid chloroplast with pyrenoid. One species shows a not completely divided chloroplast. Sexual reproduction is by conjugation of gametes (hologametes without flagella), empty cells of gametes may remain connected to the zygospore for prolonged time. The cosmopolitan genus possesses only a few aerophytic taxa.

Cylindrocystis brebissonii (Ralfs) De Bary (Figs. 3.75 and 3.76)

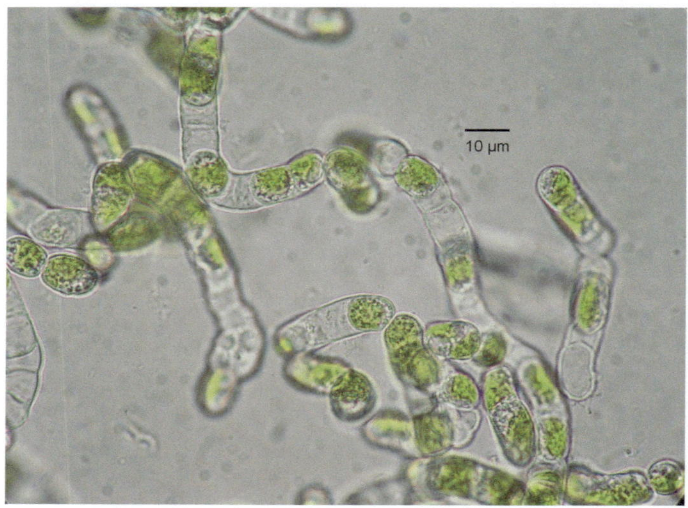

Fig. 3.74 *Klebsormidium* sp., no common features with known taxa

Fig. 3.75 *Cylindrocystis brebissonii*, some cells with different sizes of chloroplasts

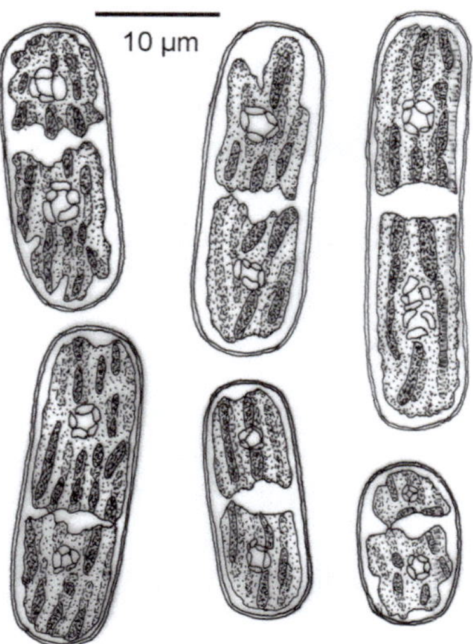

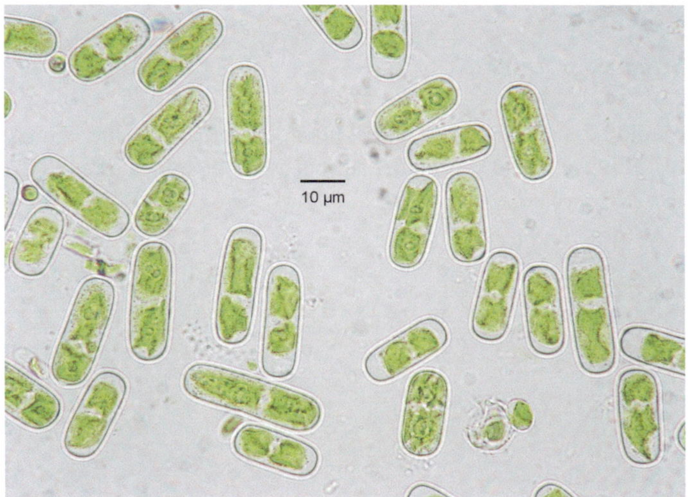

Fig. 3.76 *Cylindrocystis brebissonii* from culture

Cells were elongate, not embedded in mucilage (Figs. 3.75 and 3.76) with two asteroid chloroplasts with a pyrenoid. Sometimes chloroplasts had two pyrenoids. Conjugation and zygospores were not observed in culture (autoincompatibility?). The alga was isolated from an additional surface sample.

According to John (1942), the pH of the substrate for terrestrial populations ranges between 3.7 and 8.2.

Family **Desmidiaceae** Ralfs

Genus *Actinotaenium* (Nägeli) Teiling

Cells are spindle-shaped rounded, ellipsoidal to cylindrical, with rounded, slightly tapering poles. Median constriction is very light; the cell wall shows no special structure but is finely punctate. There is one chloroplast in each semicell and one (to several) pyrenoid(s) with starch sheath. The genus is distributed cosmopolitan with at least four aerophytic forms (Ettl and Gärtner 2014).

Actinotaenium sp. (Fig. 3.77)

Together with *Cylindrocystis brebissonii* some living cells of *Actinotaenium* sp. were detected on additional samples (Fig. 3.77), but without culture a species identification was impossible.

Genus *Cosmarium* corda ex Ralfs

The largest desmid genus has a more or less rounded cell shape; the isthmus is very narrow. The cell wall is specifically ornamented but without spines or extended projections

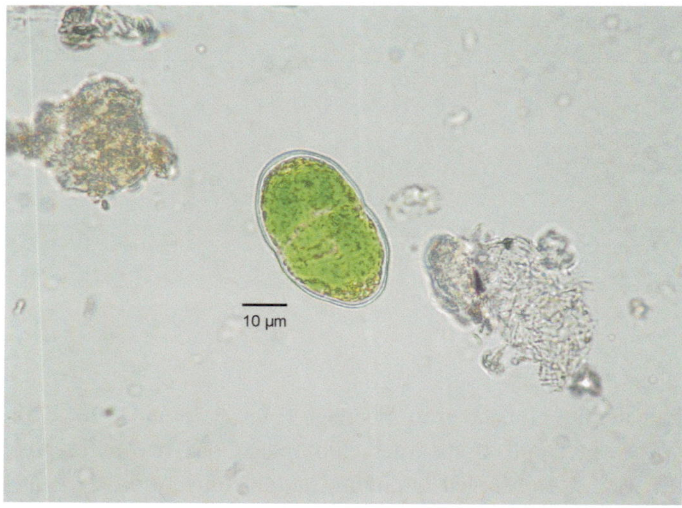

Fig. 3.77 Cells of *Actinotaenium* sp., from a frozen sample with amorphous cell content

(Bellinger and Sigee 2015). Very diverse cosmopolitan genus is known from freshwater, terrestrial and aerophytic habitats.

Cosmarium sp. (Fig. 3.78)

In addition to the former documented desmids, a singular cell of *Cosmarium* sp. was observed on one additional sample (Fig. 3.78). The alga showed some similarities to *Cosmarium holmiense* Lundell, but the overall appearance did not allow a final identification. For *C. holmiense* a pH of 8.2 is recorded (John 1942). Without culture, no further identification was possible.

Division Strepthophyta–Embryophyta (Land Plants)
Class Bryopsida–Musci

The gametophyte of the bryopsida typically forms a differentiated thallus (with only single exceptions), consisting of cauloids, phylloids and rhizoids. In the bryophytes, the gametophyte represents the main generation which bears on its body the reproductive organs. After the fertilization of the egg cell (oogamy), the sporophyte still grows connected to the gametophyte. The main part of the sporophyte is the spore capsule with its stalk (seta). In the capsule, the meiosis takes place and more or less durable spores are formed. During favorable conditions, these spores germinate and initially form a provisional stage, the protonema, prior to budding which gives rise to the real bryophyte plants. In most forms, the protonema is a short-lived branched thread that reminds strongly of branched green algae, but typically shows oblique cell walls in the main threads. In only a few forms, the protonema is a thalloid structure, and in some forms,

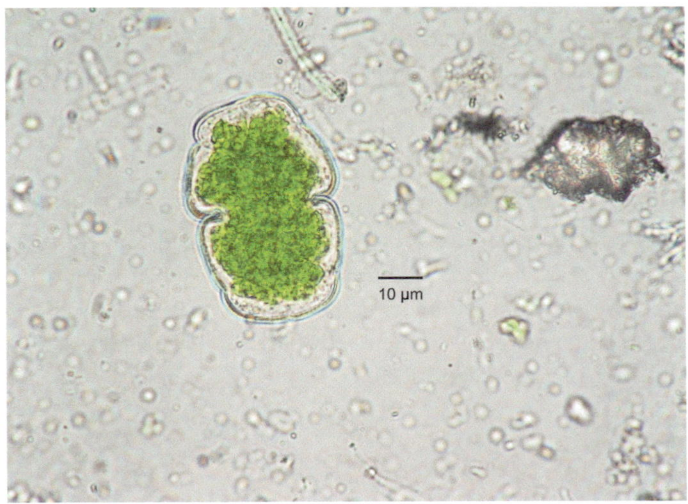

Fig. 3.78 Cell of *Cosmarium* sp. (aff. to *C. holmiense*) from an additional sample

the protonema may be perennial or a vegetative means of propagation. Many moss species are adapted to extreme locations (Bertsch 1959; Frey et al. 2006). They can stand conditions like extremely poor nutrients in the substrate, very low or very intense irradiation, and they are somehow specialists in photosynthesis. Many species can withstand complete drying for long periods without damage. After rewetting, they return to the active state very quick. Apart from some very abundant and robust species, many bryophytes require very special provisions in the underground (e.g., pH, structure and chemistry of substrate, microclimate). Different bryophytes are used in paleobotanic investigations (Dickson et al. 1996, 2019; Dickson 2000, 2003; Dickson and Hofbauer 2005). Among rock and bark inhabiting species some are able to grow on concrete surfaces, plaster or other anthropogenic substrates (e.g., Marcos Laso 2001). Also, on some specimens at the IBP, such species were recorded. Short descriptions of genera and species are given in this section. Further taxonomic information on families and orders can be found in standard works such as Frey et al. (2006).

Order Pottiales M. Fleisch.
Family **Pottiaceae** Schimp.

Genus *Tortella* (Lindb.) Limpr.

This calciphilous acrocarpous genus often pioneers on rock and gravel, and it is cosmopolitan. The midrib of the narrow lancet-shaped smooth leaflets possesses ventral and dorsal bands of stereids. At the base of the leaflet a conspicuous, V-shaped seam is developed. Often growing incomplete and dwarfed, there have been recorded *Tortella* species on masonry.

Tortella tortuosa (Hedw.) Limpr. and *Tortella fragilis* (Hook. et Wils.) Limpr. form regular elements of the colonization of old concrete or stone walls (Frey et al. 2006).

Genus *Tortula* Hedw.

This diverse genus has a worldwide distribution in very different habitats, commonly on soil. The leaflets of *Tortula* are often broad and tongue shaped, sometimes equipped with a hyaline hair tip. A section of the midrib shows only a dorsal band of stereids. Basal lamina cells are different from the upper cells, and they are bigger and translucent. The leaflets are smooth at the margin, and their cells bear papillae. The peristome teeth of the capsules usually are elongated and filiform.

Tortula muralis Hedw. (Fig. 3.79)

The cells at the leaf tip were of round shape and not different from the other papillous lamina cells. The margin of the leaflet was seamed by brighter square cells and recurved

Fig. 3.79 *Tortula muralis* at the damaged surface of a concrete wall. Adult plants with sporogons (moderately enlarged)

in its whole length. The midrib of the tongue shaped leaflets was even and not thickened in its upper third, and it was protruding as a long and smooth glass hair. Plants were usually about 3–5 mm high (Fig. 3.79).

This species is distributed worldwide on rocks and walls (Nebel and Heinrichs 2000). It may be found on monuments (Marcos Laso 2001) and has already been documented on modern façade coatings. *Tortula muralis* was recorded on masonry specimens after only 3 years' exposure.

Order Grimmiales M. Fleisch.
Family *Grimmiaceae*

Genus *Grimmia* Hedw.

The cells at the base of the leaflets do not possess wall thickenings in the shape of barbed wire (distinguishing features against *Racomitrium*). Leaflets are usually lancet shaped without pronounced longitudinal furrows. Leaf lamina are one or two cell layered with smooth margins, recurved or enrolled. The calyptra of the capsule is cap shaped, the peristome teeth are divided until the base. The capsule often is upright but may also be immersed and the columella is not attached to the lid. Mainly rock inhabiting mosses which form more or less compact cushions and are distributed worldwide.
Grimmia pulvinata (Hedw.) Sm.

The leaflets of *Grimmia pulvinata* possessed a hair tip and were curled when dry. Lamina were single cell layered and translucent in the upper part of the leaflet with more or less square cells throughout, and the leaf margin was enrolled. Setae were curved; therefore, the capsules were bend down but still lifted above the leaves. Setae could straighten when old. The egg-shaped capsules were striped with conspicuous crests, and the capsule lid was beaked.

One of the most common species of the genus growing on rock and walls (Nebel 2000b) and also documented on modern building surfaces (Hofbauer et al. 2003).

Genus **Schistidium** Brid.

This genus anatomically resembles the genus *Grimmia*. The leaflets are often two cell layered in their upper part with recurved margins. Further differentiating characters are present in the habitus. The plants are usually much more elongated and produce less dense cushions or loose mats. The capsule with a very short seta is always immersed, and the columella is connected to the capsule lid. It is a cosmopolitan and taxonomically diverse genus, which often grows on rock. Some species specialize in dry habitats, but others occcur in wet conditions, even submersed.

Schistidium apocarpum (Hedw.) B. S. G. s.l. (Fig. 3.80)

The leaflets possessed a glass hair, their midrib was smooth at the back, and the margin recurved. *Schistidium apocarpum* comprises a collective species rich in forms which have been divided into a range of sibling species (Blom 1998; Fife 2000; Holz 2000; Nebel 2000a). For their distinction, fully developed plants are necessary which are often not found on younger anthropogenic surfaces. In recent works, a barcode library was built which allows to identify poorly developed samples or even fragments of plants (Hofbauer et al. 2014, 2016).

Schistidium apocarpum s.l. belongs to the early pioneer forms of bryophytes which may emerge on masonry of modern buildings (Fig. 3.80).

Order Bryales Limpr.
Family **Bryaceae** Schwaegr.

Genus **Bryum** Hedw.

The lamina cells of the spirally arranged leaflets of *Bryum* are of rhomboidal or hexagonal form down to the base. The capsule sits on a straight seta and usually is pear shaped,

Fig. 3.80 Young plantlet of *Schistidium apocarpum* s.l. (ca. 1 mm in length) developed on the surface of a specimen exposed in Heggen/Finnentrop, Germany. Glass hairs are not yet developed and the plant is just beginning to branch

often pendulous or bent down (Ahrens 2001). *Bryum* is one of the most diverse genera of bryophytes. If underdeveloped, the species may be difficult to assess. Furthermore, some species groups are critical and may be obscure to the non-specialist. The cosmopolitan genus has many species that are present in a vast variety of habitats, in wet or dry biotopes or also on man-made substrates.

Bryum argenteum Hedw.

Plants of *Bryum argenteum* usually appear silvery or silver green, and only plants that grow in very shady and damp places may be fully green (Frey et al. 2006). The leaflets were oppressed to the stem and gave it a vermiform appearance. Leaflets were contracted abruptly to a fine tip, and the lamina cells of the upper part usually were discolored and translucent—therefore the silvery shine. The midrib ended shortly above the center of the leaflet. The capsule was developed on a short seta, pendent and usually dark red.

The species is nitrophilous (indicator for fertilizers or high nitrogen), often appears synanthropic and is comparatively rare in nature. It has been detected on very different substrates like walls, roofs, etc. (Ahrens 2001), but rarely on vertical surfaces of masonry (e.g., seen on ETHICS in Kufstein).

Genus **Pohlia** Hedw.

Plants are usually not vermiform. Lamina cells of the lancet shaped leaflets are elongated rhomboidal to narrow and linear, more than six times longer than wide. Capsules are pendent or nodding. It is a cosmopolitan genus, very similar to *Bryum*. Some species may occur in anthropogenic habitats.

Pohlia annotina (Hedw.) Loeske agg.

Typical for this critical group of species are bulbils formed in the shoulders of the leaflets, and sporogons are often rare. Species are distinguished by the form, number and color of the bulbils (Nebel 2001). If the bulbils are not sufficiently developed, only the collective species can be identified.

Some sibling species of the aggregate prefer ruderal or pioneer habitats and may be found also on masonry, as we have discovered it on façades of building parts stored in Germany.

Order Orthotrichales Dixon
Family **Orthotrichaceae** Arnott

Genus ***Orthotrichum*** Hedw.

The leaflets of the genus *Orthotrichum* are arranged spirally, usually not curled when dry and do not possess a seam with hyaline cells at the base. The calyptra of the capsule is bonnet shaped. The species inhabits rock and bark and occurs mainly in the temperate and boreal climate zones and is less abundant in the tropics.

Orthotrichum anomalum Hedw. (Fig. 3.81)

The leaflets of *Orthotrichum anomalum* showed a more or less narrow tip; they did not possess a hairtip, nor were gemmae produced on their surface. The reddish capsule was elevated above the leaflets on a 2–4 mm long seta; in the emptied stage, it was furrowed and showed longitudinal stripes. The stomata at the base of the capsules were crypto-porous (immersed); the peristome teeth were upright or horizontal in the wet stage.

Orthotrichum anomalum is abundant on concrete walls and roofs (Schäfer-Verwimp 2001) and has also been documented as part of the initial colonization of modern building surfaces (Hofbauer 2007). We discovered a first tinge on various specimens after outdoor exposure in Germany for only 3 years. On the plaster surface and sometimes connected with single shoots, the development of caulonemata was observed, and we recognized that *Orthotrichum anomalum* may propagate by aid of persistent protonemata/caulonemata like an aerophytic alga. The superficial similarity of the perennial protonema/caulonema to the green algal genus *Trentepohlia* is remarkable (Fig. 3.81).

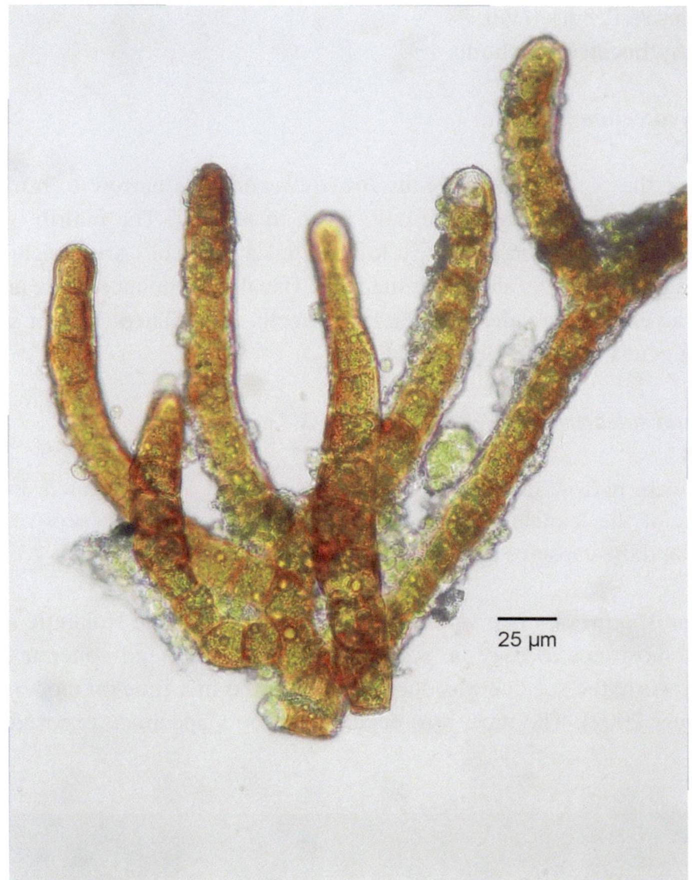

Fig. 3.81 Part of a protonema/caulonema of *Orthotrichum anomalum* showing similarity to the alga *Trentepohlia*

Orthotrichum diaphanum Brid.

The plants were appr. 1 cm high. Typical leafs showed a hyaline tip and hairpoint. Capsules were often present, pale brown and almost immersed in leafs. Stomata of the capsules were cryptoporous.

We observed the species on concrete walls in Germany, Austria and Italy, sometimes already a few years after errection of the building. The nitrophilous species is generally abundant in Europe, occuring on bark, rock and man-made structures (Frey et al. 2006).

Order Hypnales W. R. Buck & Vitt
Family **Brachytheciaceae** Schimp.

Genus ***Brachythecium*** B. S. G.

The leaflets of the pleurocarpous genus *Brachythecium* are narrow to broad lanceolate with a smooth margin, if denticulate only down to midleaf. The midrib is often weak, and the leaflets are narrowed gradually to the tip. Lamina cells are typically more than five times longer than wide, shorter at the base. Usually pronounced ears are developed, and cells of the ears are broader than the lamina cells and inflated (Frey et al. 2006). The cosmopolitan genus occurs in diverse habitats.

Brachythecium salebrosum (Web. et Mohr) B. S. G. (Fig. 3.82)

The leaflets were narrow lanceolate, strongly corrugated and elongated, but not decurrent. The tips of the leaflets were straight, and the stemlets were prostrate. The setae were smooth and the capsules bent and curved.

Brachythecium salebrosum is a variable species abundant in the Holarctic and is known from diverse substrates as may be soil, damp timber, rock and concrete (Nebel et al. 2001). In our work, the species was documented for the first time on modern plaster surfaces (Hofbauer 2007). The moss was encountered on a specimen exposed in Germany

Fig. 3.82 *Brachythecium salebrosum*, young stemlets developed between the plastic frame and the plaster of a specimen, exposed in Germany for 3 years (whole picture width ca. 2 cm)

for 3 years. Several plants developed attached to the plaster of the specimen at the margin of the surrounding plastic frame (Fig. 3.82).

Family **Hypnaceae** Schimp.

Genus *Hypnum* Hedw.

The ribless leaflets of the pleurocarpous genus *Hypnum* are bent to falcate, usually smooth at the margin or toothed only at the tip. The capsules typically are bent and curved. The genus is cosmopolitan, often forming extensive covers. The following species is the most abundant with a broad ecological amplitude.

Hypnum cupressiforme Hedw.

In some raw cultures of ETICS specimens which were exposed in Germany for 3 years, protonemata developed which finally gave rise to plantlets of *Hypnum cupressiforme*. The identification was confirmed by morphological comparison to well-developed plants in other habitats. The following features are characteristic for *Hypnum cupressiforme* (Sauer 2001; Sauer and Philippi 2001): The margin of the falcate leaflets is flat and at most toothed at the tip. Lamina cells are approximately 15 times longer than wide without plasmodesmata at the base. Leaf ear cells are quadrangular, getting bigger toward the base. The stemlets of the irregular branched plantlets are greenish and do not possess a hyalodermis. The yellowish-green plantlets are prostrate and attached to the substrate with their rhizoids. The capsule is markedly curved, and the lid is beaked.

H. cupressiforme is a widespread and variable species which grows on diverse substrates (Sauer 2001) and was also detected on modern masonry surfaces (Hofbauer 2007).

Hofbauer et al. (2003) reported further mosses which occurred on building surfaces in Germany: *Encalypta streptocarpa* Hedw., Encalyptaceae Schimp., *Ceratodon purpureus* (Hedw.) Brid., Ditrichaceae Limpr., and *Syntrichia ruralis* (Hedw.) Web. et Mohr, Pottiaceae Schimp. On older building surfaces and walls, plenty of different species can develop. In our study, we concentrated on the taxa which are encountered within the first few years after erection of modern façades.

Protonemata (Fig. 3.83)

Within the analyses of numerous specimens' surfaces, additional colonization situations as well as sedimentation probes and driving rain probes different protonemata germinated, mostly from mosses. A more detailed differentiation was not possible as no plantlets developed subsequently. Some of these protonemata could be propagated. It was remarkable that only very rarely buds and plantlets developed on BBM although the protonemata persisted. Usually, only the main filaments showed the characteristic oblique

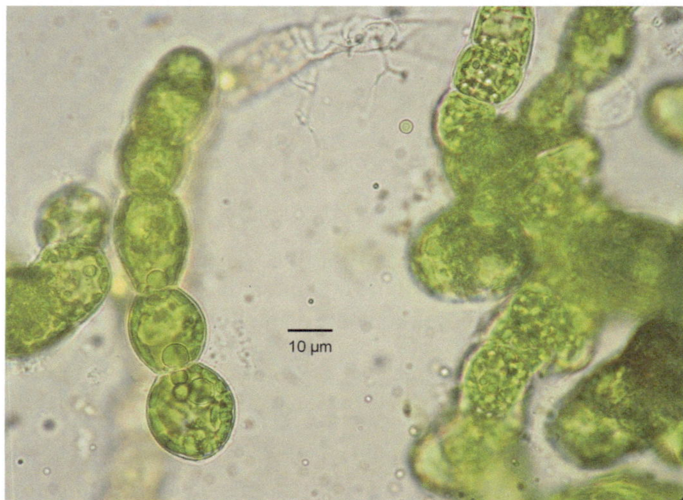

Fig. 3.83 Moss protonema, lateral branches do not show the typical oblique partitions and therefore resemble filamentous green algae

cell walls but not the lateral branches (Fig. 3.83). Such forms could also be misidentified as algae. Persistent protonemata, which may look like filamentous algae, may even spread (propagate) on building part surfaces without the development of real bryophyte plants.

Class Pteridopsida

The gametophyte of the pteridopsida (ferns) usually forms a small flat thallus (prothallium), which reminds of a simple liverwort. In contrary to the bryopsida, the main vegetation form is the sporophyte, as in seed plants. The sporophyte consists of roots, stem and leafs. Ferns, like e.g., *Asplenium ruta-muraria* L., can also grow in cracks of wall structures, especially if some humus has accumulated there (e.g., Hofbauer et al. 2003). On surfaces of modern façades in the Central European climate, a growth of ferns is unusual, but not impossible.

Prothalli of ferns

In several sedimentation probes, the germination of fern prothallia could be observed. They showed a typical single cell layer and were dioecious. When female and male prothallia were present in one Petri dish, sometimes whole fern plants developed. We therefore know that in these cases the spores derived from local fern populations. They however differed from *Asplenium ruta-muraria*. Without the development of adult fern leafs, the species remained obscure unless a molecular characterization could be performed.

Eumycota (Fungi) Divisions

The real fungi (Eumycota) represent an independent form of organization. They comprise mainly saprophytes decomposing organic substance. Among them, some metabolic specialists have evolved which can use the most extraordinary compounds for energy derivation. Apart from the saprophytic way of life, secondary parasites (especially plant parasites), pathogenic agents but also symbiotic forms of life (e.g., mycorrhiza and lichens) have evolved. Some forms may act as opportunistic pathogens, and even, predatory forms are known. In various works, fungi as colonizers of building parts have previously been investigated: e.g., coating surfaces in Florida (Reynolds 1950), frescoes in Moldavia (Ionita 1973), damp plaster and painted surfaces in Great Britain (Eveleigh 1961a, b; Hawksworth 1976), monuments made of marble in Russia (Gorbushina et al. 2002), etc. Most of these investigations dealt with monuments, historic buildings or surfaces, which are many years old, and not as in our investigations with the initial growth and development on newly made surfaces. In our diverse study, we found that certain fungi, especially black yeasts, were among the first colonizers of modern façade surfaces, if a sufficient source of organic nutrients was present (Renzl 2006; Hofbauer 2007). For identification, not only standard works (e.g., Carmichael et al. 1980; Arx 1981; Domsch et al. 2007) but also many monographs and special literature was used, which are mentioned with the respective taxon.

Some authors treat the Eumycota as an independent regnum and therefore carry the groups, which we gave the rank of classes, as divisions (e.g., Kirk et al. 2001). For practical reasons, we sticked mainly to the taxonomic system presented by Domsch et al. (2007), also with the classification of lower taxonomic ranks.

Division Archemycota
Class Zygomycetes

Zygomycetes are characterized by hyphae, which are mostly free from transverse septa, fast growth and their special mode of sexual propagation (Zygogamy). At present, 8–10 different orders are distinguished within ca. 900 species worldwide. The following order Mucorales (according to Domsch et al. 2007) is most diverse (approx. 300 species) with mainly saprophytic and parasitic species (Hesseltine and Ellis 1973; Hawksworth et al. 1995; Alexopoulos et al. 1996; Benny 2001; Benny et al. 2001). Recently taxonomic changes took place in Zygomycetes (e.g., Hoffmann et al. 2013), but we kept to Domsch et al. (2007).

Order Mucorales Schröter
Family **Mucoraceae**

Genus *Mucor* Micheli ex L.

The genus *Mucor* forms globous sporangia lacking an apophysis. The sporangiophores are simple or branched, and stem directly from substrate hyphae, stolons and rhizoids are unknown. Zygospores, if known, are produced from more or less equal opposing suspenders without attachments or protuberances (Hesseltine and Ellis 1973; Schipper 1978; Arx 1984). *Mucor* species are cosmopolitan and commonly occur on organic matter, typically lying on the ground. The genus is regarded as truly coprophilous (Domsch et al. 2007).

Mucor plumbeus Bonord. (Fig. 3.84)

The fugus represents one of the most characteristic species of the genus. Colonies on malt extract agar (MEA) were more or less dark gray to olive gray. Sporangiophores were branched sympodial or monopodial. Sporangia were hyaline initially and turned dark brown gray later, with a spiny wall that finally bursted. The columella of the sporangium was obovoid, ellipsoidal to cylindric-ellipsoidal and usually showed one to more protuberances, which were often pin-like (Fig. 3.84). Sporangiospores were densely warty-rough, typically rounded (single ones may be more irregular), 7–8 µm in diameter and yellowish brown. Sometimes chlamydospores were formed within the sporangiophores.

Mucor plumbeus is a frequent species with worldwide distribution. In our study *M. plumbeus* was documented from various surfaces (Hofbauer 2007). Previously, the species was isolated from frescoes of a cloister (Ionita 1973), but was never documented as a component of initial growth on modern building surfaces.

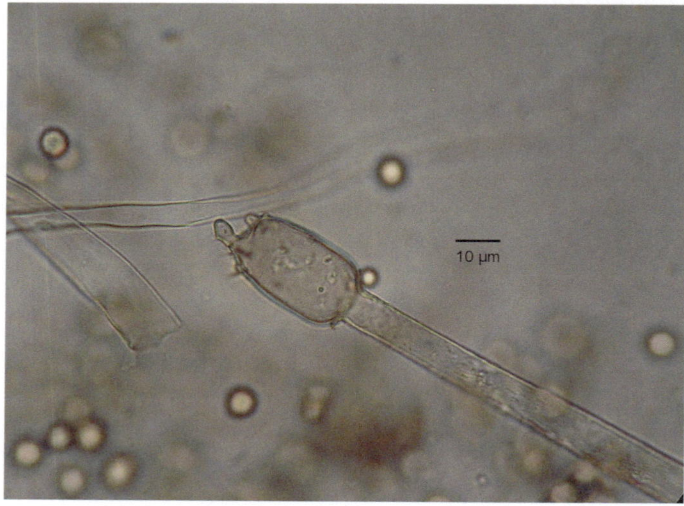

Fig. 3.84 Emptied sporangium of *Mucor plumbeus*, columella shows characteristic projections

The species is heterothallic. Zygospores are formed abundantly in the air mycelium on MEA at 24 °C, only if compatible strains are grown together (Schipper et al. 1975; Domsch et al. 2007). Unlike other species, it is frequently propagated through air because spores are released dry from the sporangium (Zycha et al. 1969; Domsch et al. 2007). The fungus can stand alkaline conditions up to pH 8 (Campbell 1938). Minimum temperature for growth lies at 4–5 °C and maximum temperature at 35–37 °C. Optimum temperature range is between 5 °C and 20 °C, while the air mycelium remains flat between 5 and 10 °C and sporulation is poor between 20 and 30 °C (Panasenko 1967; Schipper 1976; Samson et al. 2002a; Domsch et al. 2007). Minimum water potential for growth is −100 bar (Griffin 1963a).

Mucor racemosus Fres. (Fig. 3.85)

Colonies on MEA were bright to smoke gray. Sporangiophores were branched sympodial or monopodial; final branches were short and partly curved. Sporangia were hyaline initially, ripe brownish with a spiny wall (Fig. 3.85). Columellae of sporangia were egg shaped, oval or cylindric oval. Sporangiospores were gray, broadly oval to subglobose and approximately 5.5–8.5 μm long. Often, chlamydospores were found in the sporangiophores.

Mucor racemosus is a variable and frequent species with worldwide distribution. The species was documented for the first time on modern building surfaces by Hofbauer (2007). *M. racemosus* has previously been isolated from air samples (Zycha et al. 1969; Domsch et al. 2007).

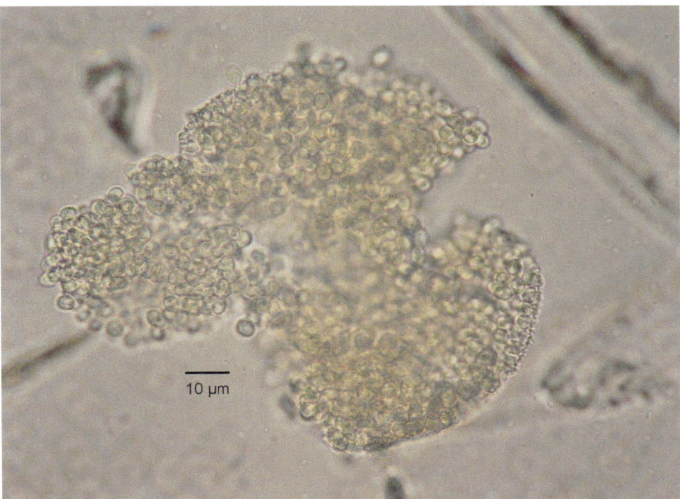

Fig. 3.85 *Mucor racemosus,* burst sporangium with characteristic delicate spiny surface of the sporangium wall (from a raw culture)

In growth substrate, pH values of 4.4–8.5 have been measured (Feher and Besenyei 1933; Campbell 1938; Apinis 1964, Loub 1967), but in culture (in vitro) pH ranges from 2.0–8.5 (Panasenko 1967). The temperature range extends from −4 to 38 °C with an optimum at 20–25 °C (Ryass and Borut 1958; Panasenko 1967; Schipper 1970, 1976; The 1972; Samson et al. 2002a). For mycelial growth, the minimum water potential is −110 bar, for sporulation −70 bar (Domsch et al. 2007). Infections of animals and man have been reported, but the pathogenicity of the species is doubted (perhaps misidentification or contamination) (Scholer et al. 1983; Hoog et al. 2000).

Genus **Rhizopus** Ehrenb.

The genus *Rhizopus* is characterized by stolons that produce rhizoids, which develop opposite to the sporangiophores (difference to *Rhizomucor*!). A further feature is the short apophysis at the sporangium. Other important characters for differentiation are the length of the sporangiophores, the growth temperature as well as the size and ornamentation of the sporangiospores (Hesseltine and Ellis 1973; Schipper 1984; Schipper and Stalpers 1984; Ellis 1985; Domsch et al. 2007). Some species grow well at high temperatures. *Rhizopus* spp. are cosmopolitan and important organisms connected to food spoilage, food fermentation (especially in Asia) and zygomycoses. They occur on plant residues, on soil and excrements, and their spores have been isolated from air (Hesseltine 1965; Schipper and Samson 1994; Weitzman et al. 1996). Monographic works on the genus have been published (Zheng et al. 2007; Abe et al. 2010).

Rhizopus stolonifer (Ehrenb. ex Link) Lind (Fig. 3.86)
Colonies were brownish to reddish gray brown and usually grew very fast. Stolons were hyaline to brown with a rigid cell wall and produced branched rhizoids (Fig. 3.86). At the nodes where rhizoids were attached, bunches of sporangiophores of different size arose. Sporangiophores were pale to dark brown and straight; the main sporangiophores grew up to 1.5–3 mm in height. Sporangia themselves were dark brown to black, had a short apophysis with a cup-like outline, a globose to oval columella and egg shaped to ellipsoidal spores (ca. 6–11 μm long) with delicate stripes (grooves).

Rhizopus stolonifer is distributed worldwide with an emphasis on warmer regions. In our investigation, we isolated the species various times from the plaster surface of specimens, which were exposed up to 2 years. Typical substrates are plant residues and feces, and the spores were regularly found in air samples (Domsch et al. 2007). *Rhizopus stolonifer* was isolated from marble statues in Russia (Gorbushina et al. 2002) and documented as component of initial colonization of modern façade surfaces (Hofbauer 2007).

Minimum temperature for growth lies at 4.5–5 °C, maximum temperature at 32–33 °C. There is no growth above 35 (37) °C which is a diagnostic character. Optimum temperature for development is at 25–26 °C (Pidoplichko 1953; Inui et al. 1965; Pierson 1966; The

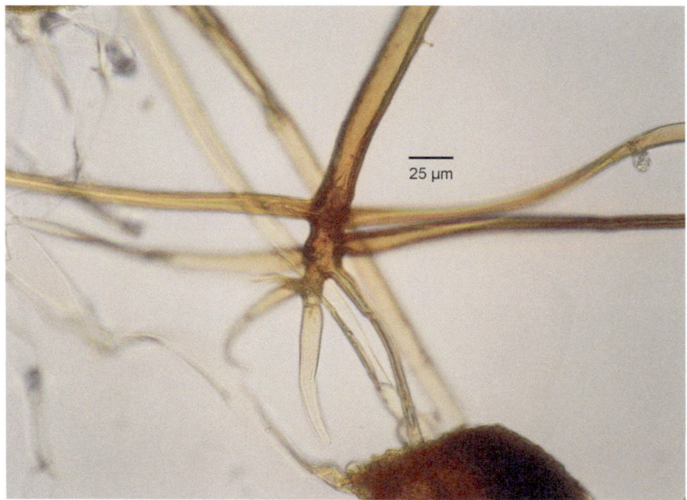

Fig. 3.86 Rhizoids of *Rhizopus stolonifer*

1972; Samson et al. 2002a; Domsch 2007). Minimum water potential is −100 bar (laboratory trial; Corry 1973); in soil, −70 bar have been measured (Chen 1966). To date, only single infections of people suffering from diabetes are known (Sandler et al. 1971; Ferry and Abedi 1983; Hoog et al. 2000). Based on the temperature range to which the species is bound (no growth at 37°C), it can be assumed that it is only capable to cause infections under exceptional conditions.

Genus *Mycotypha* Fenner

In the genus *Mycotypha*, cylindrical sporangiophores are formed on top of long chambered stalks which may be branched in age. Unicellular sporangia at the surface of sporangiophores are globose or ellipsoidal (Young 1969; Khan and Talbot 1975; Benny and Benjamin 1976; Brain and Young 1979; Benny et al. 1985; Edelmann and Klomparens 1995). *Mycotypha* spp. was isolated from feces and from soil; their distribution comprises Eurasia, North America and Africa (Mikawa 1975, 1979; Benny and Benjamin 1976; Benny et al. 1985; Ellis and Ellis 1998).

Mycotypha sp. (Fig. 3.87)

In our investigations of air spores via sedimentation, a form was isolated that fits to the morphological characters of *Mycotypha*. Globose pale-brown spores, respectively unicellular sporangia, were attached to an elongated cylindrical sporangiophore with a long chambered stalk (Fig. 3.87). This has been the first documentation of *Mycotypha* from air (Hofbauer 2007).

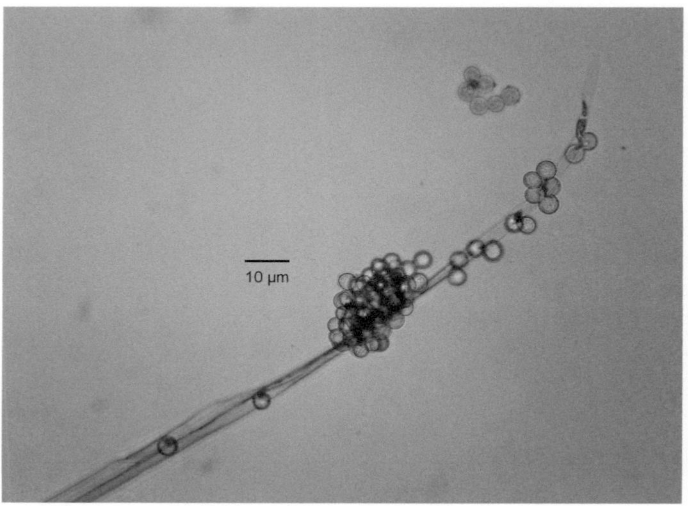

Fig. 3.87 Sporangiophore und spores/sporangia of *Mycotypha* sp

Division Ascomycota
Class Ascomycetes

Ascomycetes are characterized with mycelia showing regular septa and the development of asci (tubes) in a complicated generative cycle, typically containing eight endospores (ascospores). Secondary less or more ascospores may be released from an ascus. In many forms, typical fruiting structures like cleistothecia, apothecia or perithecia are developed, sometimes united in fruiting bodies of higher order. In some ascomycetes, the main vegetative form is a yeast stage, but even in this form, generative processes may be observed if environmental circumstances are suitable. Most known lichen symbioses have ascomycetes as fungal partner. In most lichens, the fungal partner is dominating the morphological structure; therefore, lichens usually are arranged according to the fungal taxonomy. In our work, we wanted to emphasize lichens as a special group and therefore discussed them at the end of Sect. 3.2.3 "Eumycota."

Subclass Ascomycetidae
Order **Sordariales** Chadef. ex D. Hawksw. & O.E. Erikss.

Family **Chaetomiaceae** G. Winter

Genus ***Chaetomium*** Kunze ex Fr.

Ascomata (fruiting structures) form perithecia, which are globose, egg shaped or pear shaped and to which usually characteristic hairs or setae are attached. These hairs or setae are branched, unbranched, often waved or spirally coiled, sometimes stiff and straight and provide valuable characters for identification as do perithecial tissue and morphological spore structure. Asci are cylindrical, elongated egg shaped or club shaped, usually contain eight ascospores, which are unicellular, smooth walled, pigmented and often with one or two germination pores (Arx et al. 1986). The genus is cosmopolitan, and its members are important components in the decomposition of lignocellulosic plant material in contact with soil; some species are polyphagous. Recently, the genus *Chaetomium* and further genera of the Chaetomiaceae have been revised (Wang et al. 2016, 2019). This does not affect the taxa of *Chaetomium* discussed here.

Chaetomium globosum Kunze ex Steud.

Asci of *Chaetomium globosum* were club shaped. The unbranched terminal hairs at the perithecium were waved and flexible and extended in length toward the porus of the perithecium, were dark olive brown in color and roughened by warty structures (in reflection electron microscopy [REM] cup- or bladder-like). Ascospores were regular and broad lemon shaped with an apical germ porus (ca. 8.5–11 µm in length). No aleurioconidia or phialoconidia (vegetative spores) were formed.

In our study, this species was recognized during the investigation on background germ concentration of diverse driving rain samples (Hofbauer 2007).

C. globosum has previously been documented on marble monuments in Russia (Gorbushina et al. 2002). Because of the regular occurrence in probes of driving rain, it was assumed that the species may be present on outer surfaces of building parts quite often. Indoors *C. globosum* has typically been observed as constituent of microbial damages, particularly in connection with water damage and cellulose containing materials (Piecková 2003).

Optimum pH range for growth of *C. globosum* is between 7.1 und 10.4, pH 7.3 is best for fructification (Zagulyayeva 1971; Cherepanova 1975). Minimum temperature for germination of ascospores lies at 4–10 °C (or even higher), optimum temperature between 24 and 28 °C. Above 38 °C, ascospores do not germinate. Optimum temperature range for mycelial growth is 16–25 °C (strain specific), maximum temperature is 36–37 °C (Chapman and Fergus 1975; Cherepanova 1975; Franz 1975; Dreyfuss 1976; Millner 1977; Samson et al. 2002a). At 50% salinity, 10% spores are able to germinate. Minimum water potential for growth measured in soil is −85 bar (Kouyeas 1964; Byrne and Jones 1975). The fungus is able to decompose cellulose and therefore shows an affinity to materials rich in cellulose (Domsch et al. 2007). Ellis and Ellis (1998) reported regular occurrences on dung, paper, cloth and other substrates alike. For decomposition of cellulose the optimum prerequisites are 25–32 °C in temperature and 4.5 in pH, and additionally, cellulolytic activity is enhanced by the increase of concentration of the nitrogen source (Lewi and Cowling 1969; Walsh and Stewart 1971; Sharp 1975).

Opportunistic infections in immunocompromised patients have been reported, and the overall virulence of the fungus is regarded as low (Anandi et al. 1989; Yeghen et al. 1996; Hoog et al. 2000). *C. globosum* is a potential producer of mycotoxins (e.g., Sekita et al. 1973, 1976; Domsch et al. 2007; Piecková 2003).

Chaetomium perlucidum Sergejeva (Fig. 3.88)

Fruit bodies possessed a parenchymatic structured wall (textura intricata), hairs at the perithecia were unbranched and hardly wavy. Asci were club shaped and contained eight spores. Ascospores were narrow-ellipsoidal with pointed ends, with dimensions of ca. 12×6 μm (Fig. 3.88), and they had one clearly recognizable subapical germ pore. Chlamydospores were not observed. In culture, ascomata showed bright gray color in top light. Diagnostic characters were described by Arx et al. (1986).

We gained one single isolate from a single specimen, which was exposed in Germany for 2 years. This may be regarded as a chance finding.

Rare cases of opportunistic infections of immunocompromised patients (phaeohyphomycoses) caused by *C. perlucidum* showed an affinity to clinical strains of nervous tissue (Barron et al. 2003).

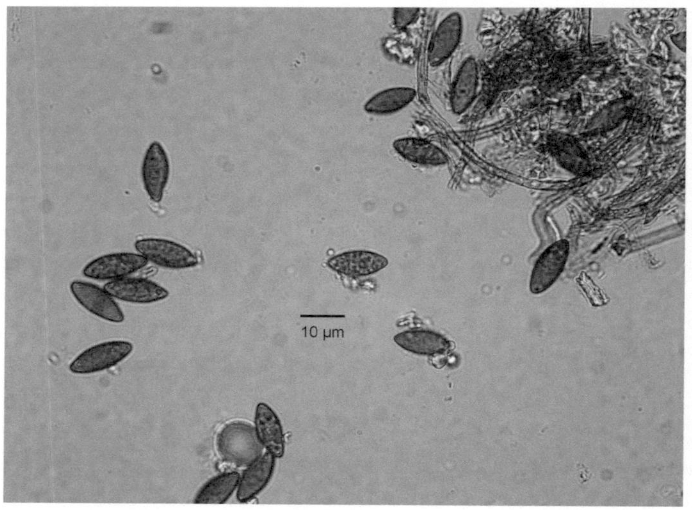

Fig. 3.88 Squeeze preparation of a perithecium of *Chaetomium perlucidum*. Ascospores with typical form and a distinct subapical germ pore. Top right: bent hairs of the fruiting body

Family **Sordariaceae**

Genus *Sordaria* Ces. et de Not.

The genus *Sordaria* forms dark perithecia without hairs on the surface and lacking paraphyses between the asci. Unicellular ascospores are dark and surrounded by a slime sheath. Spores are actively discharged (pressure built up in the ascus) and thus propagated effectively (Domsch et al. 2007; Ellis and Ellis 1998). Cosmopolitan *Sordaria* belongs to a group of fungi that prefer dung as substrate and therefore are called "coprophilous" (Domsch et al. 2007; Richardson and Watling 1997; Ellis and Ellis 1998).

Sordaria sp.

In our study, the species was regularly observed on older plaster surfaces. We interpreted the occurrence on outer building surfaces as connected to the excrement deposition by birds and insects.

Order **Eurotiales** G.W. Martin ex Benny & Kimbr.

Family **Trichocomaceae** E. Fisch.

Genus *Eurotium* Link ex Gray

Traditionally, the genus *Eurotium* is defined according to the perfect stage (teleomorph) within the section of species closely related to *Aspergillus glaucus* (anamorph), which is the type species of genus *Aspergillus*. According to changes in nomenclature, *Eurotium* has been rearranged within the genus *Aspergillus* Micheli em. Samson et al. (Samson et al. 2014). For practical reasons, we applied the system used by Domsch et al. (2007). In the group Aspergillus/Eurotium, almost always typical cleistothecia are produced beside the anamorph stage. Cleistothecia often are bright yellow in color; also hyphae surrounding the cleistothecia may produce pigments. Ascospores are lens respectively donut shaped, and their surface is ornamented species-typical. Spores often possess equatorial ledges or are winged. Vegetative spores (finely warty, subglobose, ellipsoidal or short cylindrical; bluish gray to olive) are produced by characteristic uniseriate aspergilli. *Eurotium* occurs cosmopolitan, and many species are specialized in dry and hot habitats.

Eurotium amstelodami Mangin (Anamorph: *Aspergillus amstelodami* [Mangin] Thom et Church; Syn.: *A. hollandicus* Samson et W. Gams)

Conidia bearing heads (aspergilli) were olive green. Conidia were subglobose to short cylindrical (ca. 4–5 µm long), and their surface was structured by delicate warts.

Ascospores bore two broad wings at the broad end, which surrounded a furrow of the spore; the remaining surface was covered by distinct warts or protuberances.

Eurotium amstelodami is distributed worldwide, especially in tropical and subtropical areas. The species was observed in investigations of plaster specimen exposed in Germany for one year and was isolated from air (Renzl 2006; Hofbauer 2007). Preferred habitats are soils as well as stored and/or rotten food products. The species is also known from indoor environment and has been isolated from air (Hudson 1969; Samson et al. 2002a; Domsch et al. 2007). It was the first documentation of an occurrence on modern façade coatings.

Optimum temperature for germination, growth and sporulation ranges at 33–35 °C, maximum temperature ranges at 43–46 °C (Chen and Griffin 1966; Ayerst 1969; Curran 1971; Blaser 1975; Domsch et al. 2007). Optimum water potential for germination of conidia is −100 bar, minimum water potential is −350 bar (Ayerst 1969; Corry 1973). The fungus can also grow on substrates with high osmotic activity (Chen 1964; Kulik and Hanlin 1968; Domsch et al. 2007). The species endures extreme environmental conditions and therefore belongs to groups of fungi which are called thermophilic, xerotolerant and osmophilic.

Opportunistic infections in humans have been reported (Raper and Fennell 1965; Domsch et al. 2007; Hoog et al. 2000), and however, the species is not regarded as virulent. Toxic metabolites Physcion and Echinulin may be produced (Frisvad and Thrane 2002; Domsch et al. 2007).

Order Pleosporales Luttrell ex M. E. Barr

Family **Pleosporaceae** Nitschke

Genus *Leptosphaerulina* McAlpine

Leptosphaerulina forms dark pseudothecia with a wall structure in form of a pseudoparenchym, which encloses the asci without paraphyses (Wu and Hanlin 1992). The bitunicate asci are short club shaped to bag shaped and fling off the spores actively (Graham and Luttrell 1961; Furtado and Olive 1971; Wu and Hanlin 1992). Ascospores are hyaline to brown, egg shaped to cylindrical or elongated and mostly distinctly muricate with several crosswise and one to several longitudinal partitions (Graham and Luttrell 1961; Irwin and Davis 1985; Inderbitzin et al. 2000). To date, an anamorphic stage, *Pithomyces chartarum* (Berk. and Curt.) M. B. Ellis (Roux 1986), is only known of one species, *Leptosphaerulina chartarum* Roux. The different species of *Leptosphaerulina* are cosmopolitan as parasites of leguminosae and grasses in the first place, and some species were also found on rotting plant material (e.g., Graham and Luttrell 1961; Irwin and Davis 1985; Farr et al. 1989; Inderbitzin et al. 2000; Abler 2003).

Leptosphaerulina sp.

The fungus was isolated from the surface of a specimen, exposed in Germany for one year (Renzl 2006; Hofbauer 2007). The form showed typical hyaline and muricate spores, which were elongated egg shaped. Pseudothecia were very dark, and mycelium in culture was strongly pigmented for the most part. The form could not be assigned to a certain species; therefore, further studies are necessary as this also could be a chance finding.

Genus *Pleospora* Rabenh. ex Ces. and de Not.

Pleospora forms simple dark ascostromata, and the bitunicate cylindrical asci are embedded between paraphyses. Actively discharged ascospores are muricate and pigmented, and usually, they are of irregular elliptical or egg-shaped outline (with constrictions at the cell borders). The genus is cosmopolitan, and many forms are plant parasitic. Anamorphic genera are *Alternaria* Nees ex Fr., *Dendryphion* Wallr. and *Stemphylium* Wallr. (Ellis 1971, 1976; Domsch et al. 2007). The anamorphic taxon *Stemphylium* was revised recently (Woudenberg et al. 2017). The taxonomic situation seems complicated since *Pleospora* is regarded as polyphyletic.

Pleospora herbarum (Pers. ex Fr.) Rabenh. (Anamorph: *Stemphylium herbarum* E.G. Simmons)

First the anamorphic stage was developed, and after 2–12 months, the teleomorph was produced in dependence of culture conditions. The anamorph, *Stemphylium herbarum*, was characterized by potato-shaped muricate dark conidiospores with thick cell walls and several longitudinal and transverse partitions, discharged actively. Conidia were developed on characteristic conidiophores with dark and swollen apical regions. After the formation of a conidium, the conidiophore could rupture, leading to typical structures with succeeding dark rings. Observed ascomata of the teleomorph, *Pleospora herbarum*, matched to the genus description. Ascospores were developed in a number of eight and were muricate; they were a little more regular but reminded strongly of the conidiospores.

Pleospora herbarum is a phytoparasitic fungus with worldwide distribution, especially in temperate and subtropical zones. *P. herbarum* was identified several times on diverse surfaces and from air during our investigations on modern building coatings. Living as a phytoparasite, the fungus has diverse host plants, and there has been evidence that certain strains are specialized to certain plants (Koike et al. 2001; Brahamanage et al. 2018). The species was also documented living saprophytic on plant residues, on soil or on seeds (Domsch et al. 2007). In our study, the species was documented on the surface of building coatings and from air in Germany for the first time (Hofbauer 2007).

A pH range of 3–9 is tolerated (Ellis 1931). Optimum temperature range for mycelial growth lies at 22–25 (27) °C, minimum temperature at 2 °C, maximum temperature at 30 °C. For germination of conidia, a temperature of 25 °C is optimal, minimal temperature 2.3 °C and maximal 35 °C (Perisic and Stojanovic 1967; Schneider and Crueger 1976).

Family **Sporormiaceae** Munk

Genus *Preussia* Fuckel

Ascomata of *Preussia* are smooth and possess hairs or scales in the upper part. Thread-like paraphyses disintegrate at maturity; ascospores are divided into four (or more) rounded cells by three (or more) partitions. Each cell is provided with a germination pore. Usually, the spores and their cells are surrounded by a slime sheath. Most species of the cosmopolitan genus are coprophilous and grow on different kinds of dung; some species were isolated from soil (Cain 1961; Arx and Storm 1967; Ahmed and Cain 1972; Domsch et al. 2007). Occasionally, the genus was observed on artificial substrates like rotten ropes or old sacks and decaying cloths (Ellis and Ellis 1998).

Preussia sp.

Preussia sp. was isolated from the surface of several specimens after different exposure durations and from probes of driving rain. As *Preussias* are a coprophilous species, their occurrence on plaster and coating surfaces may be connected to depositions of excrements from insects or birds.

Vegetative Ascomycetes, lacking generative stages: Hyphomycetes, Fungi Imperfecti

This artificial group of real fungi is characterized by the lack of generative processes under "normal" conditions and propagation is almost exclusively by vegetative processes (Kendrick and Carmichael 1973). Since the forms of the fungi imperfecti were assigned to their corresponding ascomycetes or basidiomycetes groups via molecular investigations or culture experiments ("mating experiments"), nevertheless uncertainties prevail. A so-called dual nomenclature was used in fungi for a long time, which meant that the anamorph and the teleomorph had different names defining the stages. This is useful if a teleomorph may develop several anamorphs or vice versa, but it also requires maintaining parallel taxonomies, which makes it more complicated. In the Melbourne Code of Botanical Nomenclature (McNeill et al. 2012), the fungal nomenclature and taxonomy was adopted to the botanical system which made clear that there is "one fungus–one name." This is continued by the Shenzhen Code (Turland et al. 2018) and will settle fungal taxonomy eventually; however, many relations still have to be clarified.

For practical reasons, hyphomyceteous genera are presented according to the system given by Domsch et al. (2007) with additional taxonomic information noted where appropriate. Dark pigmented deuteromycetes (spores and/or hyphae) are often called "Dematiaceae" or "dematiaceaous fungi" (black yeasts or black molds). Depending on authors, additionally the so-called Coelomycetes, Deuteromycetes with formation of pyknidia or derived structures, are separated. For our purposes, a distinction between dark pigmented forms ("melanised") and forms that are not or less pigmented was of interest since the former group usually is far more involved in the disfigurement of outer building surfaces.

Genus *Acremonium* Link ex Fr.

Acremonium is characterized by a hyaline mycelium; simple awl-shaped phialides attach to the hyphae or emerge from the hyphae (synnemata). If compound conidiophores are formed, the branches are restricted to the basal part. At the tip of the phialides, no or very short collarets can be observed. The single or rarely two-celled conidia are hyaline or pigmented and are accumulated in slimy heads and/or dry chains. Many species of the diverse genus are worldwide distributed. A great number are soil fungi or live on rotting plant material or comprise mycoparasites, and some are plant pathogens or have been identified in mycoses of humans (Samson et al. 2002a). Clinical trials are often connected to traumatic inoculations (Hoog et al. 2000). The big cosmopolitan genus *Acremonium* is divided into subgenera, whereby a part of the species of subgenus *Gliomastix* Guéguen is also listed under the melanogenous fungi ("dematiaceous fungi") because of dark pigmented spores and conidiophores (Hughes and Dickinson 1968; Ellis 1976; Domsch et al. 2007). Teleomorph forms of *Acremonium* are known only for a few species; examples are *Nectria* (Fr.) Fr. (Hypocreaceae, Hypocreales, Euascomycetes, Ascomycota), *Emericellopsis* van Beyma (Trichocomaceae, Eurotiales, Ascomycetes) and *Thielavia* Zopf (Chaetomiaceae, Sordariales, Ascomycetes) and others (Samuels 1976; Domsch et al. 2007; Hoog et al. 2000). According to current nomenclature, the genus *Acremonium* is therefore divided.

Acremonium strictum W. Gams

Colonies were moist to slimy and pink to orange colored. Conidiophores were single, rarely branched at the base. Phialides were very slender, ca. 20–60 µm in length. Conidia were usually cylindrical, rarely ellipsoidal (up to ca. 5 µm in length) and were accumulated in slimy heads. In culture, often shortened phialides sporulating immersed in agar were observed.

Acromonium strictum was recorded on diverse specimens and surfaces exposed in Germany, from air and from driving rain samples. The species is worldwide distributed and one of the most abundant *Acremonium* species on soil. *A. strictum* may also colonize dung and has been isolated from air and from fuel (Gams 1971; Domsch et al. 2007). To date, no teleomorph is known, and it seems likely that the ecologically very variable morphotaxon is polyphyletic.

A. strictum prefers moist substrates (Samson et al. 2002a). Accurate measurements for ecological growth parameters were not found in literature. Single cases of opportunistic infections in immunocompromised patients have been reported (Boltansky et al. 1984; Schell and Perfect 1996; Hoog et al. 2000).

Genus *Alternaria* Nees ex Fr.

The genus *Alternaria* is a typical representative of the black molds (dematiaceous fungi; Ellis 1971, 1976). Hyphae, conidiophores and conidia (porokonidia) of *Alternaria* are pale to dark brown. Conidiophores often are simple but may become complex by sympodial elongation. Porokonidia are produced single or in akropetalous (branched) chains They are built of an egg shaped or elongated ellipsoidal body with a rounded base and an apical beak or tip. They are muriform with several transverse and less longitudinal partitions (in some species longitudinal septa reduced or lacking). The diverse genus *Alternaria* contains more or less specialized plant parasites, many are host specific (Simmons 2007). Only a few species are ubiquitous and modest saprophytic. The teleomorphs are known in few species, e.g., *Lewia* M. E. Barr and Simmons (Pleosporaceae, Pleosporales, Ascomycetes), *Embellisia* E.G. Simmons (Pleosporaceae, Pleosporales, Ascomycetes), and *Nimbya* E.G. Simmons (Pleosporaceae, Pleosporales, Ascomycetes), *Pleospora* Rabenh. (Pleosporaceae, Pleosporales, Ascomycetes), *Leptosphaeria* Ces. and de Not. (Leptosphaeriaceae, Pleosporales, Ascomycetes) and others (Ellis 1971, 1976; Hoog et al. 2000; Pryor and Bigelow 2003; Domsch et al. 2007). Changes in the conception of the genera *Alternaria* and *Ulocladium* have been proposed (e.g., Runa et al. 2009; Woudenberg et al. 2013, 2015; Lawrence et al. 2016), but further investigations are necessary as both genera comprise huge and diverse groups.

Alternaria alternata (Fr.) Keissler (Fig. 3.89)

Conidia and conidiophores usually were golden brown. Conidiophores typically were developed solitary, straight or curved. Conidia showed a comparable short beak and usually were club shaped and arranged in long branched chains (Fig. 3.89). Conidia of *A. alternata* are very durable. They are discharged actively during a greater change of relative humidity in a diurnal rhythm, especially between 12.00 and 18.00 h (Machacek and Wallace 1952; Russel 1958; Meredith 1963; Hogg 1966; Pearson and Hall 1975).

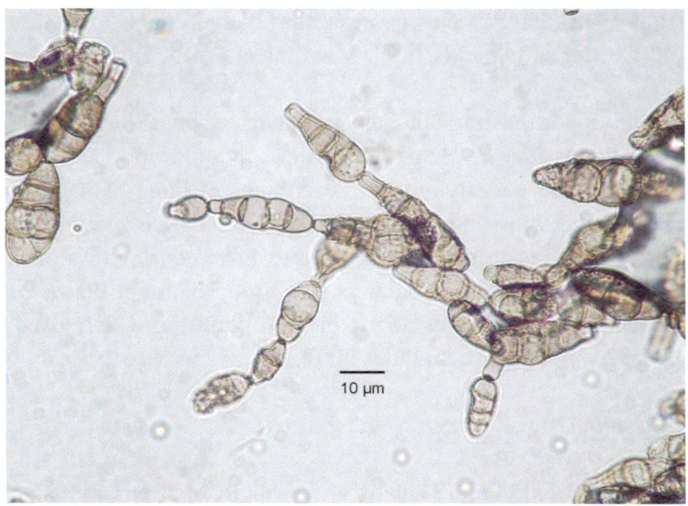

Fig. 3.89 Branched spore chains of *Alternaria alternata*

Recently, the taxonomy of the *A. alternata* species group was revised in a polyphasic approach (Armitage et al. 2015).

Alternaria alternata is very abundant and distributed cosmopolitan. In our study, it is among the most frequent fungi that were recorded from diverse specimens and surfaces. Additionally, the fungus was collected abundantly from air probes and samples of driving rain. The species can colonize diverse substrates like soil, food or textiles. It can grow on dead plant material, in various depths of soil, at different salt concentrations as well as on a great variety of living plants and seeds (e.g., Morton 1964; Domsch et al. 2007). Therefore, *A. alternata*, together with some specialized plant pathogens of the genus, has great importance in agriculture. For detection of the mycelia in barley seeds, an immunofluorescent method was developed (Warnock 1971, 1973). The occurrence of *A. alternata* on surface coatings has been reported, and it was isolated from marble monuments in Russia (Gorbushina et al. 2002).

Temperature range for optimum growth is (22) 25–28 (30) °C, maximum temperature for growth is ca. 31–32 °C, no growth is recorded at ≥ 36 °C. Minimum temperature is in the range of 2.5–6.5 °C with the exception of isolates that are cold adapted between −5 and 0 °C (Pechmann 1966; Hasija 1970; Saad and Hagedorn 1970; Årsvoll 1975). Best growth is at pH 4–5.4; between pH 2.7 and 8, growth is possible (Hasija 1970; Orynbaev and Ermekova 1973; Domsch et al. 2007; Samson et al. 2002a). Minimum humidity for growth is at −210 bar, for sporulation at −70 bar and for spore germination at −100 bar or at condensation point, respectively (Panasenko 1967; Yadav and Madelin 1968;

Pearson and Hall 1975). The species possesses a relatively high resistance against UV- and γ-radiation (Durrell and Shields 1960; Mirchink et al. 1972).

It may occur as an opportunistic pathogen in humans, especially in connection to traumata or immunosuppression. Virulence is regarded as quite low, and healing of infections was observed if the underlying disease is cured (Chung et al. 1999; Hoog et al. 2000). The relatively big muricate spores possess an allergenic potential. This may lead to hay fewer like symptoms at harvest time of cereals and during or after thunderstorms (Pulimood et al. 2007). The species belongs to a group of fungi that are more tolerant against certain compounds (biocides), as is the case with the widely used carbandazim (e.g., Paulus 2005; Avenot and Michailides 2007).

Alternaria tenuissima (Kunze ex Pers.) Wiltshire

Unlike *Alternaria alternata* conidia chains were usually shorter and also single conidia were found. The more slender conidia typically possessed a much longer beak and were often paler.

Alternaria tenuissima is a modest saprophytic species with worldwide distribution (Hoog et al. 2000). We recorded *A. tenuissima* from numerous specimens with greater scarcity on older surfaces. Possibly, it is a pioneer species that later vanishes. Furthermore, we found the fungus in air probes by filter and sedimentation method and in samples from driving rain. The fungus grows on many different substrates, e.g., soil, dead and living plant tissues or seeds, and has been isolated from air.

Bibliographic data of environmental factors (temperature, humidity) for conidia germination and vegetative growth lie within the boundaries that are known for *Alternaria alternata* (Domsch et al. 2007), although precise data on physiological boundary values lack.

A. tenuissima was reported causing phaeohyphomykoses in humans, although in general the species was only found if there was an underlying severe disease (Contet-Audonneau et al. 1991; Drouhet et al. 1991; Hoog et al. 2000).

Alternaria sp.

In some probes, forms of *Alternaria* were found that could not be assigned to a certain species because of deviating characters.

Genus ***Arthrinium*** Kunze ex Fr.

Arthrinium forms a pale mycelium. Conidiophores consist of a basal incrassate cell with a terminal thin and more or less elongated part at which blastoconidia are formed. Conidia are unicellular and strongly pigmented and they possess a germination slit. Many species are specialized saprophytes which occur on dead parts of certain plants; only a few are more commonly distributed. Some species form a teleomorph, genus *Apiospora* Sacc., Apiosporaceae, Sordariales, Ascomycetes.

Arthrinium phaeospermum (Corda) M. B. Ellis (Fig. 3.90)

Sporulation took place at irregular regions of the air mycelium which then turned sooty black. Especially in nutrient poor media (BBM, CD50) good sporulation was observed, but hyphae were hardly recognizable. According to literature (Ellis 1971; Domsch et al. 2007), the elongated part of the conidiophore is septate, but this character was hardly recognizable by microscopy. Conidia were lens shaped, dark brown, variable in size but in average ca. 10 μm in diameter with an equatorial germination slit at the broad side (Fig. 3.90).

Arthrinium phaeospermum is distributed worldwide.

We detected *A. phaeospermum* on the surface of specimens and in air probes.

In the wild, the species is observed very abundant on dead plant material, especially grasses. Hofbauer (2007) reported it for the first time on the surface of building parts.

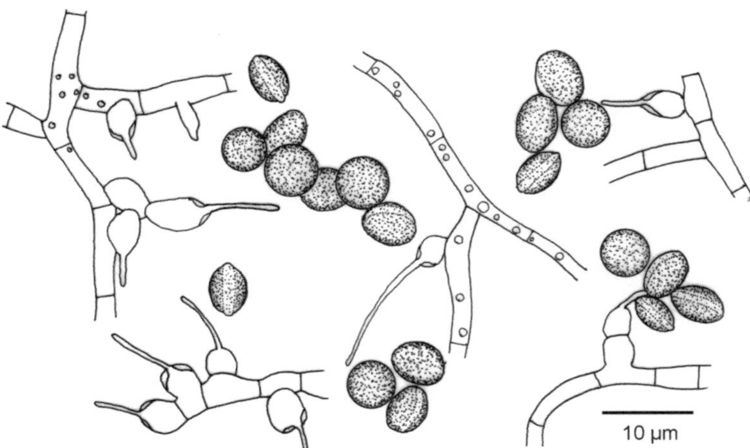

Fig. 3.90 *Arthrinium phaeospermum*, hyphae with conidiogenous structures and spores (drawn from a microscopic preparation of a raw culture on a sedimentation plate)

Little is known about ecophysiological data as ranges of temperature, humidity and pH. Sporadically skin infections in humans are reported, and nevertheless, the virulence of the species is regarded as low (Hoog et al. 2000).

Genus *Aspergillus* Micheli ex Fries em. Samson et al.

The genus *Aspergillus* is characterized by upright conidiophores that end in a vesicle which is either covered in phialides (uniseriate) or in supporting cells (metullae) that bear the phialides themselves (biseriate). Conidia are produced in basipetalous chains (more or less strongly connected by connectives); they are globose or egg shaped to elongated and often typically verrucous or echinulate ornamented. The name of the genus is derived from a holy water sprinkler, in Latin "Aspergillus." For the anamorphic form-genus *Aspergillus,* several teleomorph stages have been described, but according to current nomenclature they either belong all to the one genus aspergillus or the whole group has to be divided subject to teleomorph characters. Nevertheless *Aspergillus* forms a monophyletic group within the family Trichocomaceae Fischer (=Eurotiaceae Clem. et Shear), Eurotiales, Ascomycetes. Examples for teleomorphs are: *Emericella* Berk. and Br., *Fennellia* Willey and Simmons, *Eurotium* Link ex Fr. (see above), *Neosartorya* Malloch and Cain. Further teleomorphs are mentioned in the literature (Domsch et al. 2007; Klich and Pitt 1988). A special organization was installed which deals with the taxonomy of *Aspergillus* and *Penicillium*, the International Commission on Penicillium and Aspergillus (ICPA). According to progress in nomenclature and taxonomy, the genus has been rearranged (Samson et al. 2014), and here, we refer to Domsch et al. (2007). The genus *Aspergillus* is very diverse and of great importance in medicine (poisoning by mycotoxins through ingestion of spoiled food, especially Aflatoxins, infections in immunocompromised patients which are called aspergilloses), but also in food producing industry (for certain fermentations) and biotechnology.

Aspergillus candidus Link

Aspergillus candidus is characterized by lack of pigmentation (even whole colonies on agar are mostly pure white to yellowish cream) and smooth conidia. Aspergilli were mainly biseriate, and reduced conidiophore occurred which were uniseriate. Some strains produce sclerotia which are pink initially and eventually turn black (Domsch et al. 2007; Samson et al. 2001; Klich 2002); this was not observed in our isolates.

A. candidus is a frequent species which occurs especially in tropical and subtropical zones.

We recorded the species in air samples (filtration).

The fungus is known from diverse soils, seeds, food and dung, from air and from indoors (Mallea et al. 1972; Domsch et al. 2007; Samson et al. 2001; Klich 2002).

A. candidus shows optimum growth at a temperature range of 25–28 °C, minimum temperature range for growth is 11–13 °C; the maximum being 41/42–44 °C (Ayerst 1969; Franz 1975; Domsch et al. 2007). According to the literature, the species seems to prefer neutral to slightly base rich conditions (Domsch et al. 2007), although systematic measurements are lacking. Minimum water potential for growth under experimental conditions is −300 bar or a minimum a_w-value of 0,776 is given; optimal is −30 bar (Ayerst 1969; Wheeler and Hocking 1988). The species tolerates up to 20% NaCl or up to 40% sucrose in the medium (Chen 1964; Moustafa and Al-Musallam 1975).

That the species produces the mycotoxin Citrinin has been contested (Umeda et al. 1974; Frisvad 1989). Infections and allergies in humans have been reported (Hoog et al. 2000; Krysińska-Traczyk and Dutkiewicz 2000).

Aspergillus fumigatus Fres. agg.

Conidiophores were pigmented and possessed club-shaped vesicles with a funnel-shaped neck which is a crucial diagnostic character. Vesicles were uniseriate; phialides covered the upper half or up to two thirds of the surface whereby the lower ones were a little curved. Thus, the sharply pointed tips of the phialides were always more or less parallel to the main axis. The basal part of the phialides usually was strongly pigmented, and the heads were rather dark. The species was therefore mentioned as "dematiaceous fungi" in Ellis (1971). Conidia were shaped globose to broad ellipsoidal, usually delicate rough to delicate spiny, ca. 2–3.5 µm in diameter. Conidia chains were often united to narrow clubs; depending on culture medium, they also spread in several rows (especially if cultured on DG18). The color of the conidia respectively colonies typically was gray turquoise but can also be smoke gray or pale gray; usually, areas with rests of turquoise color remained.

Aspergillus fumigatus is commonly distributed.

We isolated *A. fumigatus* from several specimens and surfaces and frequently from air. *A. fumigatus* was documented for the first time as a component of initial colonization of modern coating systems by Hofbauer (2007). The species was documented on marble statues in Russia (Gorbushina et al. 2002).

A. fumigatus grows in a temperature range of 12–57 °C, with an optimum range of 37–43 °C (Crisam 1973; Franz 1975; Samson et al. 2002a). It is therefore one of the most thermotolerant eukaryote species known. But in a strict sense, it does not belong to the so-called thermophilic fungi because it also grows below 20 °C. At the growth place, pH 5–8.5 was measured; in the laboratory, good growth takes place at pH 3.7–7.8 (Domsch et al. 2007). *A. fumigatus* can degrade the herbicides Atrazine and Simazine (Kaufman et al. 1963; Kearny et al. 1965; Kaufman and Blake 1970). The fungus is able to produce different mycotoxins; the most important are: Gliotoxin, Verrucologen and Fumitremorgin A and B (Samson et al. 2002a).

The fungus is labeled as pathogenic, allergenic and mycotoxigenic. Infections have been recorded not only in immunocompromised patients but also in otherwise healthy persons (Hoog et al. 2000). The whole genome of *A. fumigatus* is decoded (Nierman et al. 2005). For the future, it is expected that this will enable the identification of corresponding gene sequences for virulence and allergenic and toxigenic features and the design of effective countermeasures or protecting measures, respectively. Spores are frequently present in our environment, especially in places with decomposing plant material, e.g., compost or waste heaps (Klich 2002). Many sibling species of *A. fumigatus* have been described; some of them are distinguished mainly by chemotaxonomic features (Geiser et al. 1998; Balajee et al. 2005, 2006; Katz et al. 2005; Pringle et al. 2005; Balajee and Marr 2006; Hong et al. 2005, 2007; Yaguchi et al. 2007).

Aspergillus niger van Tiegh.

Conidiophores were quite big with spherical biseriate vesicles. Club-shaped metullae covered almost the whole surface; they were relatively long and bear several tubes to bottle-shaped phialides each. Conidia were produced in chains which were not connected to one another but formed single distorted rows. Conidia were 4–5 µm, spherical and warty to spiny at the surface.

Aspergillus niger is distributed worldwide.

In our study, the fungus did not occur on outdor samples, and it was only found on not exposed specimens (0-probe).

The species was previously reported from diverse substrates (Domsch et al. 2007), indoors and on food (Samson et al. 2002a). It was also isolated from frescoes of a cloister and from marble statues (Ionita 1973; Gorbushina et al. 2002).

Minimum temperature for mycelial growth is 11–13 °C; optimum temperature range is 17–42 °C and maximum temperature 47–48 °C (Gibson 1953; Ayerst 1969; Domsch et al. 2007). Regarding the pH value of the substrate, the fungus is very tolerant; even below pH 3, growth was reported in a nutrient solution (Galbraith and Smith 1969). A water potential of −30 bar is optimal for growth on nutrient agar; values down to −310 bar are tolerated (Chen 1964; Corry 1973; Kulik and Hanlin 1968; Ayerst 1969; Corry 1973; Rai and Agarwal 1974). *A. niger* can cope with up to 5% NaCl (Mert and Dizbay 1977). Mycelia and spores are quite resistant against gamma irradiation (Tardieux et al. 1972). Because of the dark pigmentation, the species is mentioned among the "dematiaceous fungi" (Ellis 1971).

Different cases of infections in humans have been documented; mainly infections of the outer auditory canal, virulence is regarded as low (Domsch et al. 2007; Hoog et al. 2000). The fungus is an important organism for biotechnology. The genome of this species is decoded (Pel et al. 2007), which will promote an even more effective use of the species in biotechnology. The section *Nigri* has been revised, and many new species were split off or newly described (Abarca et al. 2004; Samson et al. 2004).

Aspergillus sydowii (Bain. et Sart.) Thom et Church

Aspergillus sydowii is very close to *Aspergillus versicolor* (Vuill.) Tiraboschi. Important distinguishing features are the intense blue-green color of the conidia which are also spinier at the surface (Domsch et al. 2007; Klich 2002). In future, genetic investigations will help to better understand the rank of the taxon.

A. sydowii is distributed worldwide.

The species was recorded from specimens exposed in Germany after only one year.

It colonizes various substrates outdoors and indoors and has been isolated from air (Domsch et al. 2007; Samson et al. 2002a).

Little is known on temperature reference for the species. In vitro, it can grow until a water potential of -230 bar. A minimum water activity (a_w) for growth was measured with 0,776, and a NaCl content up to 30% was tolerated (Chen 1964; Kouyeas 1964). Infections in humans are known (Domsch et al. 2007; Hoog et al. 2000; Samson et al. 2002a; Wheeler and Hocking 1988).

Aspergillus terreus Thom

Observed features in our isolates corresponded well with the descriptions in literature (e.g., Domsch et al. 2007; Klich 2002). Colonies were velvety and brown in color. The hemispherical to slightly club-shaped vesicles were biseriate; metullae were usually found only in the upper half of the vesicles. Spore rows were often united and formed long cylindrical aggregates, and therefore, the habit reminds somewhat of *Aspergillus fumigatus*.

The species belongs to the soil fungi with worldwide distribution.

A. terreus was detected on a variety of specimen variants and was isolated from ambient air.

This soil fungus is more abundant in tropical and subtropical regions (Domsch et al. 2007), and it was also isolated from the surface of marble statues in Russia (Gorbushina et al. 2002).
Optimum temperature for growth ranges from 35 °C to 40 °C; minimum temperature is 11–13 °C and maximum 45–48 °C (Pore and Larsh 1967; Franz 1975; Domsch et al. 2007). Down to a water potential of −260 bar and a pH range of ca. 2–8, growth is possible (Chen 1964; Tresner and Hayes 1971).

The fungus may produce various toxic metabolites, e.g., Patulin, Citrinin and Citreoviridin. Infections in humans have been documented (e.g., Hoog et al. 2000; Samson et al. 2002a; Baddley et al. 2003).

Aspergillus versicolor (Vuill.) Tiraboschi

Investigated isolates showed typical characters. Colonies were often multicolored; apart from green-gray portions, also yellowish or orange tones were present. Conidiophores were smooth with oval vesicles and biseriate. At the margin of older colonies, often reduced conidiophores were found; they were relatively small and short with only few metullae and phialides. In fully developed conidiophores, metullae and their phialides covered most part of the surface of the vesicle. Spore rows were not united as a whole; spore heads therefore appeared more or less radially. Conidia were spherical with delicate spines at the surface, ca. 2–3 µm in diameter. In mycelium of older colonies, so-called Hülle-cells were found which are characteristically inflated, and sometimes had distorted thick cell walls.

Aspergillus versicolor occurred irregularly on the surface of specimens exposed in Germany. It was also isolated from a driving rain probe for the first time in Germany (Hofbauer 2007).

A. versicolor is a saprophytic fungus with worldwide distribution which is found in almost all types of soil and which may colonize various substrates. It was previously isolated from air and from frescoes of a cloister (Hudson 1969; Mallea et al. 1972; Ionita 1973).

Minimum temperature range for growth is 4–10 °C; optimum temperature is 21–22 °C or 25–30 °C (dependent on geographical origin of the isolate); maximum is 40 °C (Trique 1970; Franz 1975; Samson and Mouchacca 1975; Rabie et al. 1976). Minimum water potential for growth in soil is ca. −350 bar, for spore germination −290 bar. The fungus even may grow at salt concentrations up to 30% NaCl or concentrations of sucrose up to 40%. It is therefore regarded as very xerotolerant (Griffin 1963a, b; Chen 1964; Tresner and Hayes 1971; Mislivec et al. 1975; Moustafa and Al-Mousallam 1975; Samson and Mouchacca 1975; Domsch et al. 2007).

Apart from various secondary metabolites, the fungus may produce the nephrotoxic mycotoxin Sterigmatocystin. Infections in humans have rarely been recorded (e.g., Rabie et al. 1976; Vesonder and Horn 1985; Hoog et al. 2000; Engelhart et al. 2002; Samson et al. 2002a). Progress in taxonomy of *Aspergillus* section *Versicolores* has led to the description of new species (Jurjevic et al. 2012).

Genus *Aureobasidium* Viala et Boyer

Aureobasidium usually forms slimy colonies. Initially hyphae are almost hyaline; they often turn brown and thick walled in age, at least partly. Conidiogenous cells are intercalary, lateral or terminal of the hyphae. Unicellular blastoconidia are formed at short projections or directly at the conidiogenous cell. Conidia are of various size and form and yeast like. Dark parts of hyphae often divide into portions and then function like chlamydospores or arthroconidia. Although further work on *Aureobasidium* has been presented (Zalar et al. 2008; Gostinčar et al. 2014); discussions about the rank of the different taxa are ongoing.

Aureobasidium pullulans (De Bary) Arnaud

Young colonies were slimy, pink and yeast like. Starting from the center, they often developed in concentric rings with typical dark or brown colored sections. Depending on the isolate, the tendency to develop dark parts was very different. Whereas some cultures needed several weeks to produce small and weak pigmented parts, others turned dark much faster and intense. According to the literature, this may also be a sign for the presence of different sibling species (Gostinčar et al. 2014). The dark pigmentation is due to production of incrassated cell walls and Melanin. In parts of the melanised hyphae, rounded cell ends were found; these structures then acted as chlamydospores. Conidia were produced at small projections of the conidiogenous cells which were hyaline or pigmented. Often these projections were hardly seen by light microscopy. The more or less hyaline or pink conidia were elongated ellipsoidal and slightly irregular in outline; they represented a yeast stage.

Aureobasidium pullulans is a worldwide distributed saprophyte.

The fungus was frequently detected on specimens at every outdoor location, on various surfaces, in air and driving rain probes. In general, the germ figures recorded on mineral surfaces were higher than those on other building coating surfaces.

According to the literature, the species is a saprophyte, especially on plant surfaces (Domsch et al. 2007). Furthermore, it occurs frequently on surfaces of building materials like concrete and paint and especially on timber where it is regarded as an important blue stain fungus (Reynolds 1950; Domsch et al. 2007). It was already isolated in air probes (e.g., Ellis 1971) and was documented on marble statues in Russia (Gorbushina et al. 2002). According to its characteristics, the species is a typical example of black fungi (strongly pigmented permanent cells with incrassate cell walls), black yeasts (conidia are yeast like, pigmentation) and blue stain fungi (formation of typical stain on timber), and it is often incorporated into the artificial group of dematious fungi (Ellis 1971).

The temperature range for growth of *Aureobasidium pullulans* lies between 2 °C and 35 °C, with an optimum at 25 °C (Pechmann 1966, Domsch et al. 2007). The fungus was recorded at a pH range of 3.6–6.5 (Christensen 1969; Cooke 1970; Singh 1976; Domsch et al. 2007). The species is relatively resistant against UV- and gamma irradiation (Domsch et al. 2007). A salt-tolerant strain produces MAAs which may act as sunscreen and/or compatible solute (Kogeij et al. 2006).

Infections in humans, mainly after traumatic inoculations, have been documented but virulence rate of the fungus is low (Hoog 2000).

Genus *Botrytis* Micheli ex Fries

Botrytis forms fast growing colonies, which are white initially and turn gray later due to sporulation. Frequently dark sclerotia are developed. Conidiophores are upright, branched in the upper part and bear terminal inflated conidiogenous cells which produce blastoconidia synchronously. From individual species a teleomorph, *Botryotinia* Whetzel (*Sclerotinia* Fuckel p.p.) is known.

Botrytis cinerea Pers.

Colonies grew very fast and produced a huge amount of loose air mycelium. Depending on the isolate the inclination to sporulate varied from low to strong. In the majority, also depending on isolate, black, irregularly formed and differently sized sclerotia were formed. The conidiophores were stiff and upright, the main axis was colored golden brown. The branches were hyaline and produced at the inflated terminal cells many (globose to) egg shaped to pear-shaped hyaline or bright gray-colored spores (in average ca. 12 μm long, variable).

We isolated *Botrytis cinerea* from several specimens which were exposed in different parts of Germany up to 3 years and also from air. The species was documented for the first time as component of the initial colonization of modern building coatings (Hofbauer 2007). Previously it was isolated from marble statues in Russia (Gorbushina et al. 2002).

Minimum temperature range for growth is from −2 to 5–12 °C, optimum temperature is (21) 22–25 (30) °C, maximum temperature is (28) 33–35 °C (Hennebert and Gilles 1958; Kuehn and Gunderson 1963; Årsvoll 1975; Domsch et al. 2007). pH range for growth is 2–8 with an optimum at 3–5 (Jarvis 1977). Minimum water potential for vegetative growth lies at −100 bar (Griffin 1963a).

Apart from *Botryotinia fuckeliana* (de Bary) Whetzel (*Sclerotinia fuckeliana* [de Bary] Fuckel) further forms were recorded as teleomorphs, therefore the taxon is now seen as a species complex (e.g., Samson et al. 2002a). *B. cinerea* is a typical example of the ambivalence of so-called molds. On the one hand, "gray mold" is dreaded as plant disease in agriculture and as cause of "collapsing disease" in germling plants by gardeners. On the other hand it is responsible for "pourriture noble" in the production of wine from late-gathered grapes, but only if it occurs on ripe grapes.

Genus ***Cladosporium*** Link ex Fries

Cladosporium forms velvety to flaky dark brown to black colonies, often with an olive-brown or olive-green tinge. Hyphae, conidiophores and conidia are all pigmented, therefore the genus is often placed in the artificial group of dematiaceous fungi. The straight or bent conidiophores project upright from the mycelium and are often branched in their upper part. At several conidiogenous loci acropetal and branched chains of conidia are produced. Whereas the basal conidia often are elongated and may be septated, the upper conidia get shorter and are unicellular or possess a defined number of septa. This gives the impression of the whole conidiogenous apparatus being fragmented into conidia of different sizes. Conidia outline is variable but usually there is a minimum size and form. Projecting scars are typical, with one only at terminal conidia and numerous in conidia from branching sections. The teleomorph is named *Mycosphaerella* Johanson, Mycosphaerellaceae, Capnodiales, Ascomycetes and is regarded as polyphyletic (Crous et al. 2007). Taxonomy in *Cladosporium* advanced in the last years with the description of many new species (Bensch et al. 2012). The greater part of the numerous species consists of specialized plant parasites or "plant saprophytes", respectively. Only a few are commonly distributed and within them we find the most frequent air germs (Infante et al. 1999).

Cladosporium cladosporioides (Fresen.) de Vries (Fig. 3.91)

Our isolates of *Cladosporium cladosporioides* formed dark velvety colonies, olive green to olive brown. Conidiophores did not show sympodial elongation and did not possess conspicuous swellings. Conidiophore and conidia formed dense clusters which reminded vaguely of microscopic opuntia-cacti. Conidia were comparatively slender and smooth or very delicately rough at the surface. Two to multi celled conidia were rare (Fig. 3.91).

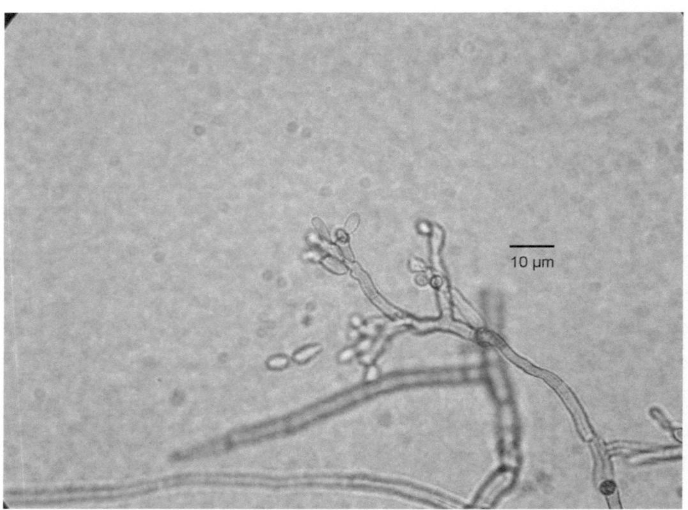

Fig. 3.91 Conidiophore and basal spores of *Cladosporium cladosporioides*. By light microscopy conidia appear almost smooth at the surface

Terminal conidia were single celled and often very small (about 2–3 µm long), narrow egg shaped with a projecting scar.

This fungus counts to the most frequent and abundant air germs worldwide with an emphasis in the temperate zone. For the British Isles seasonal abundance patterns were recorded (Harvey 1967).

In our study *C. cladosporioides* was very frequent on almost all the specimens exposed at different places in Germany, on almost all investigated surfaces and in ambient air. Only in driving rain probes it was less abundant. In our investigations the fungus can be regarded as one of the most abundant species.

It is a modest ubiquist and is found on many different substrates and in soil. Gorbushina et al. (2002) isolated it from marble monuments in Russia.

Temperature range for growth lies at −10–32 °C with an optimum at 20–28 °C (Pechmann 1966; Eveleigh 1970; Årsvoll 1975; Domsch et al. 2007). About the moisture requirements little is known. It is likely that the species needs a relatively high moisture (Domsch et al. 2007). As to pH no data was found in literature. A salt tolerant strain produces MAAs which may act as sunscreen and/or compatible solute (Kogeij et al. 2006).

The species is relatively resistant against gamma irradiation (Mirchink et al. 1972). Infections in humans are known, although virulence is regarded as low (Hoog et al. 2000).

Cladosporium herbarum (Pers.) Link ex S. F. Gray

Colonies of investigated isolates were dark, often olive brown and usually rougher than those of *Cladosporium cladosporioides*. Conidiophores were often sympodial elongated and showed conspicuous swellings at branching points or terminal. Conidia were elongated to cylindrical and conspicuously roughened at the surface. Often two-celled conidia with one transversal septum were present. Terminal conidia were also small, elongated and ca. 5–12 µm long.

This species is ubiquitous.

Cladosporium herbarum was very frequent in our study. It was isolated from a major proportion of our specimens exposed at different sites in Germany at all sample times, from probes of various building surfaces and from air. It almost lacked in probes of driving rain. Altogether, this species was slightly less abundant than *Cladosporium cladosporioides* and presented lower germ figures.

C. herbarum is among the most abundant colonizers of organic substances, especially on dead plant material, and it is frequently isolated from air (Domsch et al. 2007). For the British Isles, a seasonal abundance pattern was measured (Harvey 1967). The species has been detected on building surfaces, on frescoes (Ionita 1973), on marble monuments (Gorbushina et al. 2002) and on damp plaster (Hawksworth 1976). In some isolates, the teleomorph, *Mycosphaerella tassiana* (de Not.) Johanson was recognized.

Minimum temperature for growth is −6 °C, maximum temperature range is 28–32 °C, and the optimum is 18–28 °C, depending on the isolate (Brooks and Hansford 1923; Joffe 1962; Pechmann 1966; Årsvoll 1975; Domsch et al. 2007). A pH range for growth is set from 4.4 to alkalinity with an optimum at pH 6 (Feher and Besenyei 1933; Brown 1958; Mehrotra and Kakkar 1972; Domsch et al. 2007). Minimum water potential for vegetative growth is −210 bar. The fungus is therefore regarded as xerotolerant (Domsch et al. 2007).

The species possesses quite a high tolerance against UV- and gamma irradiation (Durrell and Shields 1960; Johnson and Osborne 1963). Infections of humans caused by *Cladosporium herbarum* are rare; the virulence of the species is regarded as low (Hoog et al. 2000).

Cladosporium macrocarpum Preuss (Fig. 3.92)

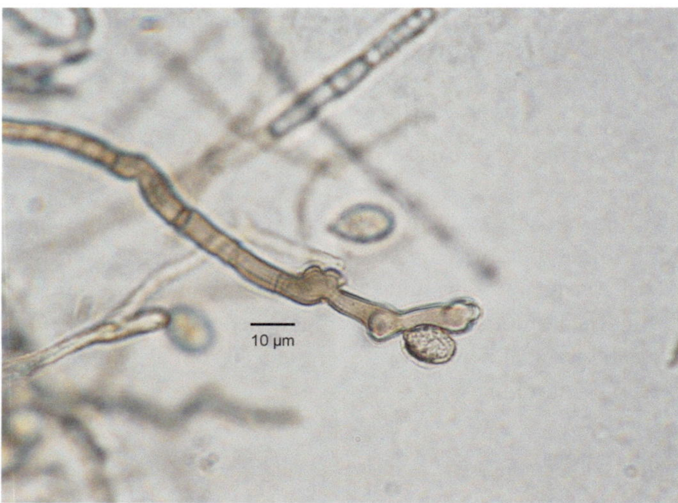

Fig. 3.92 *Cladosporium macrocarpum*, conidiophore with a single unicellular conidium. Note the conspicuous swellings from the conidiophore (microscopic preparation of a colony from raw culture of a sedimentation sample from air in Holzkirchen)

Cultures showed slower growth than other isolated species of *Cladosporium*; they were somewhat irregular in outline and dark brown in color. The conidiophores were sympodial elongated similar to *Cladosporium herbarum* and showed distinct swellings at branches or terminally (Fig. 3.92). In general, conidiophores and conidia chains were only little branched. Terminal conidia were bigger than those of the other discussed species (ca. 10–15 μm long), delicately rough and appeared swollen; mostly they were unicellular but could have up to three septa.

Cladosporium macrocarpum is commonly distributed in temperate zones.

We isolated *C. macrocarpum* only from air in Holzkirchen, although several times, using the sedimentation method.

The species is mainly known from dead plant material and from timber surfaces, but is much rarer than the previously discussed species (Domsch et al. 2007). It was previously isolated from air and seasonal abundance patterns were measured in the British Isles (Harvey 1967).

Cladosporium sphaerospermum Penz.

Colonies of the isolates were very similar to *Cladosporium cladosporioides*, but generally darker because of thicker cell walls. Conidia, including the basal ones, were usually shorter and more inflated in comparison to *C. cladosporioides*. Often terminal chains of uniform conidia were formed, each ca. 4–5 µm in diameter. Besides also smaller terminal conidia occurred, which were similar to *Cladosporium cladosporioides* and less frequent.

The species is common and cosmopolitan.

We isolated *Cladosporium sphaerospermum* only from air in Holzkirchen.
The species occurs on various organic substrates often as a secondary colonizer. It has frequently been isolated from air and from paint surfaces (e.g., Harvey 1967; Ellis 1971).

A salt tolerant strain seems to be able to produce MAAs which may act as sunscreen and/or compatible solute (Kogeij et al. 2006). Only few infections of humans have been recorded; the virulence of the taxon is regarded as low (Hoog et al. 2000; Yano et al. 2003).

Cladosporium sp.

In some analyses of our specimens, obscure forms of *Cladosporium* were found with weak deviant growth and very little sporulation. This did not allow a reliable identification.

Genus *Coniothyrium* Corda em. Sacc.

Coniothyrium belongs to a group of fungi which can form pyknidia and usually is incorporated into the Coelomycetes (Sutton & Carmichael 1973; Sutton 1980). According to current taxonomy, the position of the genus is accomodated at least in part to Coniothyriaceae W. B. Cooke, but some forms still are unresolved (Aveskamp et al. 2010; Gruyter et al. 2012). Pyknidia are pale or strongly pigmented, spherical or form an irregular folded hollow. Conidiogenous cells (phialides) are hardly different from the inner wall forming tissue, broad flask shaped with a narrow phialidic opening; paraphyses may be present. Conidia are usually unicellular, rarely two celled, with pale to dark brown color. Pigmentation of the spores is an important diagnostic feature. The genus *Leptosphaeria* Ces. and de Not., Leptosphaeriaceae, Pleosporales, Ascomycetes, forms a teleomorph stage of the form-genus.

Coniothyrium sp.

In our study, some coelomycetes with typical characters of *Coniothyrium* were isolated from older surfaces. They formed a pigmented mycelium in raw culture with sparse spherical pyknidia containing gray unicellular and ellipsoidal conidia (ca. 3–5 µm long), arising from simple phialides. Unfortunately, subcultures were not successful, and thus, further identification was not possible.

Genus *Drechslera* Ito

Some authors propose to split off taxa from the originally rather broadly understood form-genus (Shoemaker 1959; Subramanian and Jain 1966; Ellis 1971, 1976; Leonard and Suggs 1974; Domsch et al. 2007). Further to this, there are also arguments about the classification of teleomorph taxa. Conidiophores of *Drechslera* are upright or ascending, pigmented and geniculate due to sympodial overtopping. Through pores/openings ellipsoidal, tube-shaped and pigmented conidia are produced. Conidia are usually straight.

Drechslera poae (Baudys) Shoemaker (Fig. 3.93)

Investigated isolates sporulated copious on nutrient poor media (BBM, CD50). Conidiophores arose single or in groups, upright and geniculate. Conidia were up to 100 µm long, straight, subcylindrical (usually broadest below the middle), mature golden brown and with five to maximal eight septs (Fig. 3.93).

In our investigations, *Drechslera poae* (designation according to Ellis [1971]) was found only on several specimens in Holzkirchen which implied the first recording on building materials and in the air (Hofbauer 2007).

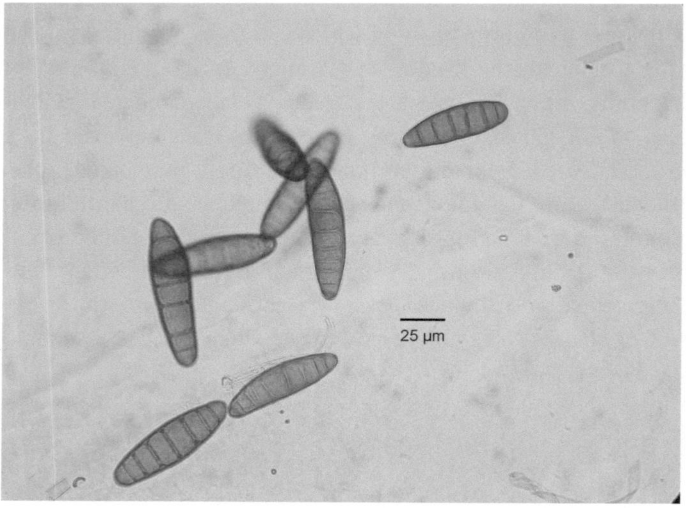

Fig. 3.93 *Drechslera poae*, several typical conidia

The species is mainly known as parasite on different grasses, especially the genus *Poa* L. (Ellis 1971).

Genus *Epicoccum* Link ex Schlecht.

Epicoccum forms are fast growing colonies with loose air mycelium. The production of yellowish, reddish and greenish pigments which are released into the agar is typical. On rich media, sporulation takes place only late or does not happen at all. On nutrient poor media, sporulation is usually copious (depending on isolate). Hyphae are colorless or pigmented brownish. Pigmented conidiophores are united in sporodochia. Conidia are black (golden brown to dark brown by microscopy), spherical, multicellular and possess a thick warty outer wall. Where the spore was attached, a typical pale projecting scar remains. According to recent taxonomy, *Epicoccum* is integrated into the family Didymellaceae (Chen et al. 2017).

Epicoccum nigrum Link (Fig. 3.94) (Syn.: *Epicoccum purpurascens* Ehrenb. ex Schlecht.)

The characters of the species matched the description of the genus (Fig. 3.94). It was remarkable that isolates from outdoor surfaces produced such copious amounts of pigment that the color of agar turned black. Some isolates sporulated poor even on nutrient poor media and under light irradiance.

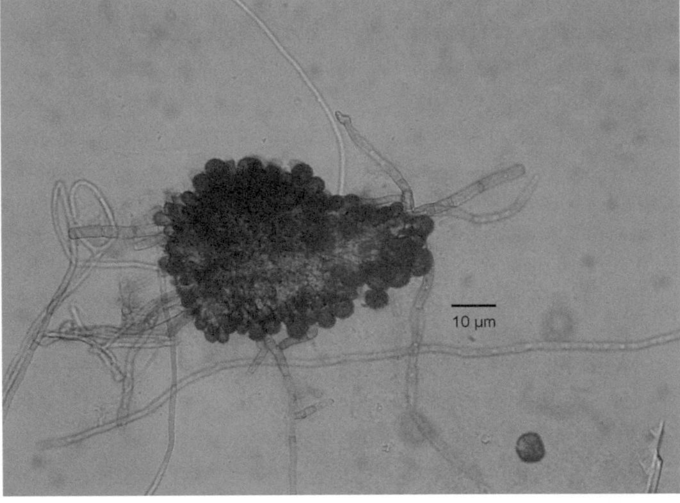

Fig. 3.94 Typical sporodochium of *Epicoccum nigrum* with a single mature spore detached

Epicoccum nigrum is distributed worldwide.

The taxon was one of the most abundant species in our study. *E. nigrum* was isolated from almost every analyzed specimen, from all surfaces and air samples and only lacked in a few probes. It was not abundant in driving rain probes.

The species occurs on a wide variety of substrates and is a frequent secondary colonizer of dead plant material (Schol-Schwarz 1959; Ellis 1971; Domsch et al. 2007). Furthermore, it is known as frequent air germ (Schol-Schwarz 1959; Meredith 1966; Ingold 1971). Riedl and Hubacek (1988) undertook trials to use *E. nigrum* as a protecting agent against detrimental microorganisms on building surfaces, but the method turned out not to be applicable.

Optimum temperature range for growth is 23–28 °C; minimum temperature is -3–4 °C and maximum temperature 45 °C (Mulder and Pugh 1971; Årsvoll 1975; Domsch et al. 2007). No growth takes place at $pH \leq 2$ and ≥ 10, optimum pH for growth lies at 5–6 (Domsch et al. 2007). For germination, a water potential of ≥ -110 is necessary, optimal growth takes place at vapor saturation (Webster and Dix 1960; Pelhate 1968).

Genus *Fusarium* Link ex Fr.

Colonies of *Fusarium* usually grow very fast, mostly pale or in pastel colors of pink, orange or yellow. Often copious air mycelium is produced. Sporulation happens diffusely, or sporodochia are formed. Conidiophores are branched at the base and can build complex three-dimensional structures. Sporogenous cells are phialides; depending on species or on location in the mycelium or at the conidiophore, they are shaped differently and may have more openings. Conidia are accumulated in slimy masses. Typical is the formation of two different types of conidia: unicellular to two-celled microconidia and transverse septate curved macroconidia which usually possess characteristic foot cells and a typical sickle shape. Between the two forms of conidia, all transitions may occur in addition. Depending on species and isolate, the emphasis is more on the production of micro- or macroconidia. Furthermore, some species form additionally characteristic chlamydospores. The taxonomy of the species rich form-genus is very complex and often demands profound investigations for final determination (e.g., Gerlach 1970; Booth 1971b, 1977; Joffe 1974; Gerlach and Nirenberg 1982; Nirenberg 1990). *Nectria* (Fr.) Fr., *Gibberella* Sacc., *Calonectria* de Not. and *Plectosphaerella* Kleb., all Nectriaceae, Hypocreales, Ascomycetes are examples of teleomorphs of *Fusarium*. The greater part of the species comprises soil fungi with worldwide distribution. Some species or strains are plant parasites. Human infections mainly occur in connection with traumatic inoculation (Hoog et al. 2000). Many species are food spoilers and important toxin producers (Domsch et al. 2007).

Fusarium sp.

In our analysis of several surfaces, air probes and probes of driving rain different types of *Fusarium* were found but not further differentiated. The most abundant form showed characters which match the species complex of *Fusarium solani* s.l.

Genus ***Harzia*** Costantin

At the beginning, colonies are colorless, later cinnamon to brown with numerous conidiophores. Colorless conidiophores are easily branched; the branches often set off in a right angle and with a fine and acute tip. Conidia are relatively big, egg shaped or spherical, golden to cinnamon, smooth to warty. They possess a double wall, the inner one thickened with a pore at the attachment. Additionally, *Acremoniella* may show a phialidic stage with much smaller and simpler spores (Ellis 1976). The small genus is cosmopolitan and comprises soil saprophytes and parasitic forms.

Harzia acremonioides (Harz) Cost. (Fig. 3.95) (Syn.: *Acremoniella atrata* [Corda] Sacc.)

The delicate mycelium spreaded on and in the agar. On BBM-agar clusters of colorless conidiophores with rectangular branches developed (Fig. 3.95). Conidia were unicellular, egg shaped and golden brown colored. The outer wall of the conidia was wrinkled in some spores. According to Ellis (1976), sometimes additionally a second type of conidiophores with phialides is formed. This could not be observed in our isolates.

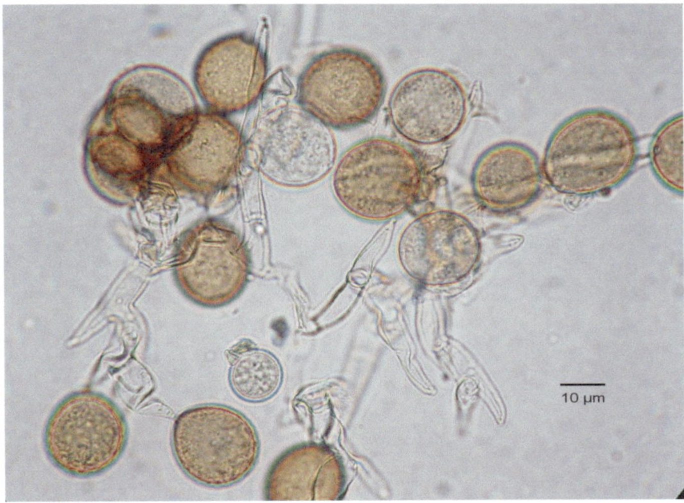

Fig. 3.95 *Harzia acremonioides*, with conidiophores in different stages of development (raw culture from a sedimentation probe)

Harzia acremonioides was only found in sedimentation probes of ambient air.

The fungus was previously recorded on various plants, dung and soil, also in the air, on frescoes and on marble monuments (Ionita 1973; Ellis 1976; Gorbushina et al. 2002).

Genus **Hormiactis** Preuss sensu Sacc. et Marchal

The genus *Hormiactis* is characterized by akropetalous branched chains (bunches) of mostly two celled hyaline and tube shaped conidia which arise from undifferentiated upright hyphae. It is distinguished from *Cladosporium* by the lack of pigmented structures in the conidiogenous apparatus as well as the lack of a differentiated conidiophore. There are still uncertainties regarding the scope of the genus, which may have to be revised (Domsch et al. 2007).

Hormiactis sp. (Fig. 3.96)

In the analysis of an older surface, a form was isolated that corresponded to the genus type of *Hormiactis* (Fig. 3.96). The observed characters showed great similarity to *Hormiactis candida* Höhn., although no microsclerotia were found. An assignment to this species remained uncertain.

Genus **Humicola** Traaen

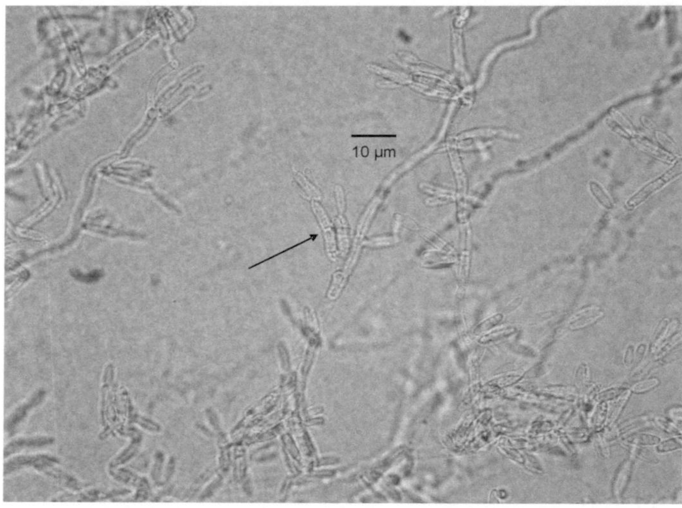

Fig. 3.96 *Hormiactis* sp. produced branched chains of conidia which arose from undifferentiated hyphae. Mature conidia were usually two celled and elongated cylindrical (arrow)

Humicola forms conspicuous dark unicellular conidia which are pear shaped, egg shaped or globose. They may be produced terminally, on side branches or intercalarily (within a hypha). Furthermore, in some species, a second form of conidia are produced via phialides; these remind of *Acremonium*. Phialoconidia are hyaline, small and obovoid. In some isolates, they may be suppressed. Vegetative hyphae of the genus are usually colorless; also the conidiogenous structures are not pigmented. Domsch et al. (2007) regarded *Humicola* as an anamorphic hyphomycete. The genus was recently revised and has been included into Chaetomiaceae (Wang et al. 2019).

Humicola sp.

In the analysis of a filtration probe of ambient air in Holzkirchen, a colony grew that showed typical characters of *Humicola*. Pear-shaped dark conidia were observed mainly on terminal side branches. Unfortunately, differentiation could not be carried out.

Genus *Oidiodendron* Robak

Colonies of *Oidiodendron* usually grow slowly. While vegetative hyphae are uncolored (colored exudate may be produced), conidiophores usually possess a conspicuous pigmentation. Conidiophores are apically branched and produce shorter or longer chains of arthroconidia. Arthroconidia are connected with sterile cell portions which may remain after release of spores as stalklets. The unicellular spores typically are cylindrical or barrel shaped, rarer subglobose. Conidia mature in a basipetalous series; they are smooth or ornamented and hyaline or pigmented. Examples for teleomorphs are: *Byssoascus* v. Arx, *Myxotrichum* Kunze and *Toxotrichum* Orr and Kuehn. The conception of the taxa and the arrangement of species has been contested (e.g., Ellis 1971, 1976; Domsch et al. 2007). In order to clarify the relations, the genus should be revised. The genus contains mostly saprophytes which occur in soil on rotting plant material (Domsch et al. 2007).

Oidiodendron sp. (Fig. 3.97)

A form with characteristics of the genus *Oidiodendron* was isolated from air and driving rain probes in Holzkirchen/Germany (Fig. 3.97). Colonies grew slowly and appeared dirty pink in top light, in view gray. Branched and pigmented conidiophores chains (partly branched by themselves) of subglobose to short cylindrical, more or less hyaline conidia with thickened cell walls were produced. In some spore chains, it was visible that they were connected with sterile portions. Since subcultures died no further differentiation was possible.

Genus *Paecilomyces* Bain.

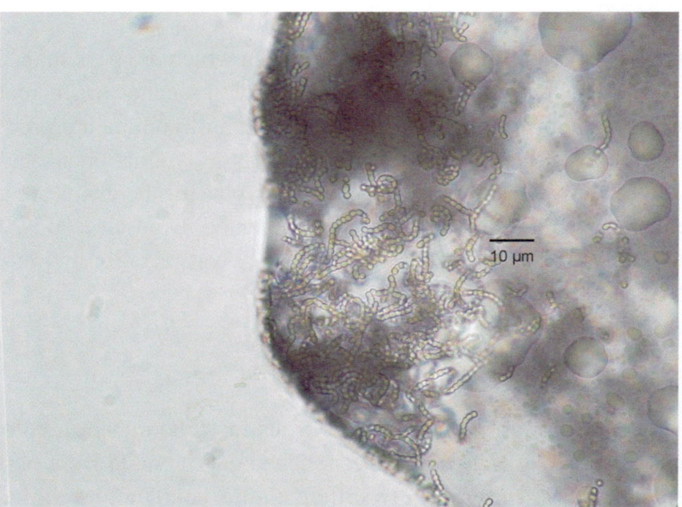

Fig. 3.97 *Oidiodendron* sp., some conidia chains were branched. Conidia were of subglobose to short cylindrical outline with thickened walls

The heterogenic anamorph genus *Paecilomyces* reminds strongly of *Penicillium* Link ex Fr., by comparison colonies are never blue green. Conidiophores are, like some species in *Penicillium*, often terverticillate (three fold branched), but in some species also branched irregularly. Phialides are compared to *Penicillium* spreading, inflated at the base and end in a long tip; no conspicuous neck is present (Samson 1974). *Byssochlamys* Westling, *Thermoascus* Miehe and *Talaromyces* C. R. Benjamin, all Trichocomaceae, Eurotiales, Ascomycetes form teleomorph stages for *Paecilomyces,* which is regarded as polyphyletic according to recent taxonomy (Samson et al. 2009).

Paecilomyces sp.

In course of our study, *Paecilomyces* sp. was isolated from various specimens and surfaces and from ambient air in Holzkirchen/Germany. Isolates formed pale pink to dirty violet colonies, and conidiophores showed the characteristic phialides for *Paecilomyces* with irregular to terverticillate structure. Conidia were oval to lemon shaped, ca. 3–4 µm long. No further differentiation was achieved; the reported characters could match several species.

Genus *Penicillium* Link ex Fr.

The heterogenetic anamorph genus *Penicillium* is characterized by broom or brush-like conidiophores which bear bunches of phialides at their terminal ends. Branches inbetween (if present) are called metulae. Depending on the presence of metulae and how intense branching is developed, conidiophores are called mono-, bi- or terverticillate or even quaterverticillate. Phialides are typically bottle shaped with a short neck and a more or less pronounced shoulder part, rarely lanceolate. Phialides produce basipetal dry chains of conidia which are linked by connectives (sterile cell portions). Often colonies show gray-green or gray-blue tones in the sporulating areas. For identification, we employed different techniques, see, e.g., Samson et al. (1976, 2002b), Pitt (1979, 2000) and Domsch et al. (2007). Most species of the very diverse genus *Penicillium* are ubiquitous saprophytes with an emphasis on temperate climate zones (Domsch et al. 2007). *Eupenicillium* Ludwig, *Talaromyces* C. R. Benjamin, *Trichocoma* Jungh. and *Hamigera* Stolk and Samson, all Trichocomaceae, Eurotiales, Ascomycetes are examples for teleomorphs according to Domsch et al. (2007) and other authors. In current taxonomic work, the very huge genus was conceived newly (e.g., with the exclusion of *Talaromyces*) and a new family of Aspergillaceae was proposed (Houbraken and Samson 2011; Visagie et al. 2014). The genus *Penicillium* s.l. has great industrial importance. Some species are very important producers of antibiotics (e.g., Penicillin). Other species are used to process food (e.g., cheese, ham and sausage production). Further species are food and feed spoilers and may produce strong mycotoxins (e.g., Mills et al. 1995; Samson et al. 2000);, a part of these species may also occur in the interior of buildings.

Penicillium brevicompactum Dierckx

Investigated isolates presented the typical characters of *Penicillium brevicompactum*. Conidiophores were terverticillate to quaterverticillate; metulae usually were somewhat swollen and approximated so that compact heads developed. Conidia were subglobose to spherical and ca. 3–4.5 μm in diameter.

Penicillium brevicompactum was isolated from various new specimens at various exposure locations in Germany at different times and from air and driving rain probes in Holzkirchen.

P. brevicompactum is distributed worldwide. The species is mainly known from soil; it is not very abundant and also is a food spoiler (Domsch et al. 2007; Samson et al. 2000).

Temperature range for growth is −2 to 30 °C, with an optimum at 23 °C (Mislivec and Tuite 1970). Good growth was observed at water potential as low as −100 bar, and on soil growth was even recorded at −200 bar (Griffin 1963b; Pitt 1973). As to pH values no data were found. It is known that the species exudes organic acids and therefore even may corrode metal (Domsch et al. 2007).

Penicillium chrysogenum Thom

Investigated isolates were in accordance with the typus (Domsch et al. 2007; Hoog et al. 2000; Pitt 2000; Samson et al. 2002). At smooth ter- and quaterverticillate penicilli blue-green slightly oval conidia were formed (ca. 3–4 µm long). Some isolates lacked the typical yellow exudate.

Penicillum chrysogenum was isolated from the surface of a specimen but also from ambient air in Holzkirchen/Germany (Renzl 2006).

P. chrysogenum is distributed worldwide, especially in temperate zones. It represents the best known fungus that produces Penicillin (Domsch et al. 2007). The species has already been isolated from surfaces of outer building parts (Ionita 1973), indoors it is common in connection with water damage.

Temperature range of *P. chrysogenum* for growth is 5–37 °C, with an optimum at 23 °C (Mislivec and Tuite 1970; Pitt 1973). Growth was reported from water saturation to a water potential of −350 bar (Chen 1966; Chen and Griffin 1966; Mislivec and Tuite 1970; Tresner and Hayes 1971; Pitt 1973; Moustafa and Al-Mousallam 1975). For physiological processes an optimum pH range is 3–4.5, data on pH limits could not be found (Domsch et al. 2007).

Apart from the antibiotic Penicillin, important toxic metabolites which may be produced are Roquefortin C and Meleagrin (Domsch et al. 2007; Frisvad 1989; Samson et al. 2002).

Penicillium citrinum Thom

The conidiophores of our isolate were smooth and biverticillate. Conidia were subglobose and up to ca. 3 µm in diameter. The back of the colonies on agar plates was bright yellow. Pigment was also released into the agar.

In our study, *Penicillum citrinum* was isolated from ambient air in Holzkirchen/Germany only once.

P. citrinum is a frequent fungus with worldwide distribution. The species can grow on various substrates (Domsch et al. 2007) and was recorded on marble monuments in Russia (Gorbushina et al. 2002).

Optimum temperature range for growth is 26–30 °C; minimum temperature is 5–7 °C. At 37 °C, only traces of growth were found (Mislivec and Tuite 1970; Domsch et al. 2007). Good growth is observed at a water potential of −100 bar; vegetative growth can prevail until a water potential of −350 bar (Griffin 1963; Chen 1966; Chen and Griffin 1966; Mislivec and Tuite 1970; Pitt 1973; Domsch et al. 2007).

Some strains of the fungus may produce Citrinin and other mycotoxins. Only few infections of humans have been reported (Frisvad 1989; Hoog et al. 2000; Kozlovsky et al. 2005).

Penicillium corylophilum Dierckx

Investigated isolates were in accordance with the typus. Conidiophores were smooth and biverticillate, mostly with metulae of different length. Conidia were smooth, subglobose, up to ca. 3 μm in diameter. The back of the colonies on agar plates was dark, single sectors showed a greenish tinge.

Penicillium corylophilum was isolated once from the surface of a specimen after 3 years' exposure in Holzkirchen/Germany

P. corylophilum is distributed worldwide and is more abundant in warmer climate zones (Domsch et al. 2007).

Temperature range for growth is ca. 5–37 °C (Franz 1975; Domsch et al. 2007). For vegetative growth, a minimum range of water potential is −70 to −120 bar (Chen 1966; Pitt 1973). At pH 2, growth was recorded, and apart from that no data on pH was available (Sinclair and Herring 1975).

Penicillium frequentans Westling (Syn. *P. glabrum* [Wehmer] Westling)

The investigated isolates showed typical characters. Conidiophores were monoverticillate, and phialides and subglobose conidia (ca. 3 μm in diameter) were smooth. Conidia were often accumulated in compact cylinders. The back of the colonies on agar plates was pale brown to yellowish brown.

In our study, *P. frequentans* was only isolated from ambient air in Holzkirchen/Germany.

Penicillium frequentans is distributed worldwide and known from various habitats and substrates (Domsch et al. 2007).

P. frequentans grows at a temperature range of 5–37 °C; optimum temperature is at 23 °C (Mislivec and Tuite 1970; Pitt 1973). Down to a water potential of −210 bar (20% NaCl), good growth is possible, and the fungus is regarded as osmophilic (Chen 1966; Domsch et al. 2007). Regarding soil reaction, a pH range of 3.8–4.4 is documented (Nicholls 1956; Singh 1976; Domsch et al. 2007).

The mycotoxin Citromycetin may be produced (Domsch et al. 2007; Samson et al. 2002).

Penicillium solitum Westling (Syn.: *P. verrucosum* var. *melanochlorum* Samson, Stolk et Hadlock)

Investigated isolates presented typical characters. Conidiogenous structures were terverticillate. Stalks were slender and distinctly roughened as were the metulae which were upright and appressed. Conidia were subglobose, ca. 4 µm in diameter.

In our study, *Penicillium solitum* was identified from the surface of specimens and from ambient air (Hofbauer 2007).

The soil fungus *P. solitum* is not very widespread; the center of distribution lies in cold-temperate areas (Domsch et al. 2007). The species has been documented on food (Pitt et al. 1991; Samson et al. 2002).

P. solitum grows in a temperature range of ca. (-2) 4–35 °C (Kuehn and Gunderson 1963; Mislivec and Tuite 1970; Pitt 1973). At a water potential down to -240 bar, growth was still recorded (Whittle 1977).

Genus ***Phoma*** Sacc.

The genus *Phoma* is regarded as a Coelomycete and produces pigmented pycnidia. According to Domsch et al. (2007), it belongs to the Sphaeropsidales, whereas Aveskamp et al. (2010) arranged it within the Dothideomycetes, Pleosporales, Didymellaceae. Pycnidia are formed single or in groups, but are not aggregated in a pycnostroma. Conidiogenous cells are little differentiated and possess narrow phialidic openings. Conidia are accumulated in slimy masses; they are hyaline, ellipsoidal and mostly unicellular. In some species, chlamydospores are present (Sutton & Carmichael 1973; Sutton 1980; Johnston 1981).

Phoma eupyrena Sacc.

The gained isolates developed typically. Chains of chlamydospores were produced in older parts of the colony. Chlamydospores typically were elongated barrel shaped, rarely globose. Conidia from the pycnidia were ellipsoidal and generally unicellular.

In our investigations, *Phoma eupyrena* was isolated from various specimens and surfaces and from probes of driving rain.

P. eupyrena is a frequent and worldwide distributed soil fungus. This soil fungus may also colonize dead plant material and has only rarely been detected on living plant tissue (Domsch et al. 2007).

Optimum temperature range for growth is 20–21 °C; maximum temperature is 28–32 °C and minimum temperature −3 °C. Optimum pH value for growth is pH 4 (Kranz 1962; Årsvoll 1975).
So far, only one report of a human infection exists (Bakerspigel et al. 1981).

Phoma exigua Desm. (numerous synonyms, see, e.g., Aa and Kesteren 1971; Boerema and Dorenbosch 1973; Boerema 1976)

The investigated isolates showed colonies with a more or less irregular margin. Vegetative hyphae did not produce chlamydospores. After addition of concentrated NaOH, the mycelium first turned blue green and subsequently red brown. Conidia were elongated, ca. 4–6.5 µm long and scarcely two celled.

In our study *Phoma exigua* was frequently isolated from various specimens exposed at different sites in Germany, from air and from probes of driving rain. The species thus was documented as component of the initial growth on modern building surfaces (Hofbauer 2007).

P. exigua is a frequent soil fungus. It usually grows on dead plant material; some isolates were also parasitic on living plants (Boerema and Dorenbosch 1973; Domsch et al. 2007; Johnston 1981). The species has previously been isolated from air (Entwistle 1972).

The temperature for optimum growth is 22–26 °C, minimum temperature is 4 °C, and maximum is 30°C. Optimum pH ranges between 4 and 6 (Kranz 1962; Logan and Khan 1969).

Phoma glomerata (Corda) Wollenw. et Hochapfel

The investigated isolates showed typical characters. Especially on nutrient poor media (BBM, CD50), copious branched chains of dictyochlamydospores were produced; these reminded of *Alternaria*. Pycnospores were ellipsoidal, unicellular and ca. 6–7 µm long.

Phoma glomerata was isolated from the surface of various specimens, old and new and from different other building part surfaces.

P. glomerata is ubiquitous. The species was typically found on dead plant material, on soil and as parasite on different plants (Johnston 1981). It has been recorded from frescoes and from marble monuments (Ionita 1973; Gorbushina et al. 2002).

Reports of infections in humans are scarce (Hoog et al. 2000).

Phoma herbarum Westend. (Syn.: *P. violacea* [Bertel] Eveleigh)

Isolates were characteristic with the exception that some submited less or no red pigment into the agar. After adding concentrated NaOH to the mycelium, the typical blueish coloration was induced. No chlamydospores were produced. Conidia were ellipsoidal, unicellular and ca. 5 µm long.

In our study, the species was documented from a specimen and an older building surface and from ambient air in Holzkirchen/Germany.
Phoma herbarum is distributed worldwide. It is known from various substrates (Domsch et al. 2007; Johnston 1981). The species has been documented on painted surfaces in Great Britain (Eveleigh 1961a,b).

Optimum temperature range for vegetative growth is 22–25 °C, maximum temperature for growth is 35°C. The fungus may grow in a pH range of 3 to<8 with two optima at pH 4.5 and pH 7.5 (Domsch et al. 2007). Conidia germinate at water potentials down to −210 bar (Eveleigh 1961b).

Only few infections in humans are known, and virulence is regarded as low (Hoog et al. 2000).

Phoma pinodella (L. K. Jones) Morgan-Jones et K. B. Burch (Syn: *P. medicaginis* Malbr. et Roum. var. *pinodella* [L. K. Jones] Boerema)

Investigated isolates were in accordance with the typus. In culture, chains of inflated chlamydospores were produced. They were mostly globose and distinctly bigger than those of *Phoma eupyrena*. Pycnospores were ellipsoidal, ca. 5–7 µm long, scarcely two celled.

In our study, the species was isolated from various surfaces of specimens, from older building surfaces and from ambient air

Phoma pinodella is a worldwide distributed soil fungus. It is a frequent soil fungus and a weak plant parasite, preferentially on legumes (Domsch et al. 2007; Johnston 1981).

Phoma pomorum Thüm.

The investigated isolates showed typical characters. Chlamydospores were produced in chains of thick-walled whole cells and also in form of dictyochlamydospores (muricate chlamydospores). Dictyochlamydospores did not form upright branched chains as is typical for *Phoma glomerata*, but developed single and prostrate. Pycnospores were ellipsoidal, unicellular and ca. 6 µm long.

In our study *Phoma pomorum* was infrequently isolated from specimens and from ambient air. It was documented on modern building surfaces for the first time (Hofbauer 2007).

P. pomorum is a frequent soil fungus. This soil fungus may grow on various dead or diseased plants (Domsch et al. 2007; Johnston 1981).

Temperature range for optimum growth is 18–21 °C. pH values from 3–10 are tolerated, optimum pH for growth is 7–8 (Domsch et al. 2007).

Phoma sp.

In course of our analyses of various specimens, old and new, of older bulding surfaces and of ambient air forms of *Phoma* were isolated that could not be assigned to a certain species because of scarce sporulation or otherwise deviant growth.

Genus **Pithomyces** Berk. et Br.

The genus *Pithomyces* produces single multicellular conidia at short stalks at otherwise undifferentiated air hyphae. Conidia are broadly attached and only liberated by rupture of the stalklet. Residues of the stalklet remain attached to the conidium and form an important diagnostic character. The genus counts about 15 species that are mainly distributed on soil, on plant litter and other plant residues (Ellis 1971, 1976; Domsch et al. 2007). *Leptosphaerulina*, Pleosporales, Ascomycetes, forms a teleomorph of *Pithomyces* (Russomanno et al. 2003).

Pithomyces chartarum (Berk. et Curt) M. B. Ellis (Fig. 3.98) (Syn.: *Sporidesmium chartarum* Berk. et Curt., *Sporidesmium bakeri* Syd.)

For information regarding systematic and synonyms see Ellis (1960, 1971), Dingley (1962) and Domsch et al. (2007). The investigated isolate of *P. chartarum* formed fast growing colonies which could overgrow a Petri dish (Ø: 9 cm) within a few days. Good spreading growth was observed on nutrient poor media like BBM and CD50. Colonies were hyaline initially and could turn gray to dark gray when aged. Note that the dark brown and muricate conidia were slightly laced at the cell borders; their outer surface was rough by warts. They were born out of inconspicuous short conidiophores. The cell wall of the connecting stalklet between spore and hypha ruptured when the spore was released and usually remained as a visible attachment (hilum) on the spore (Fig. 3.98).

In our study, *Pithomyces chartarum* was isolated only once from a specimen exposed in Holzkirchen/Germany for 3 years. Thus, it was documented the first time on modern building surfaces (Hofbauer 2007).

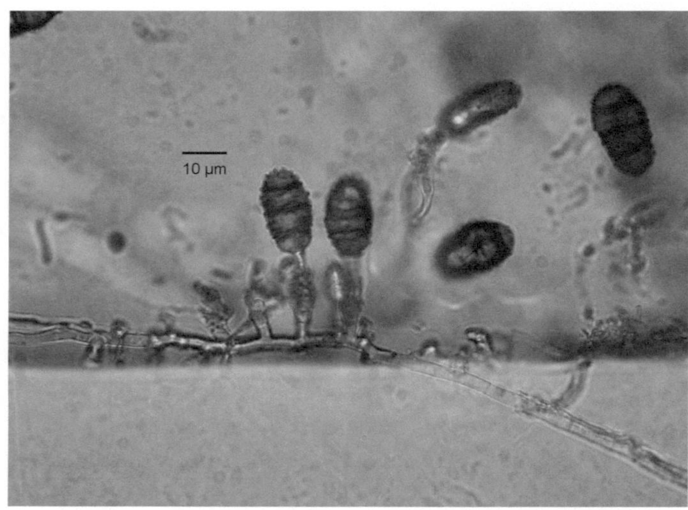

Fig. 3.98 *Pithomyces chartarum* (microscopic preparation of a raw culture derived from the surface of a specimen after 3 years' exposure in Holzkirchen/Germany)

The species is cosmopolitan. Originally, it was discovered on rotting paper in the USA (Domsch et al. 2007) and was recorded in Great Britain at the beginning of the 1960s (Lacey and Gregory 1962; Gregory and Lacey 1964). It is common on dead leafs and stalks of different plants all over the world (Ellis 1971; Domsch et al. 2007; Taber et al. 1968), prefering different grasses in pasture areas (Brook 1963). *Leptosphaerulina chartarum* Roux forms the teleomorph of *P. chartarum* (Roux 1986; Russomanno et al. 2003).

Temperature range for hyphae growth is 5–30 °C (Le Bars et al. 1990). Conidia production is high at 24 °C; huge amounts of conidia are produced in the field after rain. Conidia only germinate at water potentials ≥ -140 bar (Brook 1963).

The fungus produces highly potent mycotoxins, especially Sporidesmin (e.g., Done et al. 1961; Hodges et al. 1963; Laskin and Lechevalier 1973; Halder et al. 1979). In respect of preferred prerequisites for toxin production data differ. While Le Bars et al. (1990) gave a temperature range of 20–25 °C for toxin production, Davison and Marbrook (1965) assessed that spores show a higher Sporidesmin content at lower temperatures. Done et al. (1961) reported that toxin production is independent of the growth rate of the mycelium. di Menna et al. (1970) conducted detailed investigations and found that the amount of produced Sporidesmin clearly depended on environmental factors, especially substrate. Toxins cause different diseases in humans, especially a photosensitivity of the skin (Clare 1944; Thornton and Percival 1959; Mortimer and Taylor 1962; Rippon

1988), whereby only a part of the strains produces Sporidesmin (Ueno et al. 1974; Halder et al. 1981; Collin and Towers 1995a; Collin et al. 1996). The fungus is combated with different technical and chemical methods (e.g., Parle and di Menna 1978; Smith and Towers 1984) and also with biocompetitive methods (Collin and Towers 1995b; Fitzgerald et al. 1998).

Genus *Sporothrix* Hektoen et Perkins

Conidiogenous cells of *Sporothrix* are formed on upright or slightly branched, rare verticillate conidiophores which are hardly different to vegetative hyphae. They show numerous projections in their apical part or a rachis with stalklets at which the spores are produced singular. The conidiogenous part can be even more elongated subsequently. Conidia are small, unicellular, and obovoid to club shaped, smooth and pedunculate at the base. *Ophiostoma* H. and P. Syd., Ophiostomataceae, Ophiostomatales, Ascomycetes acts as the teleomorph of some species of *Sporothrix*. *Sporothrix schenkii* Hektoen and Perkins is a so-called dimorphic fungus and occurs besides a saprophytic life form as pathogen (Domsch et al. 2007; Hoog et al. 2000).

Sporothrix sp.

We encountered a single isolation of a *Sporothrix* sp. on one specimen exposed in Holzkirchen/Germany for one year. In a slightly wet appearing colony, pale orange-pink spores were formed which were club to tear shaped. Altogether, the observed characters allowed a designation to the genus with the exception of *Sporothrix schenkii*. Unfortunately, subcultures failed and therefore no further identification could be performed.

Genus *Stachybotrys* Corda

Many species of the genus *Stachybotrys* possess pigmented (dark) conidiophores and spores and are therefore listed among the "dematiaceous fungi" (Ellis 1971, 1976). Vegetative hyphae usually are colorless. Conidiophores are simple or branched cymous and bear a bunch of ellipsoidal to subclavate phialides which are formed successively at their top (difference to *Aspergillus*). Conidia and spores are hyaline or pigmented. Conidia are unicellular, often elongated and are accumulated in slimy masses (Domsch et al. 2007; Andersen et al. 2003).

Stachybotrys chartarum (Ehrenb. ex Link) Hughes (Fig. 3.99) (Syn.: *S. atra* Corda)

Investigated isolates were in accordance with the typus (Fig. 3.99). Conidiophores, phialides and spores were more or less pigmented. Conidiophores were formed single or irregularly branched, usually hyaline at the base and toward the top increasingly darker

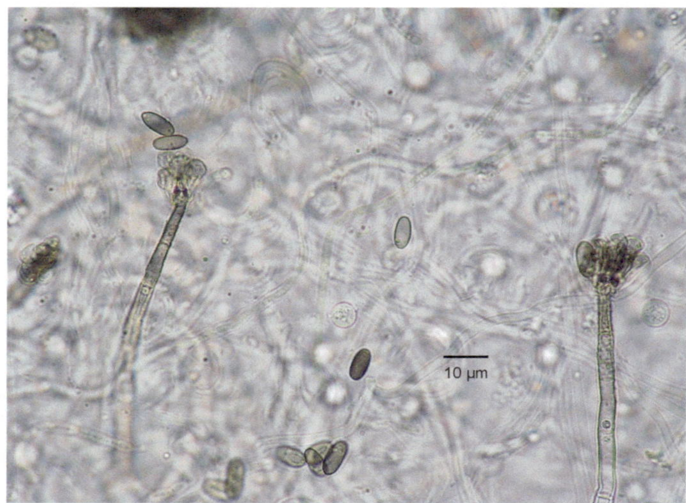

Fig. 3.99 *Stachybotrys chartarum*, conidiophores with phialides and some conidia. Background: hyaline mycelium

and rougher (warty), even the part of the vegetative hyphae that bears the conidiophore could be partially pigmented. Phialides were ellipsoidal, smooth, hyaline initially, later olive gray. Conidia were produced in slimy masses; they appeared as black globose droplets at the conidiophores (top view at weak magnification) and were ellipsoidal, ca. 7–12 µm long. Young conidia were hyaline, mature dark gray to opaque. Besides warty-rough conidia also almost smooth ones were observed.

In our study, *Stachybotrys chartarum* was isolated occasionally from specimens with mineral topcoat exposed at different sites in Germany, from other surfaces and even for the first time from probes of driving rain.

S. chartarum is by far the most abundant species of the genus and is distributed worldwide. The species occurs mainly on rotting plant material and demonstrates an affinity to cellulose rich substrates. Quite often, the species is found in connection to water damages, if cellulose-rich building materials are involved and temperature of the substrate is suitable for growth.

Optimum temperature for growth is 23 (27) °C, minimum temperature is (2) 7 °C, maximum temperature is 37–40 °C (Yadav and Madelin 1968; Ayerst 1969; Franz 1975; Jong and Davis 1976). The fungus is able to grow in a pH range of 3.6–7.7. Optimum water potential for vegetative growth is −30 bar; minimum is −100 bar (Ayerst 1969; Butt and Ghaffar 1974). It can decompose cellulose and may colonize various cellulose rich substrates like paper, textiles, methyle-cellulose, cellophane, etc. (Boneh 1953; Domsch 1960; Tribe 1960; Kanevskaya 1966). Methyle-cellulose is, e.g., added to mineral plasters. Whether this was the reason, why the fungus was recorded in our investigations only from mineral topcoats, remained unclear, because the overall occurance was too scarce to apply statistics.

Some strains of the fungus may produce various mycotoxins, especially satratoxin H, a trichothecene (Samson et al. 2002). If ingested, these mycotoxins can cause poisoning, a so-called stachybotryotoxicose and cause further illness (e.g., Forgacs 1972; Domsch et al. 2007; Harrach et al. 1988; Klinkhammer 1997; Sudakin 2000; Shariat and Collard 2007). To what extend the occurrence of the fungus on outer building surfaces may have hygienic/medical effects is not yet clear, at least ingestion is unlikely (CDC 1994, 1995, 2000; Wilkins et al. 1998; Sudakin 2000). Interesting was the finding that the fungus isolated from driving rain probes grew in raw culture in close contact to a developing moss protonema and to developing moss plantlets, respectively. Partly, the cells of the plant were dead. It was not clear if this was caused by the fungus or if it was the other way round and the fungus can only colonize dead plant matter.

Genus *Torula* Persoon ex Fries

Torula is often listed among the "dematiaceous fungi" (Ellis 1971, 1976). Conidia are formed in chains, sometimes branched. Cell borders of the conidia are constricted; they are more or less spherical, rather dark and warty. End cells of chains are often brighter in color.
Torula herbarum (Pers.) Link ex F. S. Gray (Fig. 3.100)

Investigated isolates were in accordance with the typus. Colonies were pale initially and later seemed like soot covered because of the developed conidia. Conidia chains were dark brown and warty; they showed ca. 2–7 septa. At the borders, conidia were conspicuously constricted, which gave them a spherical shape (Fig. 3.100).

Torula herbarum was isolated once from the ambient air in Holzkirchen/Germany.

T. herbarum is distributed worldwide with an emphasis on temperate regions (Ellis 1971).

The species is abundant on dead stalks of herbs, on dead leafs and on damp timber and was isolated from soil and air (Ellis 1971; Hawksworth 1976).

Little is known about the physiological values. Optimum growth was reported at 35 °C. Germination of conidia was observed at a water potential range of -100 bar up to water saturation of the air (Webster and Dix 1960; Yadav and Madelin 1968; Franz 1975).

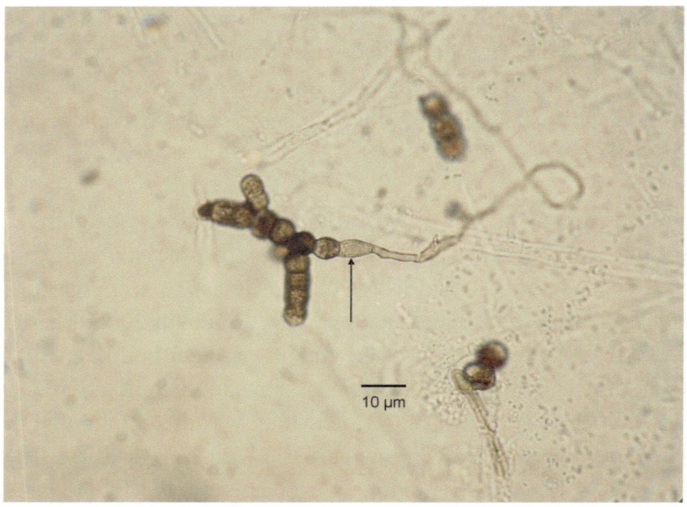

Fig. 3.100 *Torula herbarum,* conidiogenous cell (arrow) and chains of conidia (microscopic preparation from raw culture of a sedimentation probe from ambient air in Holzkirchen/Germany)

Genus ***Trichoderma*** Pers. ex Fr.

The genus *Trichoderma* is characterized by fast growing colonies which bear bunches of repeatedly branched conidiophores at which projecting and often irregularly set flask-shaped phialides are inserted. In some species, conidiophores have sterile ends and phialides are only attached at side branches. Conidia typically show green color, rarely hyaline; they can be rough or smooth. In older cultures, often hyaline chlamydospores are produced. A few species possess a teleomorph in *Hypocrea* Fr., Hypocreaceae, Hypocreales, Ascomycetes, but overall ascomata are developed only rarely (Domsch et al. 2007).

Trichoderma viride Pers. ex Gray

Many teleomorph stadia have been described: e.g., *Hypocrea rufa* (Pers.) Fr., *H. aurantiaca* P. Henn., *H. coprosma* Dingley, *H. atrogelatinosa* Dingley, *H. muroiana* Hino et Katsumoto, *H. albo-medullosa* Doi (Domsch et al. 2007). Our isolates were very similar and showed typical characters. Colonies grew very fast. At numerous places, downright heaps of conidiophores emerged which were white/yellow green initially and turned olive green when conidia ripened. Conidiophores were branched pyramidal with the basal branches more elongated. The flask-shaped phialides were in groups of two to four and irregularly projecting. Conidia were slightly uneven globose and more or less conspicuously rough at their surface (immersion lens!).

Trichoderma viride was often isolated from older surfaces and from ambient air, but was also isolated from various specimens and thus documented as a component of the initial colonization of modern building surfaces for the first time (Hofbauer 2007).

T. viride is distributed worldwide and can colonize numerous habitats, in temperate as well as tropical zones (Domsch et al. 2007).

The species was previously recorded from marble monuments in Russia (Gorbushina et al. 2002).

Optimum temperature range for growth is 20–28 °C, maximum temperature is 30–32 (37) °C, and minimum temperature is 0–6 °C (Ward and Henry 1961; Apinis 1963; Årsvoll 1975; Franz 1975; Domsch et al. 2007). In a pH range of 1.5–9, growth is possible, with an optimum pH at 4.5–5.5 (Ward and Henry 1961; Brown and Halsted 1975; Domsch et al. 2007). Minimum water potential for vegetative growth in vitro is −240 bar, and no growth was observed at a NaCl concentration of 10% (Whittle 1977; Domsch et al. 2007). The taxon is ecophysiologically very versatile. Apart from a saprophytic mode of life (many different substrates can be utilized), *T. viride* can live parasitic on plant tissue and especially on fungi. Various fungi, other Ascomycetes as well as Zygomycetes and Basidiomycetes, may be infested. Although dreaded in mushroom culture *T. viride* is used as an antagonist to reduce damage caused by fungal pathogens (e.g., Aytoun 1953; Catani and Peterson 1967; Domsch et al. 2007).

Genus *Ulocladium* Preuss

Ulocladium is very close to *Alternaria* and often listed with the "dematiaceous fungi" (Ellis 1971, 1976). According to current taxonomy, the scope of the genera has been changed, and many former *Ulocladium* species have now been incorporated into *Alternaria* (x). Conidiophores are overtopped in sympodial branching, thus often zig-zag structures are formed. Dictyoconidia usually are rounded on both ends and are only occasionally conical at one end, potato shaped and produce single or branched chains. Changes in the conception of the closely related genera *Ulocladium* and *Alternaria* were proposed (Runa et al. 2009; Woudenberg et al. 2013); however, further investigation are necessary as both comprise huge and diverse groups.

Ulocladium chartarum (Preuss) Simmons (Fig. 3.101) (Syn.: *Alternaria chartarum* Preuss)

Fig. 3.101 *Ulocladium chartarum*, muricate conidium with secondary geniculate conidiophore and secondary conidia

Investigated isolates showed typical characters. Conidia were arranged in branched chains, reminding of *Alternaria alternata*, but conidia were rounded typically on both ends, potato shaped, mostly without or rarely with stalk-like beak or secondary conidiophore, respectively (Fig. 3.101). Some isolates sporulated only late and weak in culture. In our study, *U. chartarum* was among the most frequent fungi that were recorded in our investigations. It was isolated from almost all specimen variants at all exposure locations at all probing times, from various older surfaces and from ambient air in Germany. It was conspicuous that it lacked in old specimens and probes of driving rain.

Ulochladium chartarum is a worldwide distributed saprophyte. It may colonize various organic substrates and has been isolated from emulsion paint and from marble monuments (Ellis 1976; Gorbushina et al. 2002).

In humans, only few infections caused by *Ulocladium chartarum* have been reported, virulence of the species is regarded as low (Hoog et al. 2000; Duran et al. 2003).

Ulocladium oudemannsii E.G. Simmons (Figs. 3.102 and 3.103) (Syn.: *Alternaria oudemannsii* [E.G. Simmons] Woudenb. and Crous)

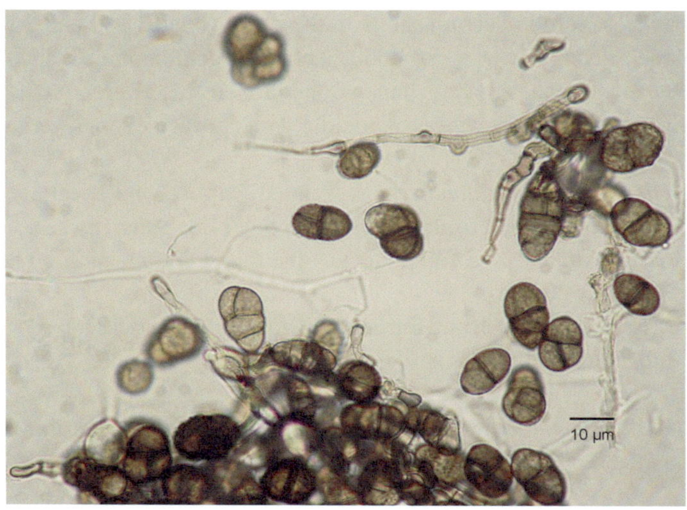

Fig. 3.102 *Ulocladium oudemannsii*, conidiophores with scars from released conidia and conidia

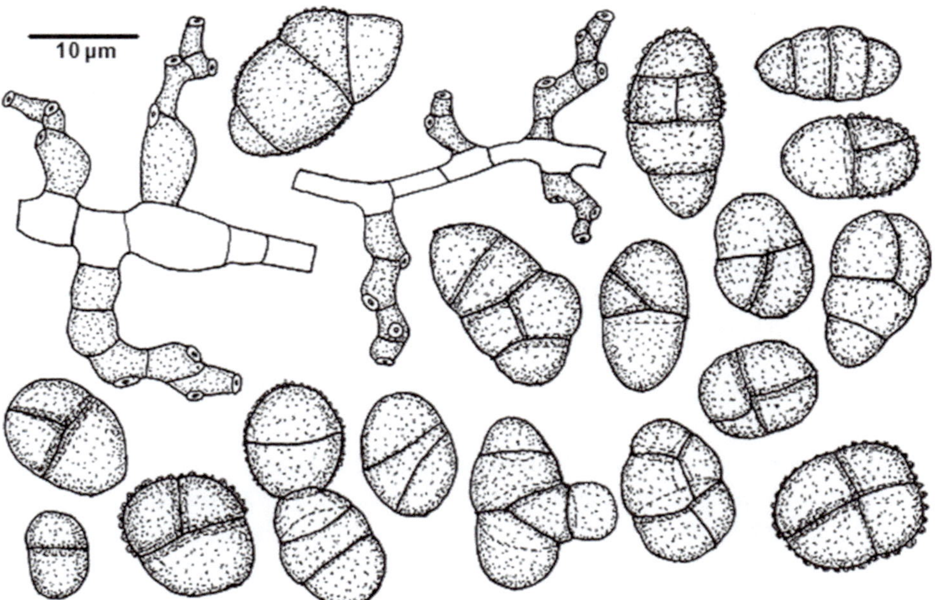

Fig. 3.103 *Ulocladium oudemannsii*, choice of characteristic conidiophores and conidia

Investigated isolates were in accordance with the typus (Figs. 3.102 and 3.103). Conidia usually developed a singular at a sympodial branched conidiophore which was in contrast to *Alternaria's* typical zig-zag shape. Conidia were of different sizes and outlines and had a warty-rough surface, spherical conidia with crosswise septa were very rare.

In our study, *Ulocladium oudemannsii* was isolated from several specimens exposed in Germany, from ambient air and from probes of driving rain. Thus, it was recorded for the first time in driving rain and as a component of the initial colonization on modern building surfaces (Hofbauer 2007).
The species is known mainly from Europe (Ellis 1971). It has regularly been isolated from organic substrates (Ellis 1971).

U. oudemannsii is a potent antagonist against *Botrytis cinerea* and is utilized in agriculture (Reglinski et al. 2005).

Division Basidiomycota
Class Basidiomycetes

Basidiomycetes are characterized by their clamp connections in the dicaryotic mycelium (in rare occasions not visible) of their persistent phase in vegetative growth. Generative spores are often produced in fruiting bodies. They are formed as exospores detached from a special organ, the basidium. In most species, always four basidiospores are produced from a whole basidium (Agaricomycetes). In the Uredinomycetes, the basidium is septate and complex modes of development are established also involving yeast stages so that affiliation to the Basidiomycota is not always recognized at once.

Subclass Agaricomycetidae (Hymenomycetidae)
Order **Agaricales** Underw.

Family **Psathyrellaceae** Vilgalys, Moncalvo et Redhead

Genus *Coprinellus* P. Karst.

The genus *Coprinellus* was formerly included in the once large genus *Coprinus* and was segregated as a result of molecular studies (Redhead et al. 2001). *Coprinellus* forms typical, in head and stalk separated fruiting bodies with the hymenium (spore producing organ) organized in lamellae. In maturity, the hymenium liquefies and drops from the fruiting body ("ink caps" in the broad sense). The spore mass ("ink") is dark brown to black. In different species, the head reaches after unfolding a diameter of less than 5 mm to several cm. Clamps at the mycelium are present or may lack. In some species also, an additional vegetative stage, *Hormographiella,* is developed. In these species, often no clamps can be recognized in the mycelium. Furthermore, in some species so-called ozoniums may be produced, which represent balls of dense mycelium which help to survive on unfavorable substrates. The different species are saprophytes, which often live on nutrient-rich substrates, on soil, dung, rotten wood and plant residues (Moser 1978).

Coprinellus sp.

Spores of *Coprinellus* sp., which were isolated from a surface probe of a specimen exposed in Holzkirchen/Germany for half a year, germinated in a raw culture once and the formation of fruiting bodies started. Unfortunately, the fruiting bodies decayed before maturity and were dissolved by bacteria, and thus, no further determination was possible. In other species or related forms (*Coprinus comatus* [Müll. in Fl. Dan. ex Fr.] S. F.Gray, *Coprinus atratmentarius* [Bull. ex Fr.] Fr.) it was observed that they like to colonize places where fresh building material was stored or worked at or places where fresh lime gravel was deposited, respectively.

Genus *Rhodotorula* Harrison

The main vegetative phase of *Rhodotorula* is a yeast stage. Colonies are pink or reddish to yellowish, slimy or dry. Budding cells are ellipsoidal or spherical with narrow scars. In culture also structures like hyphae (pseudomycelium) may arise. The genus *Rhodotorula* contains many saprophytes, of which a majority is psychrophilous, and some live in lakes, marine waters or on the surface of leafs (phyllosphere) (Hoog et al. 2000). In some species, a teleomorph is developed which is called *Rhodosporidium* Banno, Sporidiobolaceae, Sporidiales, Uredomycetidae, Basidiomycetes.

Rhodotorula mucilaginosa (Jörgensen) Harrison (Fig. 3.104)

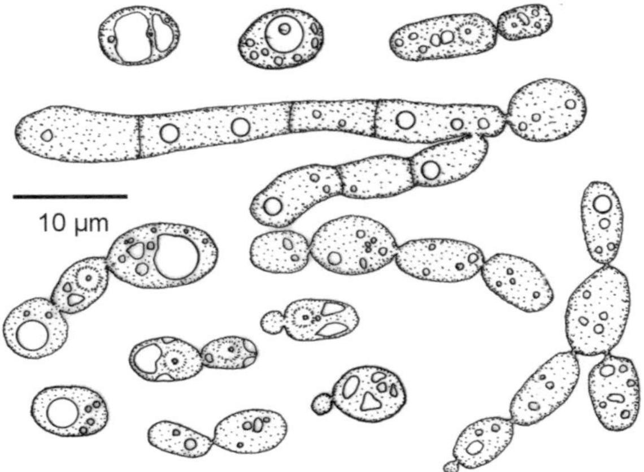

Fig. 3.104 Habitus of a strain of *Rhodotorula mucilaginosa* isolated from masonry (sketch of the microscopic observation)

So far, no teleomorph is known for this species, although according to its morphological affinities to basidiomycetous yeasts and genetic results the form is grouped here. The observed characters corresponded to the type. Colonies were purple, shiny, slimy and smooth. Budding cells were usually unipolar with a length of ca. 5–7 μm. In culture, short pseudomycelia were apparent (Fig. 3.104).

The species was identified during investigations of wall surfaces and ambient air (Hofbauer 2007).

Infections of immunocompromised patients have been recorded (e.g., Zaas et al. 2003).

Form of Organization Lichenes
Formgroup Ascolichenes

Lichens comprise a peculiar form of organization, a symbiosis of certain fungi ("mycobiont") with certain algae or blue greens, respectively ("phycobiont"). In general, the form is defined by the fungal partner who incorporates the algae usually in special areas of the vegetational body (thallus). Therefore, the taxonomy of lichens is generally aligned to the taxonomy of the main fungal constituent of this peculiar symbiosis (Hawksworth and Hill 1984; Gärtner 1985e; Wirth 1995a, b; Wirth et al. 2013; Kirk et al. 2001; Peršoh et al. 2004; Nash III 2008). There are known lichens formed by Ascomycetes, Basidiomycetes and in rare cases by further fungal taxa. In our study, only lichens with Ascomycetes as fungal partner were detected.

Diverse groups of lichens show a preference for certain phycobionts (Rambold et al. 1998). Genetic characters are more and more used for the classification of lichens (Peršoh et al. 2004), but so far no conclusive results are available.

As partner in the symbiosis, the involved algae enjoy the advantage that they are better protected against desiccation and adverse environmental conditions. Additionally, their supply with nutrients is improved by the enlarged surface through the fungus which allows to gather more nutrients than the alga could get on its own. The fungus benefits from assimilates of the algal photosynthesis which are partially shared. On the one hand, organic acids may be produced by both alga and fungus. Because these may be secreted to the substrate, they may cause shifts in ions or dissolving of minerals. On the other hand, lichens often produce oxalate crusts which protect the material beneath from negative environmental influences because of their hardness. The discussion whether lichens corrode or consolidate the colonized substrates is still controversial. Apart from organic acids lichens produce complex metabolic compounds that cannot be produced by a single symbiosis partner alone (e.g., lichen pigments, lichen acids). Lichens and bryophytes present a different sensitiveness against air pollutants and some species are suitable as indicator organisms. They are already utilized for cartographic documentation of air quality (Frahm and Solga 1999; Stapper and Kricke 2004). A standardized method is used according to VDI 3957 Blatt 13 (KRdL 2005). The role of lichens in the pioneer phase of microbial life on façades is so far only incompletely documented. Although in the existing literature many forms have been described as potential colonizers of walls (Bertsch 1964; Poelt 1969, 1977, 1981; Kremer and Muhle 1991; Purvies et al. 1992; Wirth 1995a, b), accurate definitions of the kind and age of the substrate are absent. The emphasis of research on lichens as colonizers of building parts was mostly restricted to investigations of historic buildings and monuments (e.g., Nimis et al. 1992; Marcos Laso 2001). The presented nomenclature of ascomycetous lichens and sterile stages follows Wirth (1995a, b), Wirth et al. (2013) and Kirk et al. (2001), respectively.

Class Ascomycetes, Subclass Lecanoromycetidae
Order **Lecanorales** Nannf.

Family **Lecanoraceae** Koerb.

Genus *Lecanora* Ach.

The artificial genus *Lecanora* (sensu Wirth) is very extensive and diverse and contains several distinct groups (Wirth 1995a). In *Lecanora,* species are united with crustose, partially lobed margin, with unicellular green algae, with so-called lecanorine apothecia (margin of the apothecium is built of thallus tissue including algae layer) and unicellular, typically ellipsoidal ascospores. Some of the more than 100 species of the genus live in extreme habitats as heavy metal rich or greatly eutrophicated places, and some species also colonize anthropogenic surfaces (Purvis et al. 1992; Wirth 1995a, b).

Lecanora muralis (Schreber) Rabenh. (Syn.: *Protoparmeliopsis muralis* [Schreb.] M. Choisy)

Bright gray-green thalli were attached on their whole underside; the margin was typically lobed, with flat on the underground appressed lobes. The apothecium development started in the center of the thallus.

In our study, several individuals were found on the surface of older specimens.

The distribution ranges from arctic to mediterranean vegetation zones. The lichen is abundant on different anthropogenic locations, like monuments, wall tops, tombstones, fence poles and roof tiles (Wirth 1995a, b; Marcos Laso 2001).

It is an euryoeke species, which is neutrophilic and toxitolerant.

Lecanora sp.

A further species of *Lecanora* was detected on the surface of some older specimens. It was a form with an inconspicuous thallus. The observed thalli showed already young, incompletely developed apothecia with typical lecanorine structure, Therefore, the genus affiliation was clear but further differentiation was impossible because of missing further characters (e.g., spores).

Family **Physciaceae** Zahlbr.

Genus ***Physcia*** (Schreber) Michaux

Physcia forms narrow lobed, rosettes or lawns forming thalli with *Trebouxia*-similar phycobionts which are attached to the ground with simple or forked rhizines. In some species, cilia develop at the margin. Many species possess sorales and/or isidia for vegetative propagation. Apothecia are equipped on the surface with a lecanorine margin and a brown to black disc, the surface often grayish owing to pruine. Paraphyses are entire or forked. Asci are club-shaped cylindrical with two-celled thick-walled ascospores in typical numbers. The different forms of the species rich genus colonize various, also anthropogenic habitats.

Physcia adscendens (Fr.) Oliv.

Typical characters were bright to almost white thalli with narrow lobes (breadth ≤ 1 mm) that were vaulted dome like at the tips where sorales were sheltered. At the margin, characteristic cilia were inserted.

In our study, *Physcia adscendens* was recorded from older specimens exposed in Holzkirchen/Germany (Hofbauer 2007).

P. adscendens is frequent and is distributed from boreal to mediterranean zones. The species has previously been documented on building material surfaces (Hofbauer et al. 2003).

The species is quite toxitolerant and is characterized as nitrophil and photophil. It usually colonizes eutrophic habitats rich in light like free-standing leafy trees and bird rocks.

Physcia sp. (Fig. 3.105)

A pioneer form of a *Physcia* sp., which could not be attributed to a certain species, was identified from numerous specimens after 2 and 3 years' exposure at different sites in Germany. The young thalli reminded strongly of *Physcia adscendens* and possessed cilia at the margins of their lobes. But no sorales were developed at the very small thalli.

Fig. 3.105 Young lichen thallus of a *Physcia* sp. attached by sucker like rhizines. Whole thallus (including rhizines) measured ca. 1 mm in diameter

The sorales localization as well as the final thallus color were crucial characters for the final differentiation. According to the characters presented, so far several species had to be considered in theory. The mode of attachment of these thalli was peculiar. They formed rhizines which were able to attach even to very smooth surfaces like a sucker (Fig. 3.105). It was also observed that the bigger thalli were exposed to shearing power (e.g., by precipitation) which could detach them again.

Hofbauer et al. (2003) reported *Physcia caesia* (Hoffm.) Fürnr. from established building surfaces in Germany.

Order **Acarosporales** Zahlbr.

Family **Acarosporaceae** Zahlbr.

Genus *Sarcogyne* Flotow

The genus *Sarcogyne* comprises a group of rock inhabiting species with simple crustose or endolithic thalli that contain unicellular green algae. Apothecia are attached or immersed, with a special margin and a dark red or black disk which may be pruinose bright. Numerous small unicellular and spherical to ellipsoidal spores are produced in the asci by mode of secondary cleavages. In addition, pyknidia may be formed as a means of vegetative propagation of the phycobiont (Vězda 1978; Wirth 1995a, b).

Sarcogyne regularis Körber

Typical characters were the mainly endolithic thallus, the more or less immersed and pruinose apothecia (then with blueish disc) and ellipsoidal ascospores about twice the length of the width.

In our study, *Sarcogyne regularis* was found to be an important element of the colonisation of older specimens exposed in Holzkirchen/Germany for more than 15 years. *Sarcogyne regularis* also formed a component of the initial colonization of modern building coatings and was detected in "young" specimens after 3 years' exposure. We only detected the species on mineral coating substrates.

The species is distributed from the Arctic to the Mediterranean and reaches up into the alpine zone.

Sarcogyne regularis has previously been documented on walls, plaster and artificial stone (Ariño and Saiz-Jimenez 1996; Wirth 1995a, b; Marcos Laso 2001; Hofbauer et al. 2003).

The species is termed an euryoeke pioneer but with moderate concurrent power which prefers basic, light exposed and dry (but rained) habitats.

Order **Teloschistales** D. Hawksw. et O. E. Erikss.

Family **Teloschistaceae** Zahlbr.

Genus ***Caloplaca*** Th. Fr.

Caloplaca comprises crustose lichens with simple structures sometimes endolithic to highly differentiated thalli, lobed at the margin, with yellow to red, rarely blackish apothecia. Very characteristic is the structure of the bicelled ascospores, which show an inflated septum that is run through by a fine channel. Phycobiontes are *Trebouxia* species or *Trebouxia*-like algae, respectively (Kärnefelt 1989; Purvis et al. 1992; Wirth 1995a, b). Corresponding to the great species number (more than 500 species worldwide) various habitats are colonized, some species are toxitolerant and nutrient loving and also occur regularly on anthropogenic substrates like monuments (Marcos Laso 2001).

Caloplaca holocarpa (Hoffm. ex Ach.) Wade

Thalli of the investigated forms of *Caloplaca holocarpa* were inconspicuous, the often densly arranged apothecia were orange brown. Ripe ascospores showed a typical septum.

In our study, *Caloplaca holocarpa* was recorded on various specimens (old and new) and on older surfaces from different sites in Germany.

The species is frequent and distributed from the Arctic to the Mediterranean (Purvis et al. 1992; Wirth 1995a, b; Hofbauer et al. 2003).

It is a lichen with a broad ecologic amplitude which often occurs on anthropogenic substrates like walls, artificial stones, terrazzo and washed concrete or materials from masonry.

Hofbauer et al. (2003) described *Caloplaca crenulatella* (Nyl.) Oliv. on long-established building surfaces in Germany.

Genus ***Candelariella*** Müll, Arg.

Candelariella forms crustose-grainy thalli that are conspicuous by their bright yellow-green to yolk-orange colors. Phycobionts are unicellular green algae. Apothecia are the same color as the thalli; asci are club shaped and contain 8–32 spores, which are usually unicellular. Additionally pyknidia are formed. Many species prefer nutrient-rich habitats (Hakulinen 1954; Wirth 1995a, b).

Candelariella aurella (Hoffm.) Zahlbr.

The yellow thalli were rather delicate and grainy with numerous more or less flat apothecia with separate margin and asci containing 8 spores.

We detected the species on various specimens (old and new) exposed at various sites in Germany.

The species is distributed worldwide. It prefers nutrient-rich substrates and occurs mainly in synanthropic habitats, for instance monuments, walls, washed concrete and materials of masonry (Wirth 1995a, b; Marcos Laso 2001; Hofbauer et al. 2003).

Hofbauer et al. (2003) moreover reported *Candellariella vitellina* (Hoffm.) Mll. Arg. as inhabitant of established building surfaces in Germany.

Genus ***Xanthoria*** (Fr.) Th. Fr.

Xanthoria forms leafy thalli with broad to narrow lobes, appressed to the substrate or ascending. The color of the thalli ranges from bright yellow-green to bright orange. More or less strong attachment of the thalli to the ground is facilitated by rhizines. Thalli form apothecia or soralia or isidia. Ascospores are polar diploblastic, which means that they contain two unequal cells separated by a septum. Additional pyknidia may be formed. Phycobionts are *Trebouxia* species or *Trebouxia* like, respectively. Many species prefer nutrient-rich or alkaline locations open to sunlight (Purvies et al. 1992; Wirth 1995a, b). *Xanthoria* was divided in new genera according to morphological but mainly molecular phylogenetic data (see Wirth et al. 2013), in our work we keep the previous status.

Xanthoria elegans (Link) Th. Fr. (Syn.: *Rusavskia elegans* [Link] S. Y. Kondr. et al.; see Wirth et al. 2013)

Thallus lobes were narrow and elongated, showing more or less the same breadth in their whole extend, slightly vaulted and arranged in a rosette with a typical orange color.

In our study, *Xanthoria elegans* was observed on older specimens exposed in Holzkirchen/Germany.

X. elegans has a broad arctic and alpine distribution, reaching up to the alpine zone.

The nitrophile and photopile species prefers nutrient-rich locations and also anthropogenic substrates. It has already been documented on the outer surface of building parts

Fig. 3.106 *Xanthoria parietina* on a concrete surface (Hofbauer et al. 2003). The occurrences documented in our study do not show such big and well-developed thalli (thallus here ca. 2 cm in diameter)

like concrete, tiles, walls, plaster and asbestos-concrete (Purvis et al. 1992; Wirth 1995a, b; Hofbauer et al. 2003).

Xanthoria parietina (L.) Th. Fr. (Fig. 3.106)

Thalli typical were bright yellow green and lobed. The disk-shaped orange apothecia (Fig. 3.106) form a vivid contrast to the thallus. Color and structure of the thallus are unmistakable, even in minute specimens without developed apothecia

In our study, *Xanthoria parietina* was observed on some specimens (old and new) exposed in different locations in Germany.

The species has a boreo-mediterranean distribution. It has previously been recorded from anthropogenic habitats like walls, roofs, washed concrete and materials from masonry (Purvis et al. 1992; Wirth 1995a, b; Hofbauer et al. 2003).

X. parietina prefers nutrient-rich substrates and is relatively tolerant against acid air pollution. Therefore, it is one of the few lichen species that remain in areas with high sulfur dioxide pollution but eventually may die ("lichen deserts" in the centers of cities). Because of reduction of sulfur pollution in cities, this lichen and other lichens have been able to respread in the past years (Trinkaus 2001).

Hofbauer et al. 2003 observed the occurrence of further lichens on several years old building surfaces in Germany: *Xanthoria* sp., Teloschistaceae Zahlbr., *Collema crispum* (Hudson) Weber, *Collema* sp., Collemataceae Zenker, *Gyalecta jenensis* (Batsch) Zahlbr., Gyalectaceae (A. Massal.) Stizenb., and *Pseudevernia furfuracea* (L.) Zopf, Parmeliaceae Zenker.

On older building surfaces and walls, a much greater diversity of lichen species may be encountered. In our study, we focused on the taxa that occur on modern façades within the first few years after erection.

Vegetative Thalli Fragments of Lichens

Small lichen thalli in form of little scales were found on older specimens and on some new ones after only 2 years of weathering exposure, which could not been assigned to a certain lichen taxon.

Soredious Forms

Lichen soredia or soredia like stages of lichens were observed in form of blackish crusts on almost all old specimens and on some of the investigated surfaces. At this stage, it was not possible to identify the corresponding lichen taxon. For this, the presence of the complete thalli is necessary.

Animals/Zoological Organisms

It was not originally intended to assess zoological life forms in the course of our study. Therefore, no methods and concepts for a targeted quantitative and qualitative assessment of zoological life forms were developed. Nevertheless, animals were occasionally observed in some analyses and in raw material of some locations. For the sake of completeness, the observed or co-cultured animals are discussed in brief. It is completely natural that in almost all terrestrial ecosystems, animals are part of the processes, and therefore, they also are part of the initial succession of modern building surfaces. Nomenclature of the zoological forms refers to Schaefer (1984).

Amoebae which fed on algae were observed in raw sample material and partially in raw cultures of some investigated older surfaces. These amoebae showed a great morphological similarity to genus *Vahlkampfia* Chat. et Lal. (order Amoebina, class Rhizopoda, phylum **Protozoa**; Fig. 3.107), for verification and precise determination further investigations would be necessary. Protozoa entirely covered in cilia, which therefore belong to the class Ciliata (phylum **Protozoa**), were encountered in raw material of surfaces of our other investigations.

Permanent stages from orange-reddish colored wheel animalcules (class Rotatoria, phylum **Nemathelminthes**) were found in different analyses of older surfaces and from older specimen, which immediately awakened from their dormant stage at moistening (on the slide, Fig. 3.108). The observed forms were uniform and showed great similarity with the characters of genus *Mniobia* Br.: In some individuals the content of the stomach was recognized and it was assumed that the diet of these rotatoria consisted mainly of

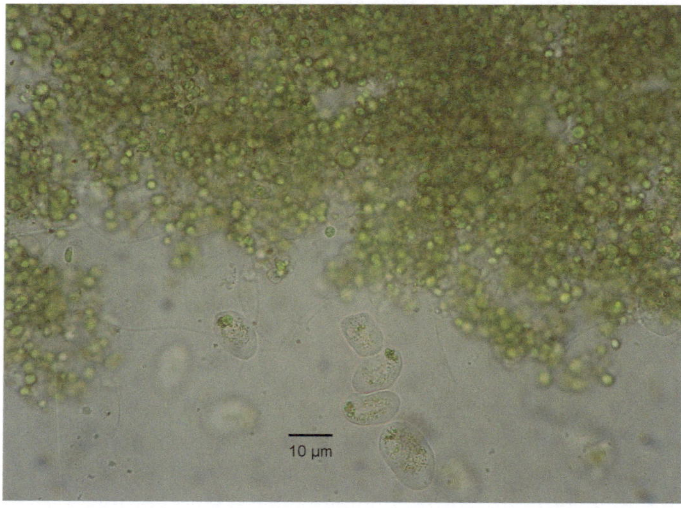

Fig. 3.107 Amoebae in the lower center of the picture (*Vahlkampfia* sp.?) with incorporated algae

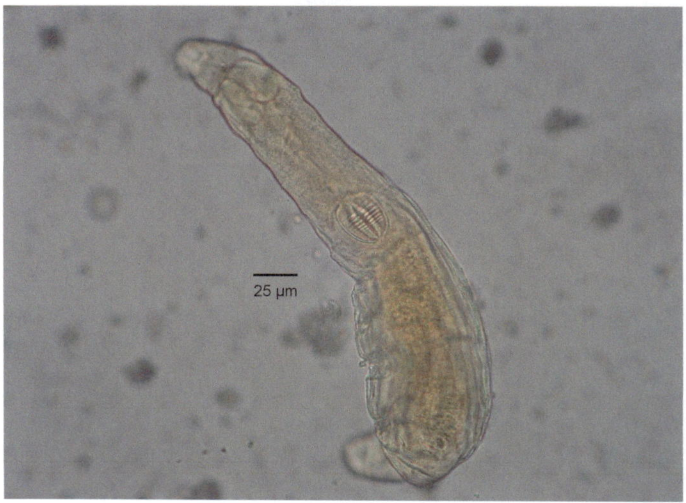

Fig. 3.108 Wheeled animalcule of the genus *Mniobia*, recovered from a building surface. The chewing stomach lays conspicuously in the center

algae. In some of the additionally analyzed surfaces also nematodes (class Nematoda, phylum **Nemathelminthes**) were detected, of which the genus affiliation could not be assessed.

Some living tardigrades (phylum Tardigrada) were discovered on raw material of one additional surface. Apart from rotatoria and ciliata, larvae of some gnats (family Chironomidae, order Diptera, class Hexapoda, phylum Arthropoda) were encountered on raw material of another additional surface. In addition, we regularly observed dust lice (order Psocoptera, class Hexapoda, phylum Arthropoda) on the surface of our specimens. Although not seen in the course of our investigations it is very likely that also mites (order Acarina, class Arachnida, phylum Arthropoda) visited the surfaces, as they are usually present on similar surfaces. Animals also influence initial succession in an indirect, nevertheless profound way via excrement deposition.

Assessment of Organisms and Environmental Factors of the Façade Habitat

Diversity of Organisms

Quantitative Growth Development of Microorganisms

Microbial growth on the specimens was continuously monitored by visual recording, and the growth development was carefully documented. Toward the end of the investigation period of 2 years' outdoor exposure, a distinct increase of the surface growth of algae on the surfaces of the specimens was observed.

For the growth of bacteria, no correlations to specific surface layers were detected; also their development in time was inhomogenous. According to the results of our study, the number of detectable bacteria varied strongly and the numbers were so low that a remarkable development of bacteria could not be recorded on the sample surfaces in the first 2 years after their production. This was additionally confirmed by analyzing the surfaces with reflection electron microscopy (REM): No "biofilm development" by bacteria was detectable. Further assessments of surface deterioration compared with untouched surface layers and surfaces in natural habitats should be performed.

NV1, NV2, NV4, NV10, NV11, NV12 and NV13 showed remarkably stronger surface growth than other systems. The development of surface growth on the same qualities of plaster was delayed by adequate paint. Samples with paint (e.g., NV14, NV15) showed less surface growth than variants without paint (e.g., NV13). A reduced particle size of the structure kernels from 0.75 to 1 mm ("smooth surface") had no effect on growth development, especially when particle size varied insignificantly.

Semiquantitative detection with a grid system is an effective method for long-time observation of primary processes of growth development on surfaces and delivers practical results. Because of the higher differentiation in the lower ranges (less surface growth) of the scale and the graphic expression of the situations, a set of data was obtained on

© Springer-Verlag GmbH Deutschland, ein Teil von Springer Nature 2021
W. K. Hofbauer and G. Gärtner, *Microbial life on Façades*,
https://doi.org/10.1007/978-3-662-54833-2_4

the visual situations of the surfaces. Contrary to chemical methods of measurements, the applied method was free of damage. Compared with methods of autofluorescence of chlorophyll (PAM-method), the former was more exact and had the same sensitivity, because the distribution pattern of microorganisms also on large areas was better detected, based on the differentiation of diverse groups of organisms (algae, fungi, mosses, animals, pollution).

Visual Assessment of Surface Growth

An important result of visual detection was the cyclic process of growth development (see also Chap. 2). During summer and at the end of winter, a decrease of optical impression was visible, while during the favorable seasons (in Central Europe mainly during autumn and early winter), an increase of surface growth was determined. Generally surface growth on specimens came in waves. During the recurrent favorable season of the following year, growth typically started on a higher level, and thus, the surface became more and more overgrown. There were exceptions only in a few specimens (NV4, NV7, NV10, NV11). Whereas the variants NV4 and NV7 seemed generally unsuitable for surface growth development, the specimens NV10 and NV11 expressed chalking after some time and part of the growth came off.

In most variants, the intensity of growth reached or succeeded level 3 (+[+]; see Table 2.4) after 5 years of exposure. This is the lowest level of growth at which complaints may occur.

Appliance of a colored top coat could reduce surface growth as seen in variants NV7 and NV8, whereas in variant NV1, it did not produce any significant effect. In general, the measurements showed a faster drying of colored surfaces.

A reduced size of structure kernel seemed to reduce growth development, but it was not clear if the effect was overlaid by chalking. A size of one mm of the structure kernels still leaves room for structures to attach microbes; mathematically, the surface area is enhanced. With the reduction by half in structure size driving rain may be more intense in its effect.

In general, coated plaster surfaces showed a tendency for better performance. With the appliance of a top coat, some depressions in the surface structure are filled up and the surface gets smoother. It is important that a suitable coating is used. If the chemical or physical properties of the plaster and the top coat do not correspond sufficiently, the coat will come off soon, as was seen in specimen NV4. Special mineral coatings (e.g., silicate coatings, NV4, NV5) stayed clean for a long time.

Silicone resin systems showed a rather intense growth development as recognized on specimens NV13 and NV14 and less drastic on specimen NV3.

Growth development of a different intensity, but markedly above the lowest complaint level, was also apparent on super-hydrophobic surfaces. The explanation for this could be that surface condensation is a main source of growth by supplying moisture, as shown in the base project (Hofbauer et al. 2006). If super-hydrophobic surfaces are regularly washed by driving rain, they can stay clean for a long time, but if the main source of moisture is only condensation, there may be not enough run-off energy to clean it completely.

Visual detection showed that melanogenic fungi grew primarily in habitats with organic contamination whereas the growth of algae proceeded autonomously.

With the visual measurements of the surface growth development valuable insight was gained. Several parameters influenced the results as is often the case in outdoor observations. Such questions can be addressed in newly designed experiments using the gained observations.

With the additional exposure sites, we showed that in a moister climate surface growth development is more intense (both in Ernsthofen and especially in Heggen, the relative humidity is usually higher than in Holzkirchen). This confirmed the result that on ETICS surfaces high humidity and condensation were very important sources of growth promoting moisture.

Figures 4.1, 4.2, 4.3, 4.4, 4.5, 4.6, 4.7, 4.8, 4.9, 4.10, 4.11, 4.12, 4.13, 4.14 and 4.15 show the sketches of the monthly/two monthly growth development on the new specimens at the main site in Holzkirchen. For a comparison between the exposure sites on a quarterly/half year basis, see Figs. 4.16, 4.17, 4.18, 4.19, 4.20, 4.21, 4.22, 4.23, 4.24, 4.25, 4.26 and 4.27, according to Hofbauer (2007). For growth development up to 2 years, see Hofbauer et al. (2006).

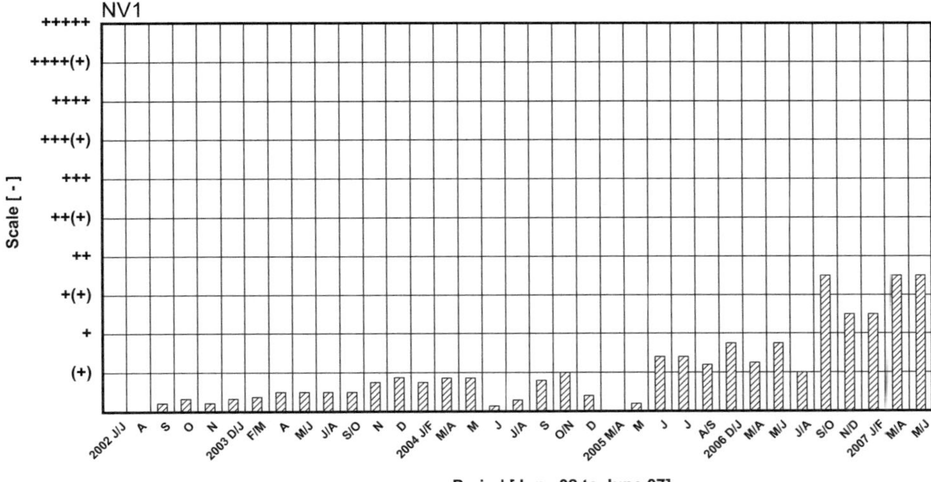

Fig. 4.1 Development of initial growth of microorganisms on series NV1 in Holzkirchen, Germany. Assessment of growth according to the visual rating scale for assessing growth intensity on building surfaces (Table 2.4) (Hofbauer 2007)

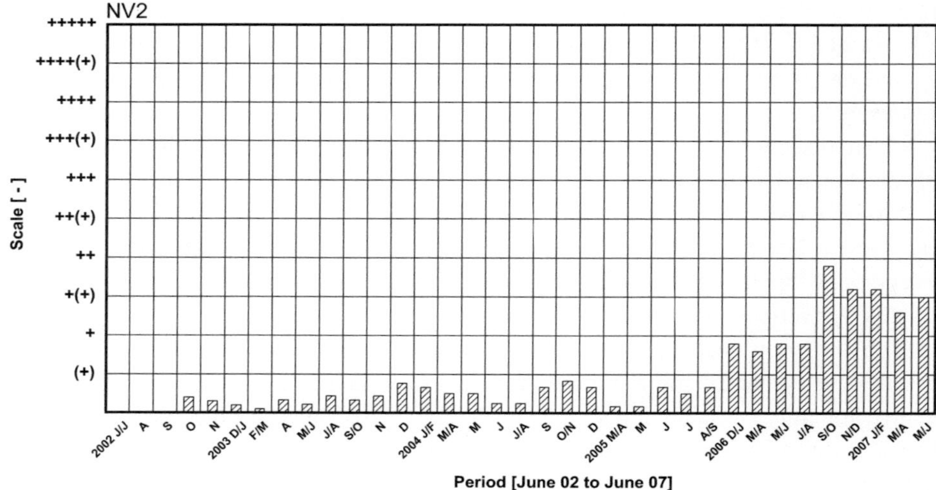

Fig. 4.2 Development of initial growth of microorganisms on series NV2 in Holzkirchen, Germany. Assessment of growth intensity according to the visual rating scale (Table 2.4) (Hofbauer 2007)

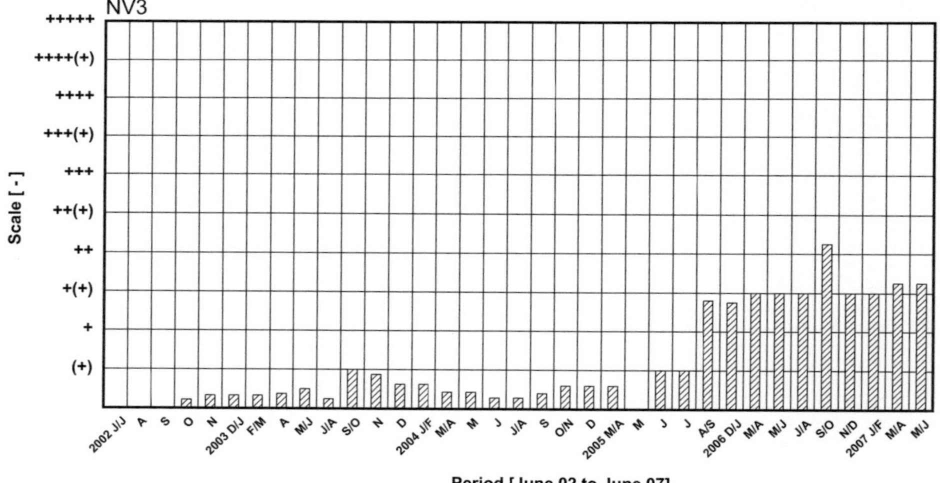

Fig. 4.3 Development of initial growth of microorganisms on series NV3 in Holzkirchen, Germany. Assessment of growth intensity according to the visual rating scale (Table 2.4) (Hofbauer 2007)

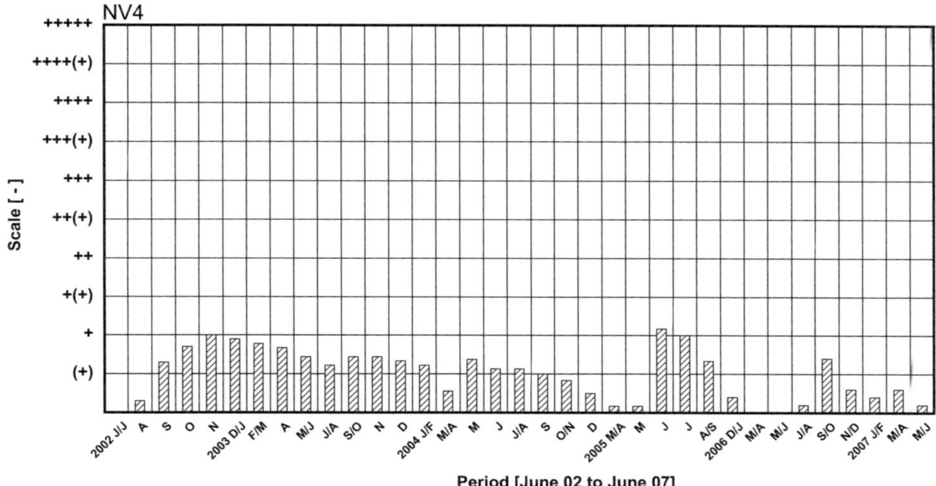

Fig. 4.4 Development of initial growth of microorganisms on series NV4 in Holzkirchen, Germany. Assessment of growth intensity according to the visual rating scale (Table 2.4). A partial reduction of growth is visible because the paint did not stick well to the plaster and came off (Hofbauer 2007)

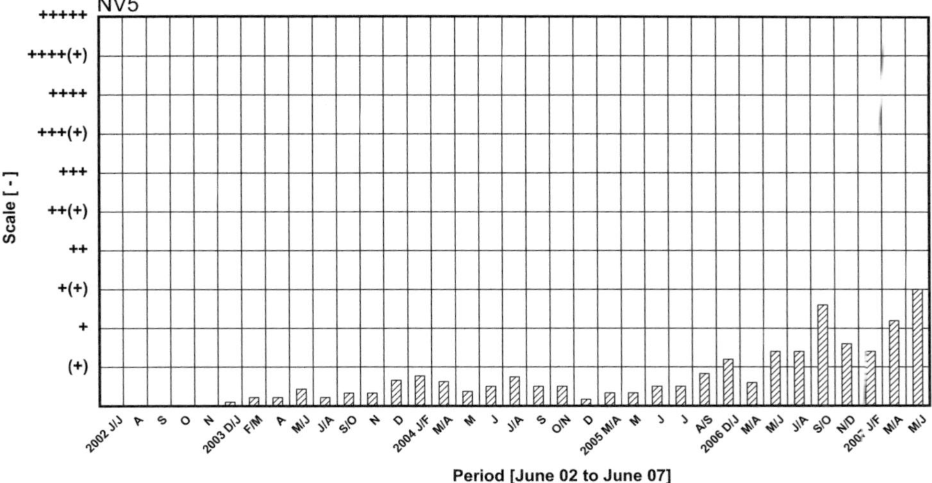

Fig. 4.5 Development of initial growth on series NV5 in Holzkirchen, Germany. Assessment of growth intensity according to the visual rating scale (Table 2.4) (Hofbauer 2007)

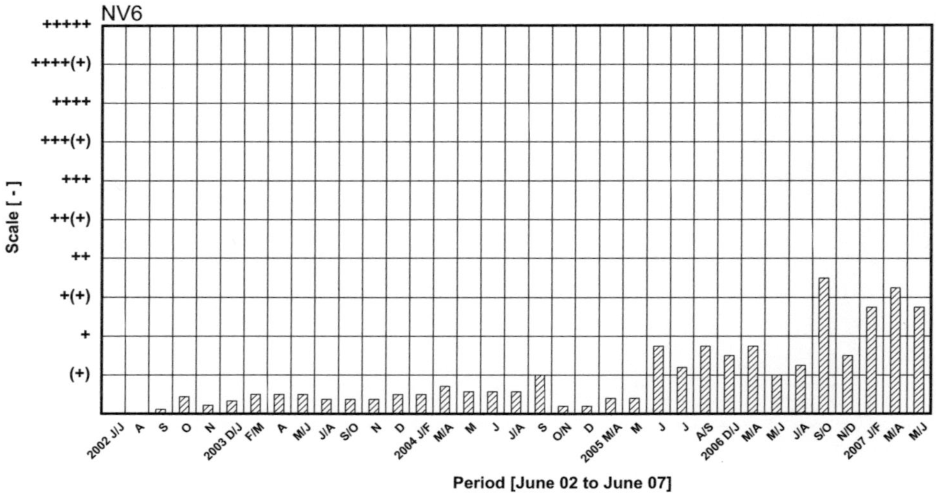

Fig. 4.6 Development of initial growth on series NV6 in Holzkirchen, Germany. Assessment of growth intensity according to the visual rating scale (Table 2.4) (Hofbauer 2007)

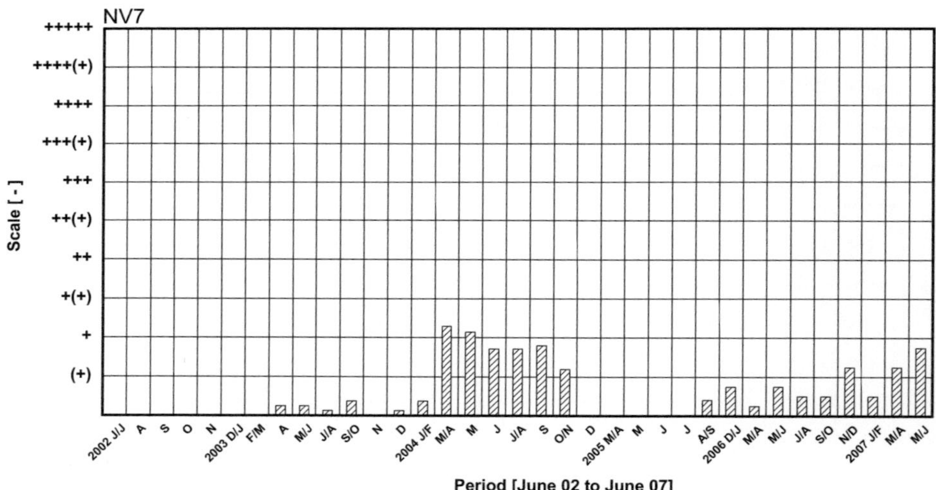

Fig. 4.7 Development of initial growth on series NV7 in Holzkirchen, Germany. Assessment of growth intensity according to the visual rating scale (Table 2.4) (Hofbauer 2007)

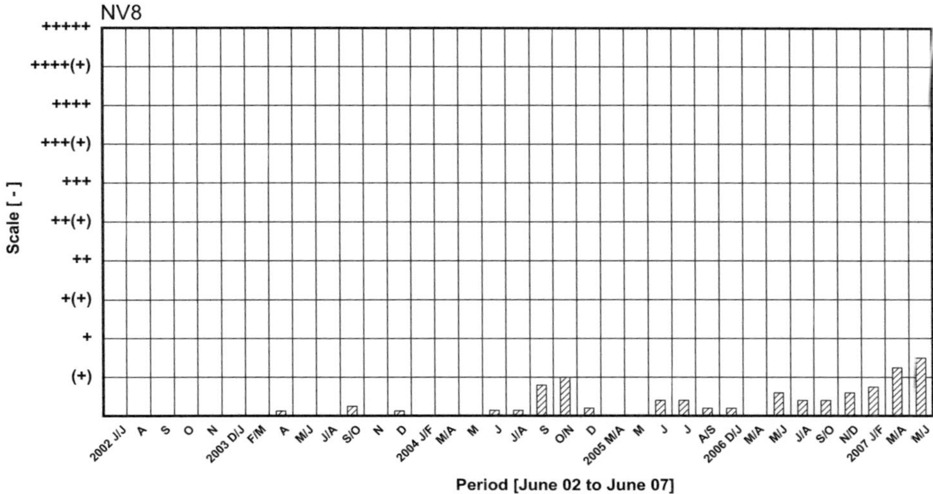

Fig. 4.8 Development of initial growth on series NV8 in Holzkirchen, Germany. Assessment of growth intensity according to the visual rating scale (Table 2.4) (Hofbauer 2007)

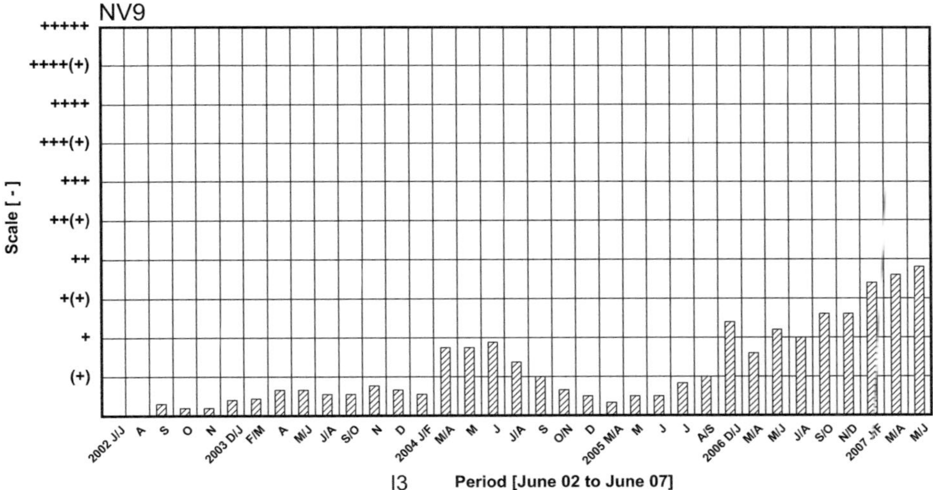

Fig. 4.9 Development of initial growth on series NV9 in Holzkirchen, Germany. Assessment of growth intensity according to the visual rating scale (Table 2.4) (Hofbauer 2007)

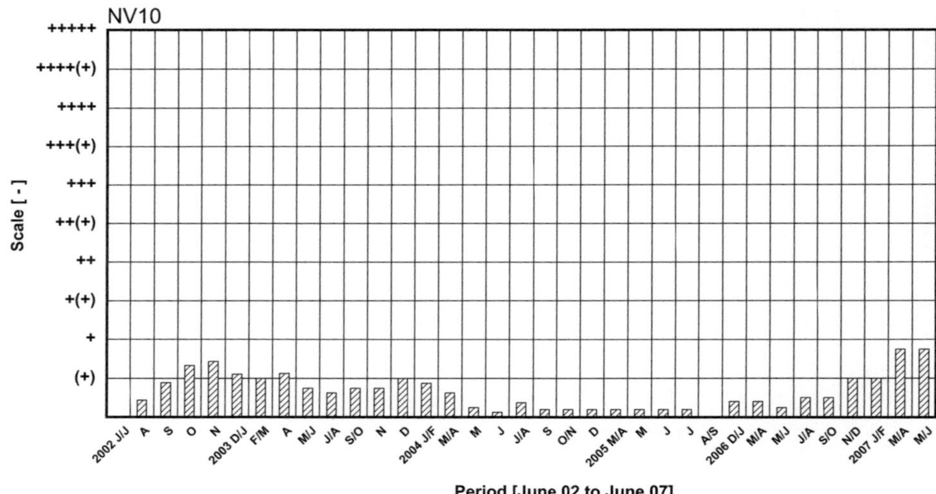

Fig. 4.10 Development of initial growth on series NV10 in Holzkirchen, Germany. Assessment of growth intensity according to the visual rating scale (Table 2.4). Interim reduction of surface growth because of chalking (Hofbauer 2007)

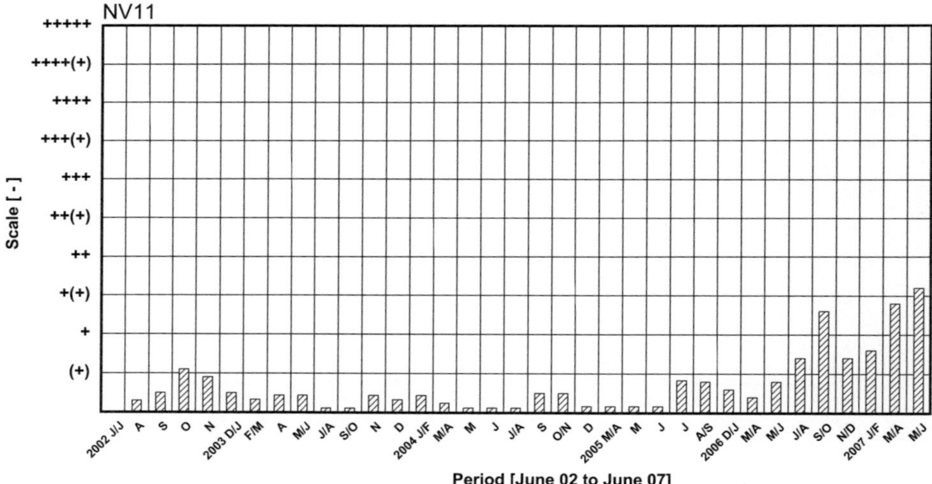

Fig. 4.11 Development of initial growth on series NV11 in Holzkirchen, Germany. Assessment of growth intensity according to the visual rating scale (Table 2.4). Interim reduction of growth because of chalking (Hofbauer 2007)

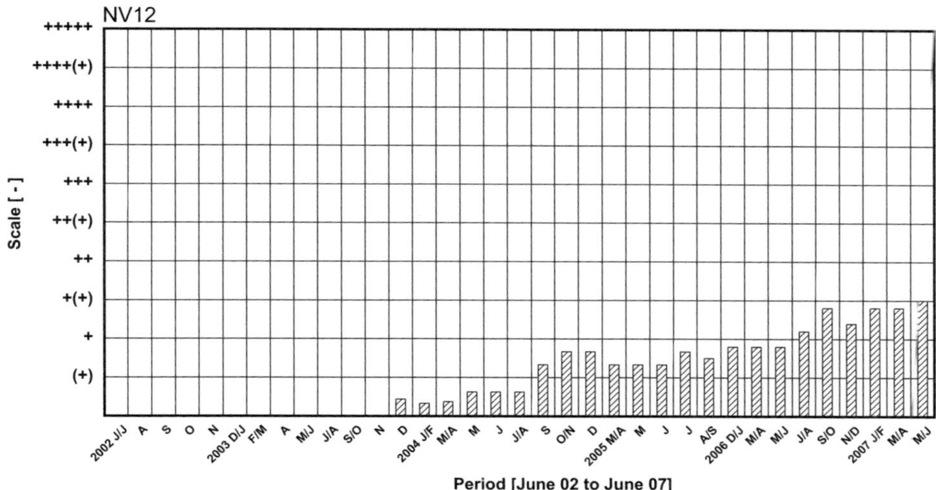

Fig. 4.12 Development of initial growth on series NV12 in Holzkirchen, Germany. Assessment of growth intensity according to the visual rating scale (Table 2.4) (Hofbauer 2007)

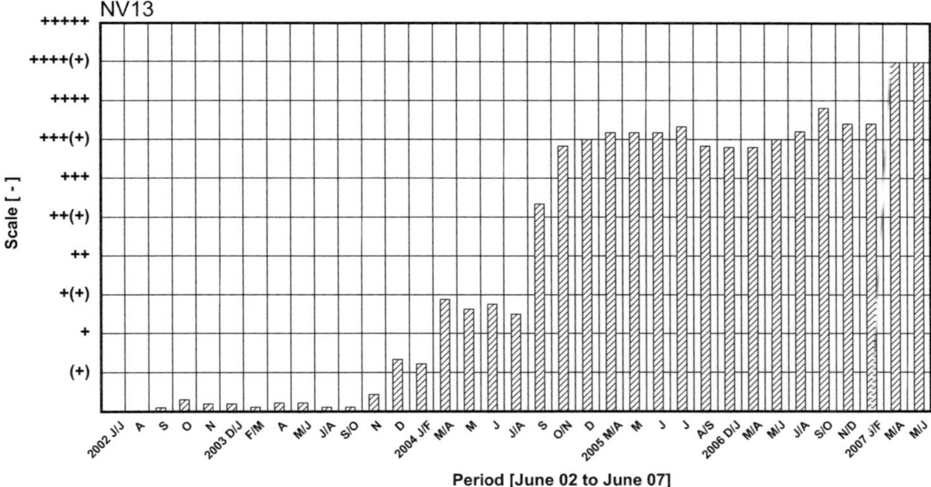

Fig. 4.13 Development of initial growth on series NV13 in Holzkirchen, Germany. Assessment of growth intensity according to the visual rating scale (Table 2.4) (Hofbauer 2007)

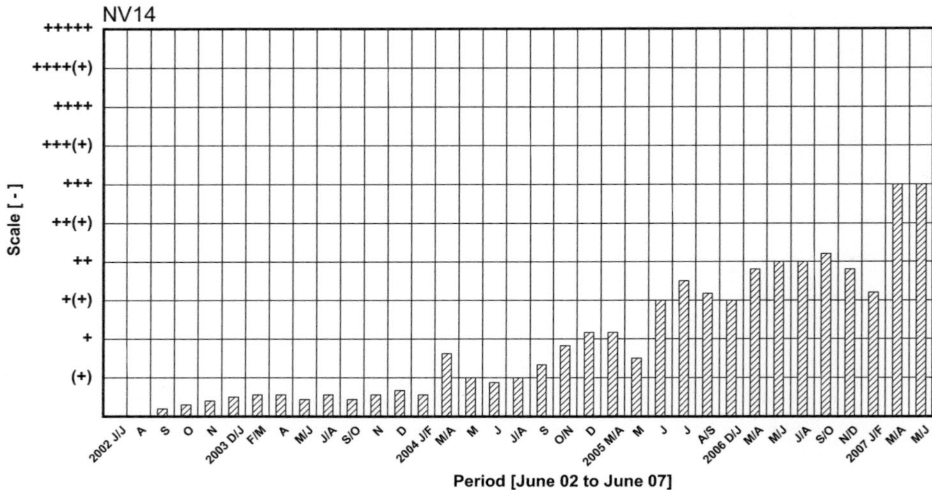

Fig. 4.14 Development of initial growth on series NV14 in Holzkirchen, Germany. Assessment of growth intensity according to the visual rating scale (Table 2.4) (Hofbauer 2007)

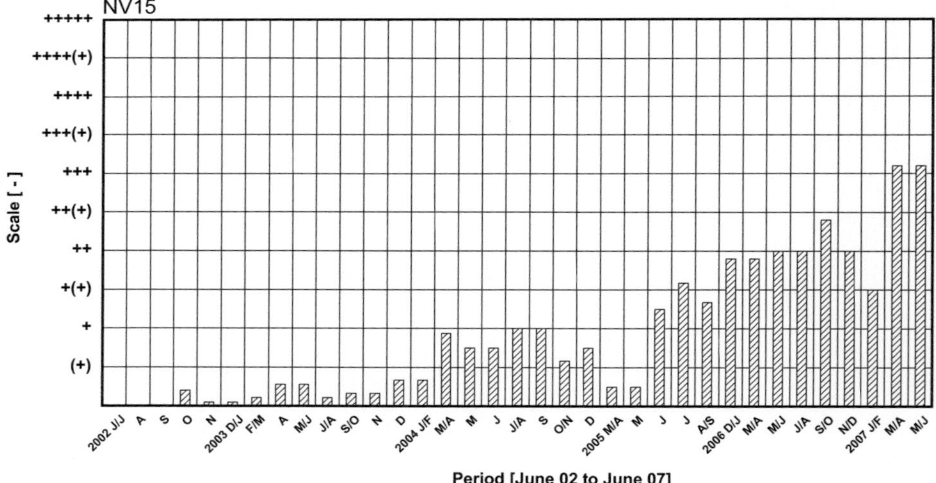

Fig. 4.15 Development of initial growth on series NV15 in Holzkirchen, Germany. Assessment of growth intensity according to the visual rating scale (Table 2.4) (Hofbauer 2007)

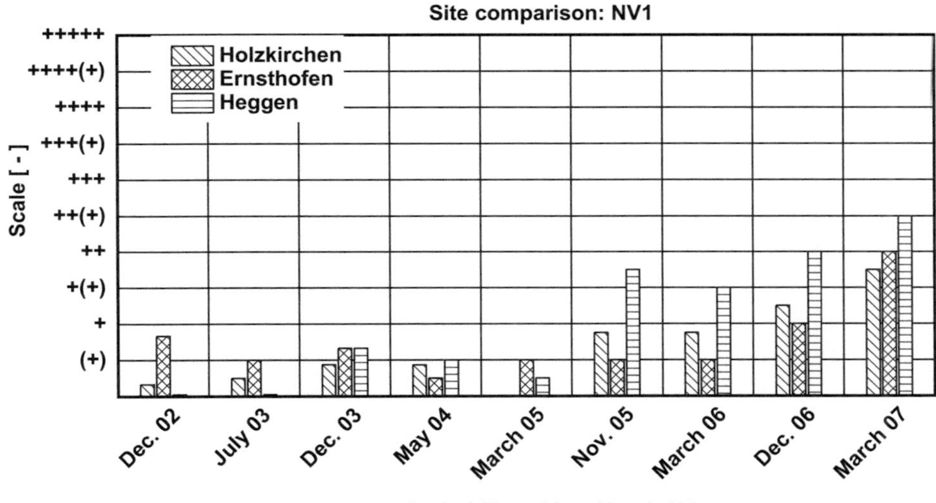

Fig. 4.16 Comparison of overgrowth development of microorganisms on series NV1 at different sites, according to the visual rating scale (Table 2.4) (Hofbauer 2007)

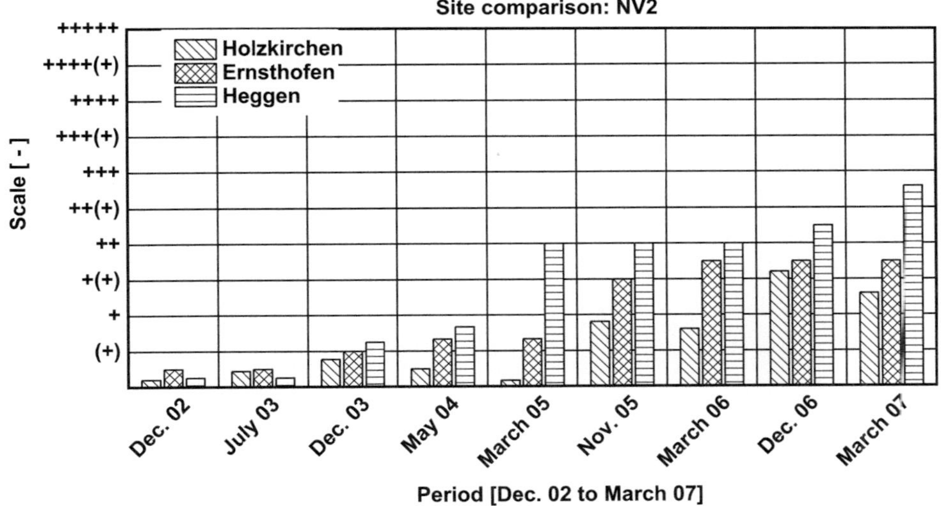

Fig. 4.17 Comparison of overgrowth development of microorganisms on series NV2 at different sites, according to the visual rating scale (Table 2.4) (Hofbauer 2007)

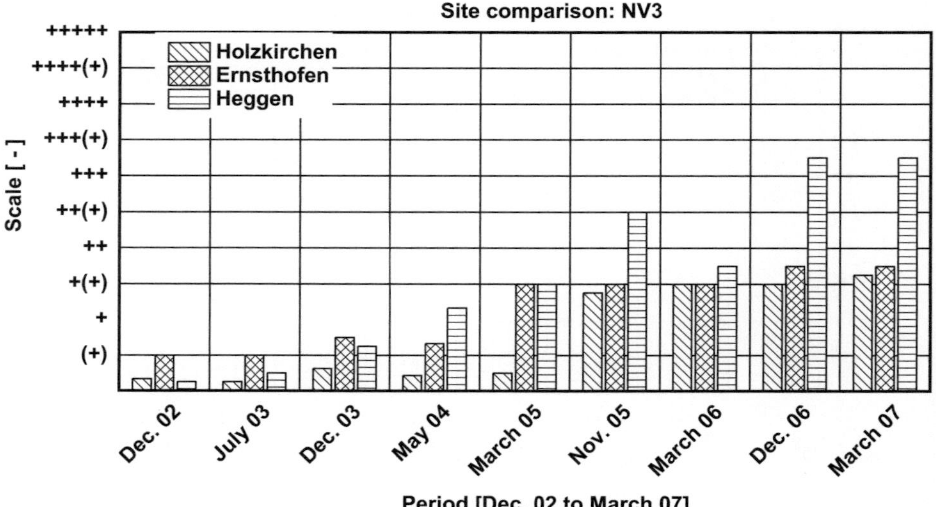

Fig. 4.18 Comparison of overgrowth development of microorganisms on series NV3 at different sites, according to the visual rating scale (Table 2.4) (Hofbauer 2007)

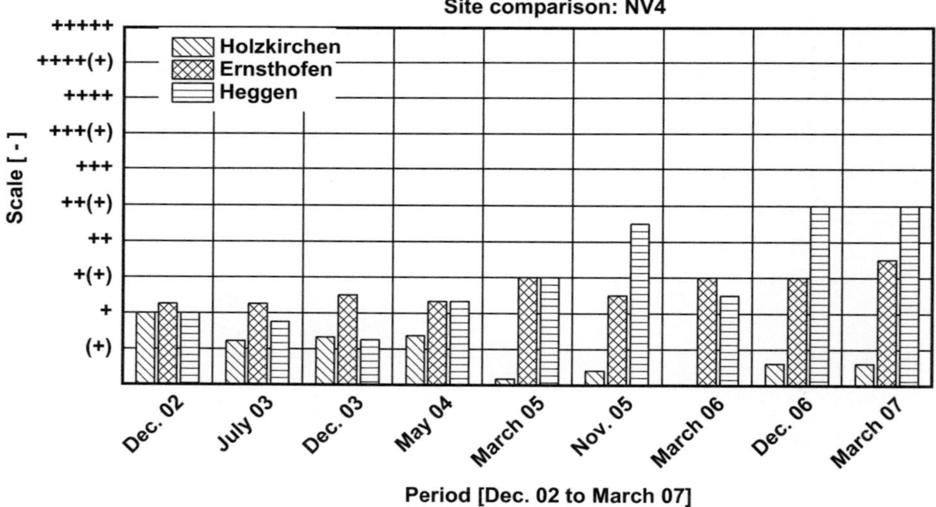

Fig. 4.19 Comparison of overgrowth development of microorganisms on series NV4 at different sites, according to the visual rating scale (Table 2.4) (Hofbauer 2007)

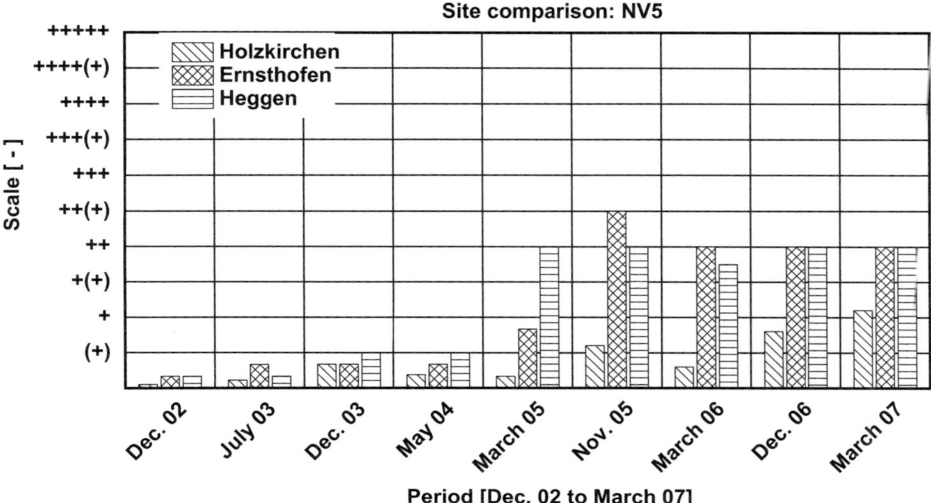

Fig. 4.20 Comparison of overgrowth development of microorganisms on series NV5 at different sites, according to the visual rating scale (Table 2.4) (Hofbauer 2007)

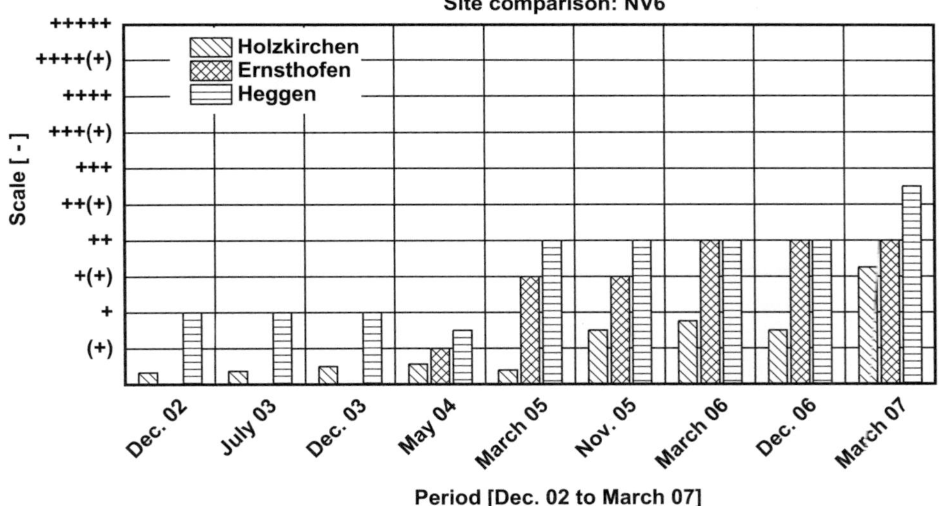

Fig. 4.21 Comparison of overgrowth development of microorganisms on series NV6 at different sites, according to the visual rating scale (Table 2.4) (Hofbauer 2007)

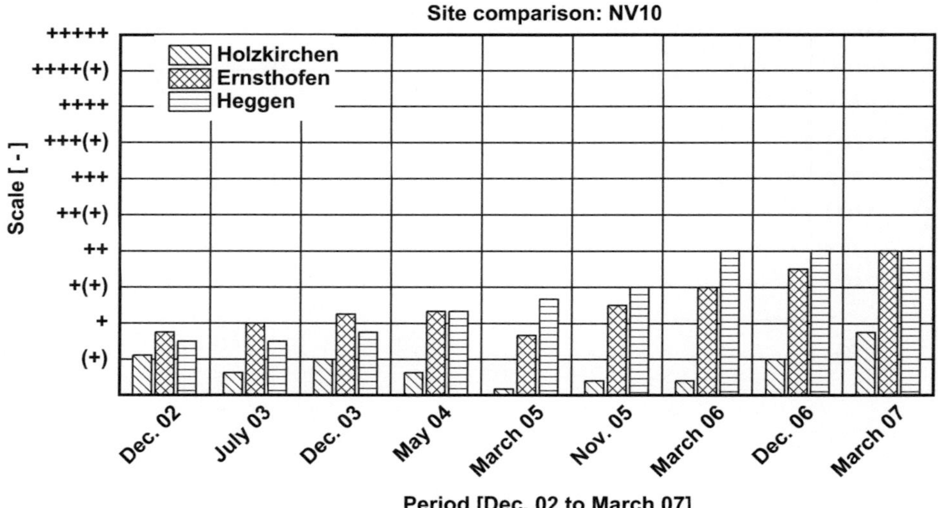

Fig. 4.22 Comparison of overgrowth development of microorganisms on series NV10 at different sites, according to the visual rating scale (Table 2.4) (Hofbauer 2007)

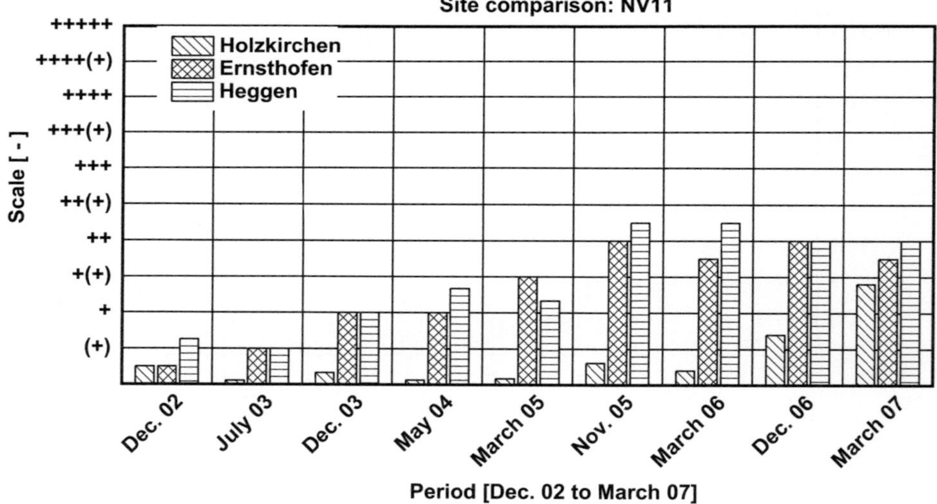

Fig. 4.23 Comparison of overgrowth development of microorganisms on series NV11 at different sites, according to the visual rating scale (Table 2.4) (Hofbauer 2007)

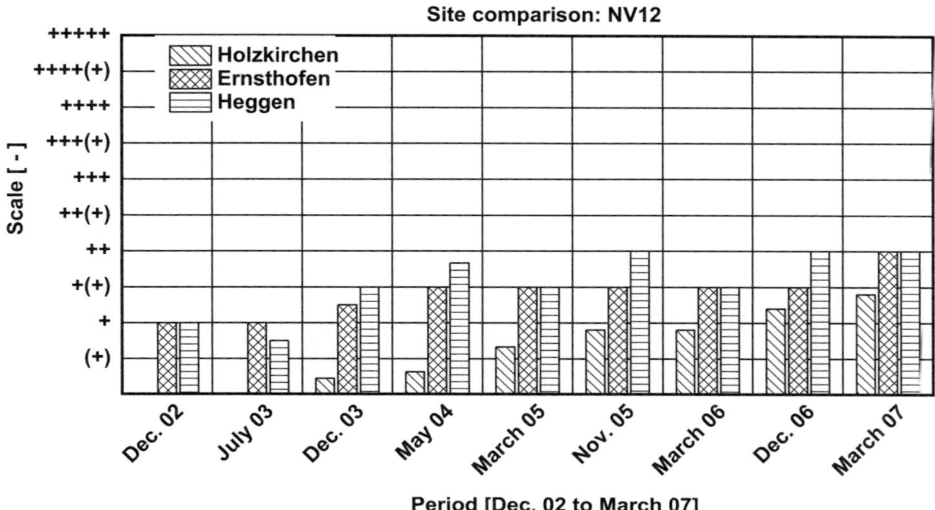

Fig. 4.24 Comparison of overgrowth development of microorganisms on series NV12 at different sites, according to the visual rating scale (Table 2.4) (Hofbauer 2007)

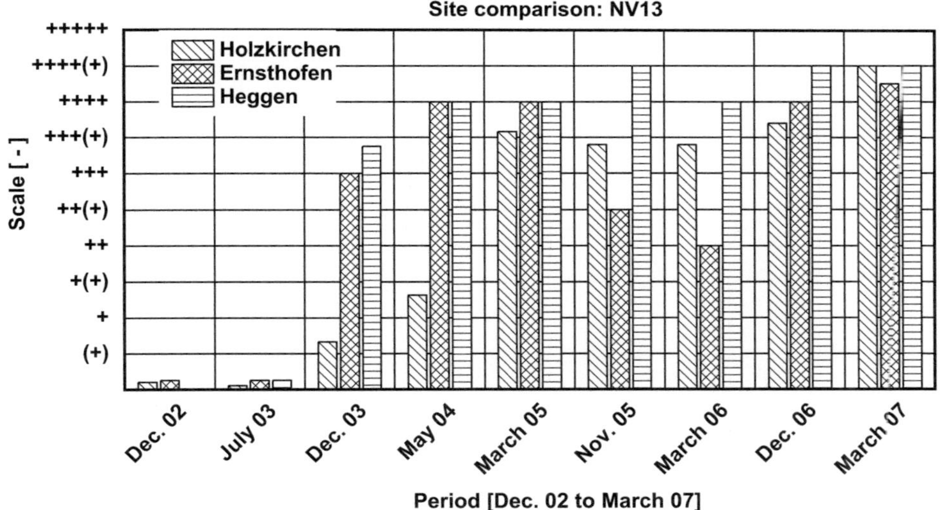

Fig. 4.25 Comparison of overgrowth development of microorganisms on series NV13 at different sites, according to the visual rating scale (Table 2.4). (Hofbauer 2007)

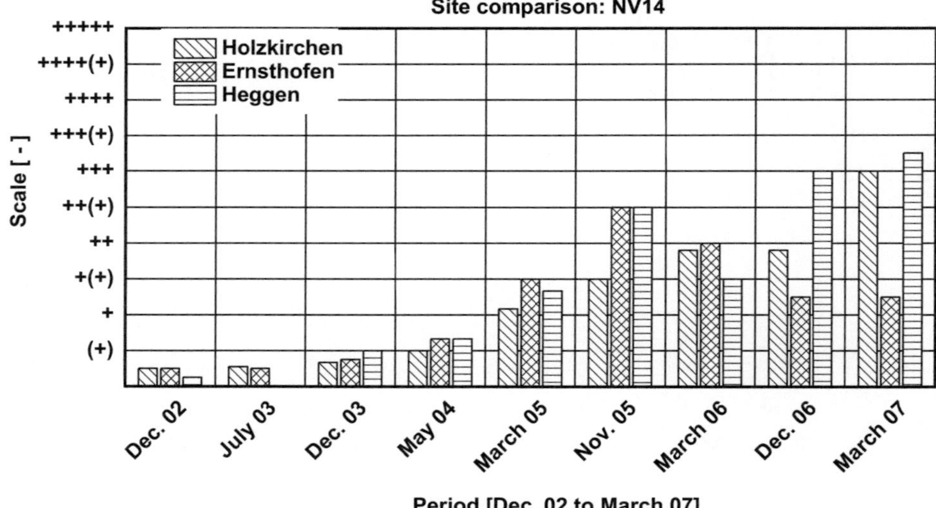

Fig. 4.26 Comparison of overgrowth development of microorganisms on series NV14 at different sites, according to the visual rating scale (Table 2.4) (Hofbauer 2007)

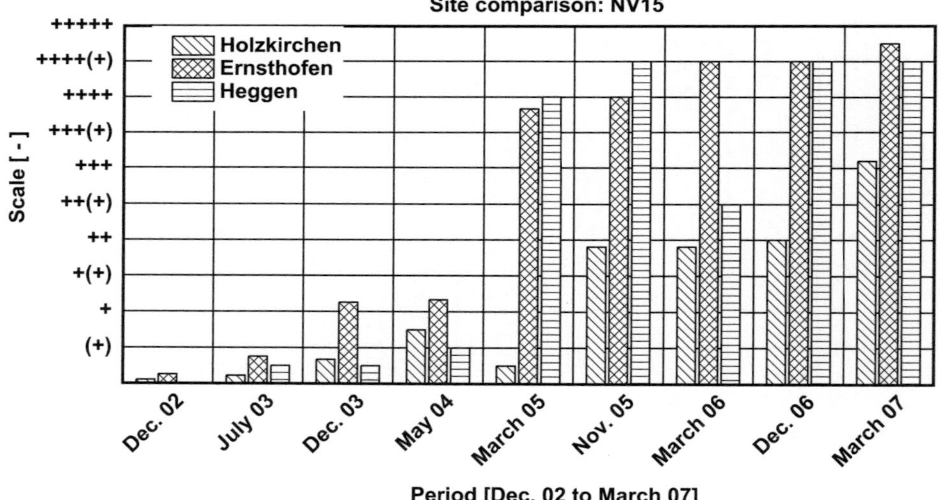

Fig. 4.27 Comparison of overgrowth development of microorganisms on series NV15 at different sites, according to the visual rating scale (Table 2.4) (Hofbauer 2007)

Growth Development of Organisms on Building Surfaces

So far, it was assumed that algae are predominant components of the initial succession of growth on façades (Sand 1995). Contrary to this assumption the investigated surfaces of the new samples were inhabitated by fungi, especially melanogene fungi ("demati-aceae"), during the first year. During this phase of outdoor exposure, no algal growth was detected. To be able to compare our collected data on the development of identified groups of taxa on the specimens and at additional locations, the results are shown in diagrams (Figs. 4.28, 4.29, 4.30, 4.31 and 4.32). Deuteromycetes were arranged in two groups: (1) melanogene Deuteromycetes (dematiaceae) together with Coelomycetes to unite all dark forms and (2) all further Deuteromycetes without dark pigmentation. The first group is of crucial importance for processes of biodeterioration of façades because their dark pigmentation makes their occurrence very obvious and may lead to complaints and damage claims.

Figure 4.28 shows the relative dominance of different groups of fungi in the number of taxa, particularly intensely pigmented forms (melanogene fungi and Coelomycetes) on new samples within the first year. Later in the course of our study, other groups of organisms appeared. Strongly pigmented fungi prevailed in rather high species numbers also on new samples in older stages, on old samples and additional variants. A relatively strong presence of Trebouxiophyceae was also remarkable. Many species of this algal class are known from lichen symbiosis but free-living taxa were dominant. A relatively constant increase was recognizable from Chlorophyceae. After 2 and 3 years of development an increase of bryopsidae and lichens according to habitats was detectible. All other groups of organisms seemed to fluctuate. Further investigations are necessary, which could result in important facts with respect to local climate and material properties. The comparison between the different exposure localities showed that the diversity of organisms developed in different ways although the main groups of Trebouxiophyceae and melanogene fungi/Coelomycetes coincided. On old samples and additional surface growth situations, a still higher diversity was visible. On the plaster of old samples, partially mosses and/or lichens were immanent contrary to the additional surface growth situations. A relatively high diversity of Trebouxiophyceae and melanogene fungi was also present, in spite of a slight decrease in comparison with other groups.

Background Load

A certain number of spores, cells and thallus particles of fungi, bacteria and aeroterrestrial algae are constantly floating in the air (McElhenney et al. 1962; Ettl 1980). Brown et al. (1964) counted up to 3000 algal cells per m^3 of air. In our study, filtration measurements (see Appendix, Table A.9), sedimentation samples (Table A.10) and investigations of driving rain (Table A.11) were performed to document the background concentrations of viable microbial particles (especially fungi and algae) present in the environment

Fig. 4.28 Development of biodiversity (detected organisms) on the new specimens after half a year (**a**), one year (**b**), 2 (**c**) and 3 years (**d**) of outdoor exposure in Holzkirchen (Hofbauer 2007)

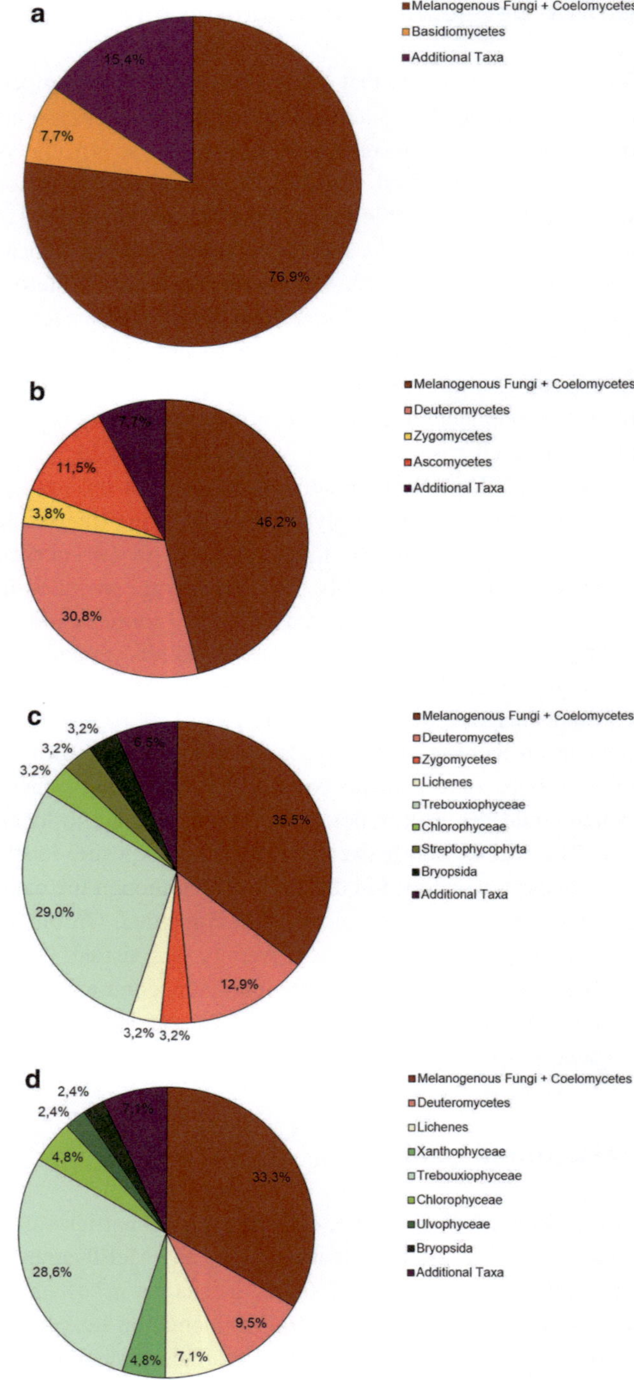

Fig. 4.29 Development of biodiversity (detected organisms) on the new specimens after 2 (**a**) and 3 (**b**) years' outdoor exposure in Finnentrop/Heggen (Hofbauer 2007)

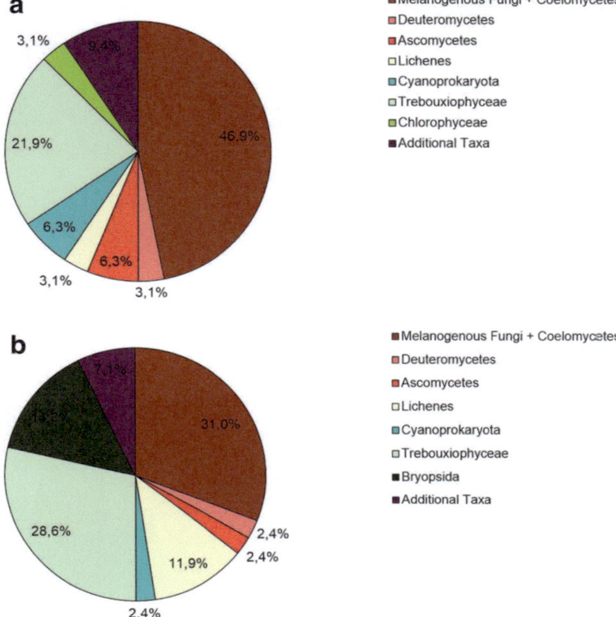

Fig. 4.30 Development of biodiversity (detected organisms) on the new specimens after 2 (**a**) and 3 years' (**b**) outdoor exposure in Ernsthofen/Oberramstadt (Hofbauer 2007)

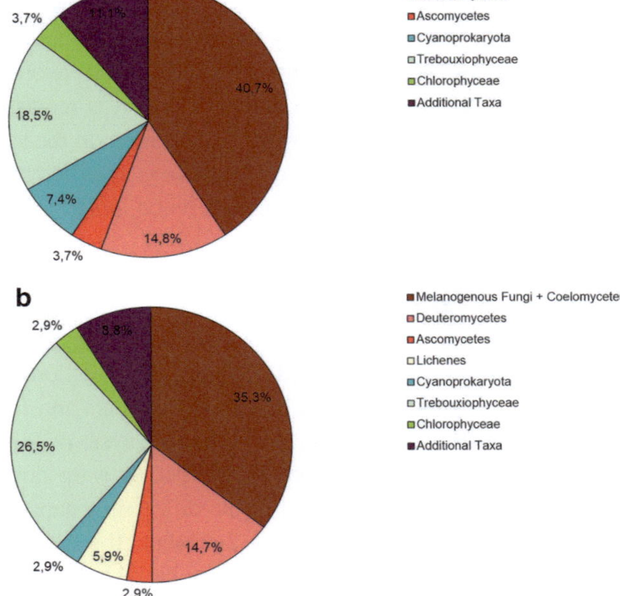

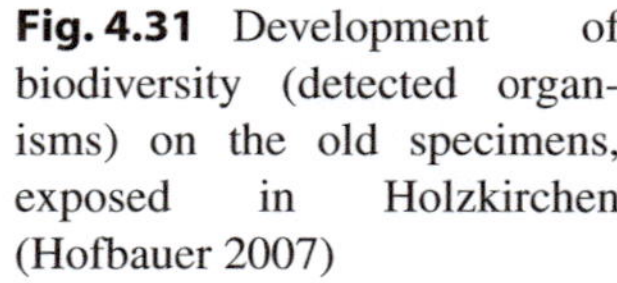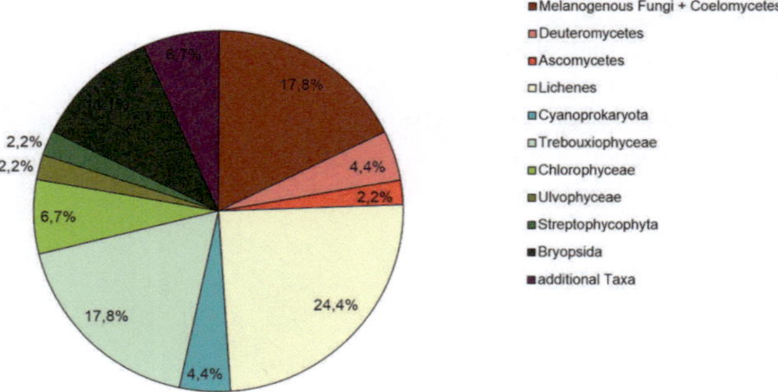

Fig. 4.31 Development of biodiversity (detected organisms) on the old specimens, exposed in Holzkirchen (Hofbauer 2007)

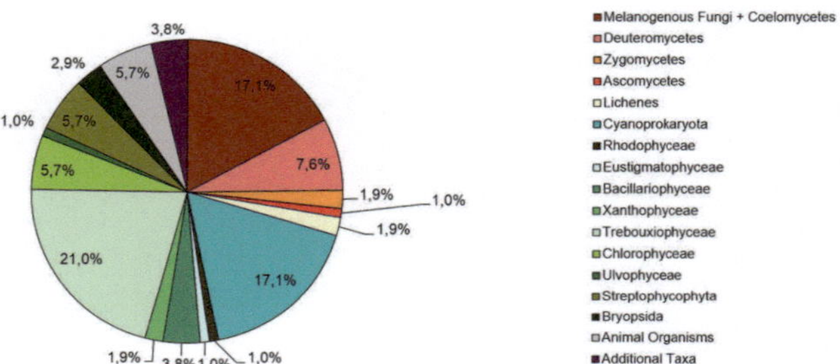

Fig. 4.32 Summary of results on the qualitative analyses of additional variants and surfaces (Hofbauer 2007)

of the exposure sites. About 1000 algal cells per ml driving rain were reported by Karsten et al. (2007). Hofbauer (2007) presented new basic data on background concentrations of diaspores of microorganisms inhabiting building parts in air and driving rain in the area of Holzkirchen. Figure 4.33 shows their taxonomic diversity.

Analyzing the background concentrations, it is remarkable that the results of driving rain matched the microbial growth development on the samples after 2 and 3 years' outdoor exposure. This shows that driving rain is an efficient and important medium for distribution and growth of microorganisms on buildings.

Diversity of Taxa

As commented in Chaps. 1 and 3, many investigations have been conducted on algae and cyanoprocryota inhabiting historical buildings and monuments. The initial development of organisms on modern façade coating systems (ETICS) is presented in our study in a wide-ranging investigation. Many species were documented on surfaces of building materials for the first time. For the different groups of specimens and locations, see Table A.1 (old specimens). Table A.2 presents the surface growth analysis of specimens after half a year of outdoor exposure in Holzkirchen. In Table A.3, surface growth of specimens after one year of outdoor exposure in Holzkirchen is analyzed. Table A.4 and

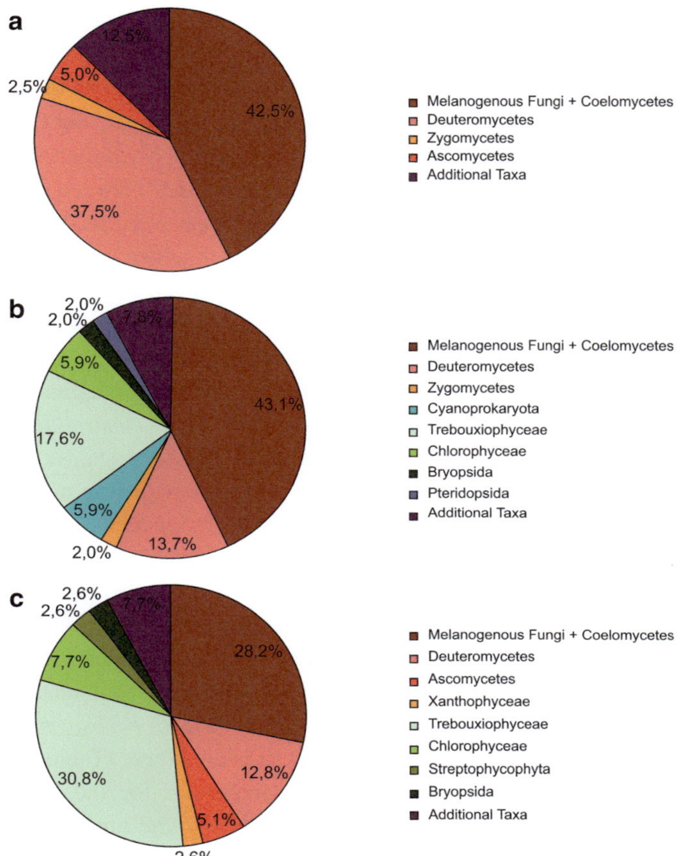

Fig. 4.33 Summary of groups of organisms found in air germ measurements (**a**), sedimentation samples (**b**) and measurements of driving rain (**c**) of the ambient air at the IBP Holzkirchen (Hofbauer 2007)

Table A.5 state the surface growth analysis of specimens after 2 or 3 years' weathering exposure in Holzkirchen. Table A.6 and Table A.7 present the surface growth analysis of specimens after 2 and 3 years' exposure in Finnentrop/Heggen or Ernsthofen/Oberramstadt, respectively. Table A.8, part one and two, shows the surface growth analysis of the additional variants/surfaces.

Apart from the quoted publications in Chap. 3, further investigations on fungi on buildings and building surfaces have been published, e.g., on façades in Sao Paulo (Shirakawa et al. 2002, 2004), on painted walls of a historical building in Germany (Gorbushina et al. 2003), on growth of Dematiaceae on marble and calcareous stone (Sterflinger and Krumbein 1997) and on biogenic coloration of concrete (Escadeillas et al. 2006). All these studies dealt with historical buildings or monuments, and they only documented genera but not species.

An unexpected diversity of more than 220 different taxa has been documented in the course of our study. Peak values were reached by eukaryotic algae and Cyanoprokaryota (ca. 78 taxa) and fungi (eumycota, ca. 70 taxa). Lichens and bryophyta were represented with 11 species each. The remaining organisms were divided between other groups like bacteria and animals. Surprisingly, lichens and mosses appeared already in the first years.

Additional fern spores were isolated from air. Alltogether, more than 400 different strains of organisms were transferred from building surfaces into clone cultures. Investigations of organisms on façades with genetic methods assumed a much lower diversity of taxa (Mudimu et al. 2007). Our investigations clearly showed that a basic differentiation with practical background is possible, if a morphological and ecological concept of species is used consequently. The example of *Coelastrella/Graesiella* even demonstrated the superiority of a morphological differentiation against a molecular and statistical classification for certain single cases. In forms with less morphological features for classification as well as for interpretation of phylogenetic relations, molecular investigations are an important and indispensable instrument.

In our study, certain taxa were mainly dominant among the algae with accessory species intermingled. Most prominent as initial colonizators were different species of genera *Stichococcus* Nägeli, *Diplosphaera Bialosuknia em.* Vischer and *Chloroidium* Nadson/*Chlorella* Beijerinck. The former two genera coincided with analyses of the "old samples" and the additional surface growth situations, but also *Klebsormidium flaccidum* (Kützing) Silva, Mattox and Blackwell commonly appeared there (Table 4.1).

Aerophytic green algae (Chlorophytes) were important for the very first surface growth, whereas other groups of algae were quantitatively and qualitatively less detectable. Eukaryotic algae and fungi dominate on substrates in the Central European geographical region. They are supplied by moisture through air humidity and dew point gradients because many of the detected forms can start their physiological activity already at a relatively high humidity and do not always need liquid water supply. Cyanoprocaryota mainly appeared on surfaces which were wetted by precipitation, e.g., splashing water or water from gutter.

Table 4.1 Most common algae forms detected on study samples. (Modified from Hofbauer et al. 2006; Hofbauer 2007)

Species	Comment
Stichococcus spp. (e.g., *S. minutus, S. bacillaris*)	Mainly coccal type, derived from thread-like organization type
Diplosphaera sp. (incl. an unknown species)	Sarcinoidic life form
Chlorella spp. s.l. (e.g., *C. ellipsoidea* agg., *C. treboux-ioides, C. angusto-ellipsoidea, Chlorella vulgaris,* etc.)	Organized coccal
Klebsormidium spp. (e.g., *S. flaccidum, S. sterile,* etc.)	Thread-like organization, mainly in surface growth more than 2 years old

Table 4.2 Fungi of high abundance and CFU numbers isolated from "old" and "new" samples (after half a year, one year, 2 and 3 years). (Modified from Hofbauer et al. 2006; Hofbauer 2007)

Species	Comment
Alternaria alternata	Melanogenous fungi (Dematiaceae), high resistance against biozides
Cladosporium spp. (e.g., *C. cladosporioides, C. herbarum,* etc.)	Melanogenus fungi (Dematiaceae)
Epicoccum nigrum	Melanogenous fungi (Dematiaceae)
Ulocladium chartarum	Melanogenous fungi (Dematiaceae), high resistance against biozides
Phoma spp. (e.g., *P. exigua, P. glomerata, P. herbarum,* etc.)	Melanogenous fungi (Coelomycares)

Stichococcus bacillaris Nägeli is generally considered one of the most common aerophytic algae and is documented in different habitats. Also, isolates from limnic, marine and thermal locations exist, but since a long time the heterogeneity of such "morphospecies" is discussed (see Ettl and Gärtner 2014) and moleculargenetic studies are necessary.

Table 4.2 summarizes fungi which appeared in analyses of surface growth on "old" and "new" samples in Holzkirchen in high abundance and relatively high colony-forming unit (CFU) numbers (dominance).

In summary, algae which appeared as initial colonizers on modern building components needed a longer starting period than fungi and bacteria. In their initial period, fungi and bacteria were often concentrated on places which are influenced by organic materials (e.g., excrements of insects and birds). No correlation between fungal or bacterial growth as a precondition for the development of algae was detectable for the initial phase (the first three years). It cannot be excluded that in later development several bacteria (e.g., nitrogen fixing bacteria) may function as nutrients for algae or that bacteria and fungi digest nutrients for other organisms, as these mechanisms work in soil formation. For the first years of development, such processes are, as we know today, of secondary relevancy.

Synsystematics

Our study used the terms "crust," "growth," "overgrowth" for all investigated biocenoses, and the validating term "infection" was avoided. Different authors use the term "biofilm" or "algal biofilm" for overgrowth dominated by aerophytic algae (e.g., Jahnke 2001). According to definition, a biofilm is a biocenosis of mainly bacteria in a film of excreted mucilage with accessory microorganisms in aquatic systems (Burzlaff 2007). A transfer of the term from aquatic to terrestrial systems seems plausible but cannot be used for biocenoses of aerophytic algae, fungi, lichens, etc., because these are independent from a mucilaginous bacterial matrix and often no mucilage is formed at all. A transitional position could be found in biocenoses with aerophytic Cyanoprokaryota, whereas in taxa like *Nostoc* spp., the formation of mucilage is part of the thallus itself and cannot be interpreted as biofilm.

Aerophytic biocenoses living on the soil surface are named "biological soil crusts," because several members of the association stabilize soil particles and work against erosion (Evans and Johansen 1999; Türk and Gärtner 2001; Souza-Egipsy et al. 2004). On tree barks and rocks, microbial overgrowth is defined as "algal overgrowth," "algal coverage" and "epiphytic communities" (Barkman 1969; Jahnke 2001).

Biogenic discoloration on historical buildings is called "Patina" (Sabbioni and Zappia 1991; Caneva 1993). For surface growth of algae in illuminated caves, the term "maladie verte" is used (Lefevre and Laporte 1969; Lefevre 1974; Kováčik 2000). This term is related to obviously discoloring features (biodeterioration) of the surface growth and cannot be used as a common scientific term for characterization of surface growth.

From the perspective of vegetation ecology, the façade is not an isolated habitat but a synusia (Walter 1990), comparable to the synusia of forests (e.g., lichens and mosses on tree bark or the coverage of organisms on rock). Biocenoses of cryptogams are in analogy to higher plants named associations particularly in ecosystems like high mountains or wetlands. Microbial overgrowth associations dominated by aerophytic algae were differentiated by Barkman (1969) and by Ellenberg (1986), but in our study only intensely investigated biocenoses like in habitats of *Prasiola crispa* (*Prasioletum crispae*, AV15) *Apatococcus lobatus* (*Apatococcetum lobati*, AV17) and of *Porphyridium purpureum* (*Porphyrietum purpurei*, AV10, AV24) could be assigned to established associations. All other investigated surface growths were early succession stages which could not be interpreted as associations.

Physical Properties of Building Materials

Measurements of Climate Parameters

Measurements of climate parameters showed a slightly higher average temperature and atmospheric humidity in the localities Heggen/Finnentrop and Ernsthofen/Oberramstadt than in Holzkirchen and therefore provided better conditions for overgrowth. During our investigations, all localities presented a relatively high development of overgrowth. Comparison of local climate parameters with data of the surfaces revealed that condensation water formed the essential source of humidity on the exposed coating systems.

Measurements of Temperature

Different types of façade coatings possess different thermotechnical characteristics. Reduced emission of thermal radiation during night and enhanced heat capacity can reduce the formation of dew on the surface of a coating, which means that less water is available for microorganisms and their physiological processes. Measurements on experimental walls showed that the use of a thick plaster system reduced the time of dew water on the surface of façades between 10 and 20%. Measurements of the surface temperatures explained why during autumn months a stronger growth on the surface coatings took place: at this time of year, the temperatures are most convenient for dew formation.

Properties of Materials Against Humidity

It has to be taken in account that overgrowth of microorganisms on coating systems raises the water uptake and the surfaces stay longer wet. In our study, different hydrophobic materials were used and the coatings acted differently to exposure. In mineralic systems dew water formed a water film on the surface, whereas strongly hydrophobic surfaces showed water drops and dried more slowly. Additionally, a fine layer of dark dust was visible on very hydrophobic surfaces which formed lines of non biological origin.

Building Surface Structures

After weathering exposure, the old samples showed about two times higher values in relation to roughness of their surface in comparison to unexposed samples. It seems that microorganisms settle easier on rough surfaces.

In a study on susceptibility of different mortars to the establishment of the alga *Klebsormidium flaccidum* roughness played an important role in biological colonization (Tran et al. 2012). A rougher surface facilitated the attachment of algal cells and so favored the development of algae. A further study documented that porosity and roughness are important factors as they favored the development of algal growth in the test situation (Fassier et al. 2013).

Measurements of the contact angle (Hofbauer et al. 2006) on the new samples showed heterogeneous results: Partly, the contact angle increased during exposure (e.g., silicon resin colors, silicate colors), partly it decreased (e.g., silicate colors). After 2 years' exposure, the slope angle of the silicon resin color NV9 measured only 43°. On all other coating systems, the droplet adhered even in 90°. Changes of contact angle and slope angle determine important characteristics of surfaces. In our study, surfaces with high contact angle seemed to support overgrowth. Also, specialized organisms were found like *Physcia* sp. which were able to stick to very water-repellent surfaces.

Recent studies showed that contact angle influences the establishment of growth of algae on building materials. A rather low contact angle means that the water is better spread on the surface and in the test situation growth of algae developed faster and more intense (Fassier et al. 2013).

Measurements of the brightness values of surface coatings showed that this factor is very suitable for temporal and comparative investigations, but surface moisture had a bigger impact on the results.

The chalking of different surfaces changed only slightly within the first 12 months but was of significance in different ways after 24 months. Particularly, in mineralic systems, an increased chalking was visible which reflected a temporary decrease of the surface growth. Our investigations on chalking showed two important facts:

1) Within the first years, the mechanical chalking of the surfaces was so significant and in dimension bigger than any influence of microorganisms on the surfaces and
2) a proper degree of chalking retarded the development of surface growth significantly.

Ecophysiological Features

Knowledge of the ecophysiological characters of the involved taxa allow us to adjust the environmental parameters like humidity, light or chemical composition of the surface, etc. (see, e.g., Albertano 1990) and thus to provide a tailored management of the biological surface colonization.

According to external factors (like fluctuation of temperature and humidity, solar radiation, etc.), the surface of façades has to be classified as an extreme habitat. The investigated processes of growth on modern building surfaces within the first months and years (guarantee period) were characterized by two important circumstances regarding the provision of nutrients:

(1) The availability of organic nutrients and anorganic mineral nutriment. Easily metabolizable organic substances were available if included in the formula of the coating or provided by external application, e.g., excrements from insects, birds or by long-distance aerial transport or precipitation.
(2) The growth of autotrophic organisms (especially algae and lichens) produced organic material for saprophytes.

Caneva and Salvadori (1987) and Kováčik (2000) presented a scheme which shows the influence of different factors on the surface growth of stones, historical buildings and monuments. The same fundamental factors function during surface growth on coating systems of modern buildings.

Cyclic Growth

An important result of the visual detection in our study was the cyclic growth development on the surfaces. During the dryer seasons, which in Central European climate are

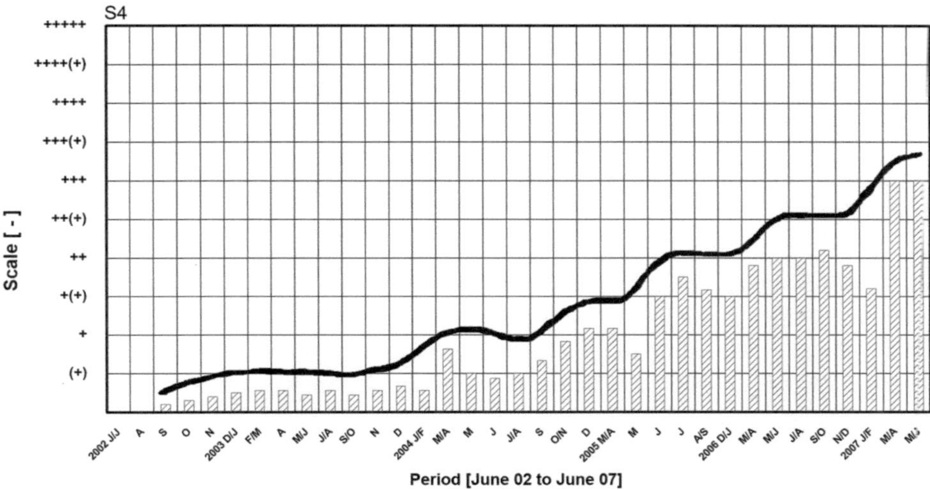

Fig. 4.34 Cyclic growth waves of microorganisms, based on original data. (Hofbauer et al. 2006; Hofbauer 2007)

high summer and end of winter, a stagnation or even slight decrease of optical impression was observed. Even if an optical decrease of surface growth was recognized, surface growth in most cases did not vanish completely, but started subsequently on a somewhat higher level. During moist periods of the year, especially at the beginning of autumn and winter, a more or less intense increase of surface growth was determined (Fig. 4.34).

Temperature

Eukaryotic algae and fungi share in life about the same extreme values of temperature. Physiological activity runs until $-15\,°C$ (exceptionally even down to $-24\,°C$ minimum temperature) (Lange 1965; Culberson 1970). The upper temperature limit for eukaryonts is ca. $60\,°C$ because of the denaturation of their proteins (e.g., Seckbach and Oren 2004). Surprisingly, several of the extremophile species (with regard to their upper and lower temperature limits) were documented on surfaces of façades. For example, *Nostoc commune* and *Prasiola crispa* showed a positive photosynthesis until $-15\,°C$ (Becker 1982; Vincent 1988; Kappen 1993), in *Aspergillus fumigatus* growth was documented until $57\,°C$ (Crisam 1973; Franz 1975; Samson et al. 2002a). High temperature limits for fungi are between 60 and $70\,°C$ ("thermal death point"). Several cyanoprocaryota of thermal springs can survive vegetatively in high temperatures up to $70–74\,°C$ (Castenholz 1969; Huber 1993; Seckbach and Oren 2004). In pressurized habitats, like the deep sea, archaea and bacteria can reach even higher temperatures for growth, up to $113\,°C$ (Seckbach and Oren 2004), of course such conditions cannot be reached on normal building surfaces. These temperature values are the maximum for metabolic active organisms. In a phase of dormancy, several pioneer forms on façades can stand even more extreme temperatures. In dry condition, it is known that *Nostoc commune* can survive up to 80 °C and *Klebsormidium flaccidum* up to 100 °C (Mattox 1971).

Humidity

The described initial colonizers belong to the poikilohydric organisms, which can keep water only with the capillarity of their surface or swellable colloids. This amount of water evaporates according to the actual temperature and the existing gradient of vapor pressure. Vacuolized cells are able to take up water also in their cell sap, which enlarges the accumulated water. Vacuoles and their included cell sap form a main difference of higher cell organization versus cells of cyanoprocaryota.

Some cyanoprocaryota have a high tolerance against desiccation but need mainly liquid water for their physiological processes, only some forms are able to use gaseous water at a high relative humidity (about 95%) (Potts 1994; Shirkey et al. 2003). *Nostoc commune* is able to take up moisture from dew in the morning and accumulate it in its mucilaginous envelope. In this way, a daily short metabolic activity is possible even in xeric regions.

Bertsch (1966) documented net photosynthesis in the green alga *Apatococcus lobatus* already at a relative air humidity of 68% which is the lowest measured value known for aerophytic algae so far. It is remarkable that *Apatococcus lobatus* is among the most common algae in longstanding established biogenic crusts on building surfaces. Also in our actual investigations, this taxon was constantly documented within older surface growth in high dominance. Cyanoprocaryota developed on surfaces regularly wetted, e.g., by precipitation, whereas several green algae developed on surfaces mainly exposed to high air humidity. This difference is due to the physiological characteristics of cyanoprocaryota and green algae. Conforming results from lichens with different photobionts (cyanobionts or phycobionts) exist (Lange 2000). Also on modern isolated façades, mainly eucaryotic algae were observed, especially in situations with dew point gradients until dew water accumulation, as presented in our study.

For growth of fungi, different results about the lowest data for relative air humidity are given, but seemingly no big differences exist to eucaryotic algae. Among pioneer forms on façades, some species were found which have a high tolerance against drought, like *Eurotium* spp. Some species of this genus survive at a relative air humidity of 73.3% (Heintzeler 1939), or even lower. In conclusion, from our results, it is preferable to reduce not only liquid water (precipitation) on the surface of façades but to achieve a desiccation until a relative humidity below 70%. The lowest measured relative humidity for growth of any fungus is ca. 61%, e.g., *Xeromyces bisporus* (Seckbach and Oren 2004). According to the ecology of this fungus, which prefers nutrient rich environments with high sugar content, growth on building surfaces would not be possible.

pH Value

Building materials especially of façade coatings have a high initial pH value. It is supposed that fresh concrete has a pH value of about 12 (Grübl et al. 2001). According to an overview by Gimmler and Degenhardt (2001), no natural habitats exist with a pH value > 11.5, in contrast Seckbach and Oren (2004) reported that soda lakes may reach a pH of 12. Many algae prefer a neutral to basic pH value, but several species can tolerate very acidic conditions until pH 0 or high basic conditions until pH 11.5 (Gehl and Colman 1985; Gimmler 2001; Gimmler and Degenhardt 2001; Sitte et al. 2002). As discussed in Chap. 3, there are several species among the cyanoprocaryota and eucaryotic algae of initial succession on modern façades which grow in a wide range of pH value. Especially fungi can live in a wide pH range. *Penicillium* Link ex Fr. spp. or *Aspergillus,* Micheli ex Fries spp. appear between a pH < 2 and 12 (Sand 1995; Gimmler and Degenhardt 2001). Additionally, fungi are able to change the pH around their hyphae by excretion of acids.

From an ecophysiological view, it is remarkable that the substrate pH has an influence on the pigmentation of *Gloeocapsa sanguinea* (Jaag 1945), in which a pH change provokes a color change of the mucilage envelopes from reddish to blue violet to blue (Fig. 4.35). The critical value for the change in color is pH 6.5. Forms with red mucilage envelopes are only found at a pH < 6.5.

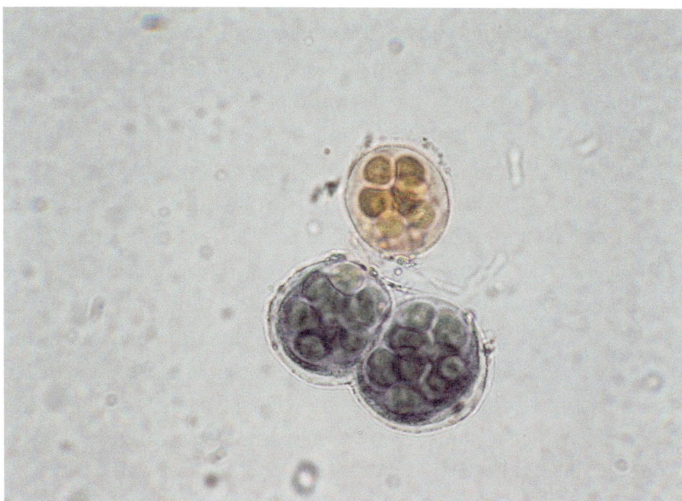

Fig. 4.35 Colonies of *Gloeocapsa sanguinea,* at the critical pH 6.5 when mucilage color change occurs. Top: red color at pH 6.5. Bottom: blue color at pH 6.5 (Hofbauer et al. 2006; Hofbauer 2007)

In a study on susceptibility of different mortars to the establishment of the alga *Klebsormidium flaccidum,* the surface pH proved to be the most important parameter (Tran et al. 2012). A lower surface pH accelerated the development of algae on the sample surfaces considerably.

Light

Light is normally available for photosynthetic organisms on building surfaces, but with different intensity according to exposure, etc., Façades exposed to strong insolation (mainly south exposed in the Central European geographical zone) are in most cases not or only slightly covered by overgrowth. Not only intensive insolation but also higher temperature cause rapid drying. On shadowed surfaces, e.g., behind shrubs, etc., a more intense development of algae is given.

Many plants form light and shade modifications. Several algae are able to change between both forms of adaption, not immediately but after a phase of accommodation. A sudden exposure of shade adapted plants to high photosynthetic active radiation can lead to irreversible damage (Asada 1994). Radiation which is too high ("light intensity" or photon flux density) can decrease photosynthesis and induce photoinhibition (Kok 1956; Björkman and Holmgren 1963). In light adapted plants, photoinhibition is mostly reversible (Baker and Bowyer 1994; Schlensog 2000).

Light of short wavelength in the blue and ultraviolet spectral range is very dangerous for organisms. It causes damages of proteins, and in very short wavelength, it also causes damages of nucleic acids. A known phenomenon is the light protection of plants by increased synthesis of carotenoids. Cyanoprocaryota like, e.g., *Nostoc* spp. additionally contain Scytonemin, a pigment absorbing between 300 and 450 nm spectral range (Krumbein and Villbrandt 1993). Many cyanoprocaryotes produce MAAs (mycosporine-like amino acids), which provide an additional protection against excess UVR (ultraviolet radiation) in exposed situations (Castenholz 2004). Some green algae in exposed habitats also show tolerance against UVB (ultraviolet B) irradiation. Their enhanced tolerance seems to be accompanied by the possession of comparably thick cell walls containing sporopollenine, which hypothetically scatters the noxious irradiation. Some green algae are also capable of production of MAAs (Xiong et al. 1996, 1999; Holzinger et al. 2009). Furthermore, some fungi, especially the black yeasts, are capable of producing MAAs, which may act either as sunscreens or as compatible solutes when salt concentration is high (Kogej et al. 2006).

While species like *Nostoc* spp. and *Trentepohlia* spp. grow on modern building surfaces and protect themselves with the abovementioned strategies against high insolation, the main part of algae which are building relevant are specialized in weak light and survive with a minimum of light (up to 0.01% of solar radiation) (Sitte et al. 2002). Therefore, shading of façades is not feasible, but an increase of insolation, e.g., with mirror systems, can reduce overgrowth.

Nutrients

During our investigations, the content of nutrients in coating systems was also measured (Renzl 2006; Hofbauer 2007). The results showed that in all coating systems the content of nutrients was relatively low. The quantity of soluble nitrogen compounds in variants NV13, NV14 and NV15 was slightly higher compared to all other variants. It is remarkable that intensive algal growth was comparatively early visible in variant NV13. Algal growth started later in the variants NV14 and NV15 because of coloration of the topcoat. It was observed that colorations retard the development of growth.

All variants were influenced by external nutrient supply. Nutrients, e.g., from agriculture and traffic, are transported notably with precipitation (Wolseley et al. 2006). Aerophytic algae are usable for biomonitoring of ammonium as documented by Abe et al. (2004). The color intensity of the culture interferes with the amount of ammonium in the culture medium in Abe's test.

Our study also showed an important influence of excrements of birds and insects on the initial succession of microbial growth on building surfaces. This influence was higher in relation to the roughness of the surface, as in NV10 and NV11. On smooth surfaces and surfaces with chalking processes such deposits (including their microorganisms) were ablated after some time.

Autecology

Several aeroterrestric algae are euryoecious organisms with a wide ecological amplitude and in some cases a worldwide distribution (Ettl and Gärtner 1995). In contrast, some algae are restricted to special habitats (e.g., *Porphyridium purpureum*). On building surfaces, generally euryoecious forms predominate in the pioneer phase.

Among fungi several species like *Trichoderma viride* and *Ulocladium oudemannsii* have antagonistic influence on other fungal taxa. Some isolates of such fungi are used in agriculture to control pests. The species, which were identified on façades so far, are only useable with reservations, because their spores and/or sporogene structures are quite intensely pigmented and could therefore result in undesired biodeterioration.

Effect of Overgrowth on Façades

Different environmental processes have an influence on the surface of modern façades. With the settlement and growth of certain bacteria, algae and fungi, the processes of soil formation start. Subsequently other species become active than those which dominate the initial succession, because other ecological strategies act. Biological influences together with abiotic factors are of significant importance for the degradation of rocks in nature (Wakefield and Jones 1998). These long-range processes are less significant on building surfaces, but it is different with damaged localities and with places with splashing water or water from eaves. In experiments melanogenous fungi ("Dermatiaceae"), especially so-called MCF (micro colonial fungi), showed their ability to drill into marble and limestone (Sterflinger and Krumbein 1997). How fast these processes run in nature is yet not known. In laboratory conditions, these fungi are slow growing even with good nutrient supply. A heavy biological attack on surface/material is under normal conditions not to be expected for façades in Central Europe. In the first years, the physical and chemical weathering is indeed more distinct. This can be desirable as the "cleaning" effect of chalking. Nevertheless an undesired biodeterioration (discoloration) may appear with microbial colonizing of façades, supported by irregularities of the surface, deposition of nutrient particles and physical and chemical properties of the surface like hygrothermic parameters and initial values of nutrients.

The described processes concern primarily the optical characteristics of the surface, the material itself is not attacked, but during a sustaining renovation of the façade, it could be necessary to refurbish parts of the coating and/or mount a new coating.

Several authors postulated that biological surface crusts on building materials lead to a damage of the building material (e.g., Hoppert et al. 2005; Karsten et al. 2005a, etc.). On the contrast, geochemical analyses showed that endolithic organisms in limestone often have a protective impact (Pohl and Schneider 2005). Algae and other microorganisms are able to protect themselves against stress factors (like predators, abrasion, high radiation, water deficit, etc.) by immigration into the surface of rocks (active gaining of space or immigration into existing cavities). To preserve such favorable ecological niches, these organisms secrete extracellular polymeric substances (EPS), which preserve the surface of rocks. This is important as the compounds of mineralic coating systems generally represent a calcareous substrate and in systems of synthetic resins at least the structure kernels often consist of calcareous material. It is to consider that modern plaster differs in its structure from natural limestone and that the above-noted interminable processes have no influence on the persistence of a modern façade system. The popular argument that algae damage building materials has therefore to be modified.

The initial colonization of external building systems is probably not a fundamental sanitary problem. Contrary to the interior of houses where spores of microorganisms can produce health problems by potential allergic effects (Gravesen 1979), a comparable exterior concentration of spores is not expected because of the enormous volume of air and the minute overgrowth on the surface. Seldom allergic reactions to aerophytic algae or to aerosoles containing algae, respectively, have been documented (McElhenney et al. 1962). Several cyanoprocaryota are known to produce toxins, but the main diversity lives in water and is to be found in blooms (Briand et al. 2003). On modern building façades, only a low amount of biomass of cyanoprocaryota is expected and ingestion is improbable. In our study, cyanoprocaryota were represented only in very low numbers during the pioneer phase of overgrowth.

Bacteria were recorded only in total germ numbers according to the used methods. Beside cyanoprocaryota, which were identified to species level, if possible, mainly chemoorganotrophic bacteria were recorded. Within the first several years, no development of damaging acid producing lithotrophic bacteria was observed. In historical buildings, chemical and physical processes are as a rule more damaging to the material than biological activities (Mansch et al. 1999). In regions with air pollution, higher numbers of germs were observed on building systems than in regions with clean air. Air pollution stimulates several microorganisms which supposedly become damaging factors to building materials (Mansch et al. 1999; Papen et al. 2001). On the one hand, the mere presence of an organism on some decayed material does not necessarily imply that it has caused the damage observed. It may be present in form of resting stages (spores, propagulae) or it may have arrived after the damage. On the other hand, there is a continuous deposition of pollutants on surfaces of buildings in urban areas and, at the same time, a biotransformation of these pollutants by biocrusts and biofilms leading to their removal through degradation or detachment (Saiz-Jimenez 1995, 1997). Although knowledge is far from complete, biotransformations, which occur on building surfaces, are assessed to be important processes for the natural removal of pollutants.

Cutler et al. (2013) described that algal patches, which developed on building materials, were associated with less weathered surfaces and with additional protective influence on moisture levels in building stones. This indicates that algal cover may have bioprotective effects. The authors concluded that a more detailed examination by scanning electron microscope (SEM) is required to definitively establish whether algae are associated with weathering. However, in the work performed by Cutler et al. (2013), it seemed likely that the main factors driving the weathering at the study locations were abiotic processes. This is in accordance with our own observations during our trials and with further experience with additional investigations of surface growth on buildings in Central Europe: There is hardly any evidence of significant damage caused by algae whereas abiotic processes have much more profound effects on the materials used.

Reducing Microbial Growth

Microbial growth on highly insulated but weather exposed building parts is, apart from a potential favorable substrate property of the surface layer, due to enhanced moisture strain and therefore improved growth conditions. Outer walls are particularly affected if they are constructed with ETICS, where the insulation layer is only covered by a thin plaster layer. The rather thin outer layer is thermally uncoupled from the rest of the construction by the insulation material, possesses only a small thermal storage capacity and has to bear major temperature variations. Notably in clear nights, the temperature of the top layer may fall below the surrounding air temperature facilitated by thermal radiation, whereby surface dew results if the dew point is undercut. This leads to a moisture supply in addition to precipitation, which gives better conditions for microbial growth.

The substrate properties of the material result from the composition and chemistry of the top coat (carbon, nitrogen and phosphorous sources, kind of binder, proportions of artificial resin, pH value, etc.) on the one hand and on the other from the depositions on the surface, which can be utilized by microorganisms (mineral and organic input by, e.g., dust and dirt particles, excrements of birds and insects, pollen, etc.). Options to influence this lie in some degree in a careful choosing of the coating recipe and further in a surface structure, which reduces deposition.

The moisture condition of the surface, which is determined by climate and building physical parameters, is regarded as a central criterion. Apart from the surface orientation, thermal insulation and thermal storage capacity of the building parts particularly the physical surface properties are of significance. As already small differences in the mid-term to long-term moisture level of the surface may cause significant differences in growth intensity, an adjustment of the parameters according to the following aspects will gain a substantial reduction in biological colonization.

Generally, in the long term, sooner or later microbial growth will establish to some degree on almost any surface. There are several measures, which can be taken in order to

© Springer-Verlag GmbH Deutschland, ein Teil von Springer Nature 2021
W. K. Hofbauer and G. Gärtner, *Microbial life on Façades*,
https://doi.org/10.1007/978-3-662-54833-2_5

avoid or reduce growth development on building surfaces (Hofbauer et al. 2006). These structural and physical measures comprise

- a roof projection,
- adjusted hygrothermic properties of the system composition with minimal water absorption and water retention (moisture absorption of the top layer),
- hydrophobic additives to the top layer (avoidance of capillarity), a high water repellent effect of the surface (short duration of moistening),
- a reduced surface roughness,
- pigmented (darker) surfaces (enhanced mean temperatures, intensified drying),
- an enhancement of the surface layer thermal capacity (thick plaster, latent warmth effects) and phase change materials (PCM),
- infrared (IR) coatings with reduced long wave radiation.

IR-active pigments, which show a reduced long wave radiation and therefore a reduced hypothermia during night, combined with lower dew accumulation on the surface are specialized solutions, which help to reduce the problem of surface growth under special conditions.

Surface Characteristics and Nanostructure/Nanomaterials (Incl. Photocatalysis)

Silver nanoparticles have been used in formulations against microbial growth; whereas the action against bacteria and some Cyanoprokaryota seems promising, the results for eukaryotic aerophytic algae and fungi are still controversial. Rather high concentrations have to be employed in comparison with applications tested for aqueous environments (Nowicka-Krawczyk et al. 2017).

Photocatalysis as a principle for growth prevention is often discussed together with nanostructure and also with biocidal measures. Usually anatase, a metastable mineral form of titanium dioxide, is used for its catalytic production of free radicals in sunlight and availability of water (vapor). Action against bacteria has previously been demonstrated; however, the influence on fungi and algae is discussed controversially (Fonseca et al. 2010). It has to be considered that the produced free radicals do not only harm microorganisms, they also act against most organic materials, including binders of coatings. The catalyst can be doped with further inorganic compounds in order to enhance its action.

A special biomimetic form of nanostructure is realized with the so-called super hydrophobicity (Cerman et al. 2003, 2004). A combination of surface structure and hydrophobic compounds affects the very high contact angle for liquids. This means that water cannot form a film or attach to the surface, but is forced into droplets which run off and take also dust and dirt with them. This principle has also been successfully applied to building coatings. The only disadvantage may be, that, if the surface is not regularly washed by rain and dirt is removed, the occasional droplets formed by condensation may lead to brighter runoff traces. Generally, it is now accepted that an equilibrated adjustment of hydrophobicity and hydrophilicity of the different layers of coating should be achieved in order to get the best results in preventing unwanted growth. A further property to combine with is the surface roughness (D'Orazio et al. 2014).

Biological Treatment

There are two different options for biological treatment to mitigate damage by surface colonization: (1) application of antagonists and (2) application of microorganisms for surface recovery. Particularly in agriculture, a number of antagonists are already used against several pests, but antagonists are also used against some domestic insect pests. The special challenge for the use of antagonists in the management of surface colonization of building surfaces is the vast diversity of organisms which are encountered. A possibility could be to inoculate the surface with preferred microorganisms so that it is already occupied and unwanted growth is therefore suppressed. We (the authors) are deeply convinced that approaches utilizing antagonists in different forms will lead to the sustainable management of surface colonization of building surfaces.

Microorganisms were also successfully applied to recover deteriorated surfaces of buildings and monuments (Ranali and Sorlini 2003). Special organisms or groups of organisms can either be used to remove black stains from surfaces or/and to recover damaged parts of the structure (Tiano 2003). The latter is hitherto limited to calcareous stonework.

Further Approaches

Another method to remove the growth from masonry surfaces is to clean it with solid CO_2 particles. Giovagnoli et al. (2009) reported that solid CO_2 particles can mechanically remove surface growth and patina involving another kind of a disinfecting principle.

Biodeterioration vs. Bioprotection

The debate is still ongoing if the biological colonization of building surfaces has more negative (biodeterioration, biodegradation) or more positive (bioprotection) effects. In general, it can be assumed that most organisms colonizing a surface tend to remain there and therefore take some action to protect their place of living at least for some time. There is a huge consent that air pollution has the greatest detrimental impact on building surfaces (e.g., Zanardini et al. 2002). In Central Europe, which has a temperate climate, climate conditions are very straining for materials. Whether the building material will additionally be affected by the activity of microorganisms depends on the kind of material, the environmental conditions and the type of microorganism involved. Grondona et al. (1997) reported a bioprotective effect of a special lichenized biocrust on building stone by establishing a constant moisture level and thus preventing further weathering (clay swelling).

Surface Greening with Algae and Moss

A complementary approach to the prevention of surface growth is the deliberate enhancement of the growth of certain organisms, like algae, fungi or mosses. Often it is not the surface growth on masonry itself, which induces complaints, but the uneven pattern and ugly colors, which are a nuisance. If it was possible to manage a uniform distribution of hygienically uncritical organisms as are, e.g., algae or mosses in a pleasant color, the acceptance could be great. Recently, in Leistner et al. 2018 it was shown that it is possible to attach a homogenous layer of moss fragments to a building surface, which is able to grow subsequently.

Filmconservation

In our study primarily, structural–physical approaches to reduce overgrowth on surfaces were included. Biocidic agents were excluded from our study to reduce the number of varying parameters. Chemical agents against algal growth mainly interfere with photosynthesis and therefore have an effect on all photosynthetic active organisms (Paulus 1993, 2005). So far, several mutants (mainly aquatic forms), which are resistant against the applicated agents, have been discovered or experimentally developed (Couderchet et al. 1995; Srivastava et al. 1995; Schmalfuß et al. 1998; Matthes 2000; Koenig 2001). It cannot be excluded that also terrestrial forms develop resistance. To compensate existing gaps in active substances, mixtures of agents are generally used against undesired surface growth of fungi. Resistances are already known notably from genera *Alternaria* and *Urocladium* (Paulus 2005).

Against surface colonization of façades by algae and fungi (molds, mildews), often biocides are used. Usually, a combination of different compounds is employed (Paulus 1993; Bagda et al. 1999; Lindner 2000). Within the European Community, the biocide product directive (directive 98/8/EG) regulates the use, the investigation and registration of products equipped with biocides. This directive means rigorous restrictions and outlines conditions, especially for the marketing of new active substances. Therefore, it has great influence on product development and product composition.

In order to unfold their action, biocides have to be water soluble to some degree so that they are transported to and into the organisms. Thus, the compounds are washed out after some time, and the time of action is limited. In own studies we found that the binder has some influence on the mobility of biocides and that the surface biocide concentration is crucial (Breuer et al. 2011; 2012, Schwerd 2011; Krueger 2020). Additionally we reported some wild strains of Diplosphaera that exhibit an increased tolerance against the investigated algaecide (Krueger 2020; Krueger et al. 2013).

Prospects

The requirements of many taxa of the initial colonizers on modern building surfaces can be specified by ecophysiological measurements. For a better understanding of the interrelationships, further overgrowth associations with reference to actual damage events should be investigated. In addition, a comparison with unobjected surfaces should be made. Beside the taxonomic diversity, it will be necessary to detect the composition and properties of the building material and all other influencing parameters. In relation to the initial succession on modern surfaces of façades, more questions exist. The efficiency and longevity of chemical arrangements for the protection of coating systems from unwelcome overgrowth currently was investigated at the Fraunhofer Institute of Building Physics with the public and private partnership (Breuer et al. 2011; 2012, Schwerd 2011; Krueger et al. 2013; Krueger 2020), additionally the influence of formula (binder material and additives) is under investigation. A deeper knowledge of the diversity of organisms of initial succession will give us the potential to use them increasingly as bioindicators. In the future, it will be more and more possible to adopt the measures in the management of microbial overgrowth in a specific and differenced way.

Appendix

Dichotomous Key for Identifying Surface Microbial (especially Algal) Layers on Building Surfaces

1 Surface layer dark brown, black, grayish-black to gray, gray-brown to whitish (if single layers of green algae are densely surrounded by colorless or brown fungal hyphae: soredious state of a lichen) ………………………………………………………………. **fungal layers**

1* Surface layer of other colors, when black with odor of mud………….………………………..2

2 Surface layer black, blackish- green to dark –gray or dark-bluish (like ink) with a characteristic odor of mud when moist ………………………………………… **Cyanoprokaryota (blue-green algae)**

2* Surface layer red, rust-red, orange-red or of different green colors ………………….. 3

3 Surface layer orange-red, of filthy structure or interwoven branched filaments (use hand lens 10 x) ………………………………………………………………… ………………**green alga *Trentepohlia***

3* Surface layer red, purple-red, rust-red or green ……………………………………………… 4

4 Surface layer purple-red or grayish-red of slimy consistency, in nutrient-rich situation (e.g. dog urine)………………………………………….. **red alga *Porphyridium*** not exposed to excess nutrients …………………………………….**blue-green alga *Gloeocapsa***

4* Surface layer rust-red (not slimy) or green ……………………………………………… 5

© Springer-Verlag GmbH Deutschland, ein Teil von Springer Nature 2020
W. K. Hofbauer and G. Gärtner, *Microbial life on Façades,*
https://doi.org/10.1007/978-3-662-54833-2

5   Surface layer rust-red, on horizontal building structures especially in water-filled depressions, which periodically dry out, when dry in brown-red parti-cles .... **green flagellate** *Haematococcus*

5*   Surface layer green, bright green to yellow-green or dark-green 6

6   Structure of surface layer smooth (seen by the naked eye), seen through a hand-lens (10 x) rough, granular .... **green/yellow-green algae pro parte**

6*   Structure of surface layer thread-like, dark green 7

7   Cells of threads/filaments always with rectangular cross-walls and one to several chloroplasts (use microscope!) .... **green/yellow-green algae pro parte**

7*   Cells of filaments with oblique cross-walls at least on the main axis and always many lens-shaped chloroplasts (use microscope!) .... **moss protonemata**

Table A.1 Organisms identified on specimens of "old" variants (OV)

Taxon	OV1	OV2	OV3	OV4	OV5	OV6	OV7	OV8	OV 9	OV 10	OV 11
Apatococcus lobatus (Chodat) J. P. Petersen			+	+	+		+		+		+
Chlorella ellipsoidea Gerneck agg.	+						+			+	
Chlorella sp.		+	+						+		+
Chlorella vulgaris Beijerinck	+			+							
Chlorococcum infusionum (Schrank) Meneghini								+			
Chlorococcum sp.									+		
Deasonia multinucleata (Deason & Bold) Ettl & Komárek				+							
Diplosphaera sp.	+		+	+	+		+				
Gloeocapsa sanguinea (Agardh) Kützing	+										
Gloeocapsa sp.									+		
Klebsormidium flaccidum (Kützing) Silva, Mattox & Blackwell	+	+	+	+	+	+		+	+	+	+
Stichococcus bacillaris Nägeli	+	+	+						+		
Stichococcus minutus Grintzesco & Péterfi		+		+	+				+		+
Stichococcus sp.		+					+				
Trentepohlia iolithus (Linnaeus) Wallroth				+	+			+			
Further green algae and bluegreens	+	+	+		+	+				+	+
Grimmia pulvinata (Hedw.) Sm.			+								
Orthotrichum anomalum Hedw.	+		+						+	+	
Schistidium apocarpum (Hedw.) B. S. G. s.l.	+		+				+	+	+		
Tortella sp.								+			
Tortula muralis Hedw.			+								
Alternaria alternata (Fr.) Keissler	+		+	+	+		+	+	+	+	+
Alternaria sp.			+								
Aureobasidium pullulans (De Bary) Arnaud					+			+			
Cladosporium cladosporioides (Fresen.) de Vries	+		+		+	+		+		+	
Cladosporium sp.	+		+		+					+	
Drechslera poae (Baudys) Shoemaker					+						

Epicoccum nigrum Link			+	+	+	+				+	+
Paecilomyces sp.									+		
Phoma glomerata (Corda) Wollenw. & Hochapfel						+					
Phoma sp.		+	+	+	+	+	+	+	+	+	+
Sordaria sp.	+		+	+							
Trichoderma sp.			+								
sterile mycelia	+					+		+	+		+
Caloplaca holocarpa (Hoffm. ex Ach.) Wade			+								
Candelariella aurella (Hoffm.) Zahlbr.	+		+	+	+					+	
Lecanora muralis (Schreber) Rabenh.			+								
Lecanora sp.	+		+				+			+	
Physcia adscendens (Fr.) Oliv.	+										
Sarcogyne regularis Körber			+				+	+			
Xanthoria elegans (Link) Th. Fr.	+						+			+	
Xanthoria parietina (L.) Th. Fr.			+								
Lichen scales (vegetative)				+							
Sorediose crusts	+	+	+	+	+	+	+	+	+	+	+
Further lichens			+	+	+	+	+			+	
Bacteria	+	+	+	+	+	+	+	+	+	+	+
+ Taxon identified as component of surface growth											

Table A.2 Overview of organisms detected on specimens of "new" variants (NV) after half a year of outdoor exposure (series of variants in Holzkirchen)

Taxon	Variant														
	NV 1	NV 2	NV 3	NV 4	NV 5	NV 6	NV 7	NV 8	NV 9	NV 10	NV 11	NV 12	NV 13	NV 14	NV 15
Alternaria alternata (Fr.) Keissler	+	+	+								+	+	+		
Alternaria tenuissima (Kunze ex Pers.) Wiltshire	+	+									+	+	+		
Aureobasidium pullulans (De Bary) Arnaud				+								+			
Cladosporium cladosporioides (Fresen.) de Vries	+	+	+							+	+	+			
Cladosporium herbarum (Pers.) Link ex S. F. Gray	+	+	+							+	+	+			
Coprinus sp.										+					
Epicoccum nigrum Link	+	+	+		+				+	+	+	+	+		
Phoma glomerata (Corda) Wollenw. & Hochapfel	+	+									+				
Phoma herbarum Westend.				+											
Phoma pinodella (L. K. Jones) Morgan-Jones et K. B. Burch	+	+	+		+				+	+					
Ulocladium chartarum (Preuss) Simmons	+	+	+								+		+		
Sterile mycelia	+	+	+							+	+		+		
Bacteria	+	+	+	+	+	+				+	+	+	+	+	
+ Taxon identified as component of initial surface growth (primary growth)															

Table A.3 Overview of organisms detected on specimens of "new" variants (NV) after one year of outdoor exposure (series of variants in Holzkirchen)

Taxon	Variant														
	NV1	NV2	NV3	NV4	NV5	NV6	NV7	NV8	NV9	NV10	NV11	NV12	NV13	NV14	NV15
Acremonium strictum W. Gams		+													
Alternaria alternata (Fr.) Keissler	+	+	+	+	+	+	+		+	+	+	+	+	+	+
Alternaria sp.	+														
Alternaria tenuissima (Kunze ex Pers.) Wiltshire	+	+		+		+					+				
Aspergillus sydowii (Bain. & Sart.) Thom & Church					+		+	+	+						
Aspergillus terreus Thom	+		+							+	+	+	+		
Aspergillus versicolor (Vuill.) Tiraboschi			+		+	+	+	+	+	+	+	+		+	+
Aureobasidium pullulans (De Bary) Arnaud											+				
Cladosporium cladosporioides (Fresen.) de Vries	+	+	+	+	+		+	+		+	+				+
Epicoccum nigrum Link		+	+	+		+				+	+	+	+	+	+
Eurotium amstellodamii Mangin					+										
Leptosphaerulina sp.							+								
Penicillium brevicompactum Dierckx							+	+							
Penicillium chrysogenum Thom								+							
Penicillium sp.									+						
Phoma eupyrena Sacc.	+		+			+					+				
Phoma glomerata (Corda) Wollenw. & Hochapfel	+														
Phoma pinodella (L. K. Jones) Morgan-Jones et K. B. Burch			+	+		+				+					
Phoma pomorum Thüm.	+			+											
Phoma sp.		+	+	+		+				+	+	+			
Preussia sp.					+										
Rhizopus stolonifer (Ehrenb. ex Link) Lind							+								
Sporothrix sp.					+										
Ulocladium chartarum (Preuss) Simmons	+	+	+	+		+	+		+	+	+	+		+	+
Sterile mycelia	+		+							+	+	+			+
Bacteria		+		+				+		+	+		+		+
+ Taxon identified as component of initial surface growth (primary growth)															

Table A.4 Overview of organisms identified on specimens of "new" variants (NV) after 2 years' outdoor exposure (series of variants in Holzkirchen)

Taxon	Variant														
	NV 1	NV 2	NV 3	NV 4	NV 5	NV 6	NV 7	NV 8	NV 9	NV 10	NV 11	NV 12	NV 13	NV 14	NV 15
Chlorella minutissima Fott & Nováková						+									
Chlorella trebouxioides Punčochářová				+						+					
Chlorella vulgaris Beijerinck													+		
Chlorococcum sp.	+			+											
Diplosphaera chodatii Bialosuknia em. Vischer											+	+			
Diplosphaera sp.	+	+		+		+				+	+	+			
Klebsormidium sp.											+				
Stichococcus bacillaris Nägeli												+	+		
Stichococcus chlorelloides Grintzesco & Péterfi					+										
Stichococcus exiguus Gerneck											+				
Stichococcus minutus Grintzesco & Péterfi	+	+								+	+	+	+		
Bryum sp.												+			
Alternaria alternata (Fr.) Keissler	+	+		+		+	+			+	+		+		+
Alternaria tenuissima (Kunze ex Pers.) Wiltshire				+							+				
Aspergillus sp.											+	+			
Aspergillus versicolor (Vuill.) Tiraboschi					+										
Aureobasidium pullulans (De Bary) Arnaud												+			
Botrytis cinerea Pers.	+														
Cladosporium cladosporioides (Fresen.) de Vries	+	+	+	+						+	+	+			
Cladosporium herbarum (Pers.) Link ex S. F. Gray				+						+					
Cladosporium sp.												+			
Epicoccum nigrum Link	+	+		+	+							+	+	+	−
Penicillium brevicompactum Dierckx												+			
Penicillium sp.		+		+						+		+	+		
Phoma pomorum Thüm.	+														
Phoma sp.	+														
Rhizopus stolonifer (Ehrenb. ex Link) Lind											+				
Ulocladium chartarum (Preuss) Simmons	+	+	+	+		+				+	+	+	+	+	+
Sterile mycelia	+	+	+	+		+	+			+	+	+	+		
Lichen thalli		+										+	+		
Bacteria	+	+	+	+		+				+	+	+	+	+	+

+ Taxon identified as component of initial surface growth (primary growth)

Table A.5 Overview of organisms identified on specimens of "new" variants (NV) after 3 years' outdoor exposure (series of variants in Holzkirchen)

Taxon	Variant														
	NV1	NV2	NV3	NV4	NV5	NV6	NV7	NV8	NV9	NV10	NV11	NV12	NV13	NV14	NV15
Apatococcus lobatus (Chodat) J. P. Petersen		+								+		+			
Botrydiopsis intercedens Pascher											+				
Bracteacoccus sp.										+					
Chlorella cf. *angusto-ellipsoidea* Hanagata & Chihara					+								+		+
Chlorella luteoviridis Chodat			+									+	+	+	
Chlorella saccharophila (Krüger) Migula							+				+				
Chlorella trebouxioides Punčochářová			+	+	+					+	+				
Chlorella vulgaris Beijerinck			+							+	+		+		
Chlorococcum infusionum (Schrank) Meneghini				+											
Choricystis minor (Skuja) Fott var. *gallica* (Bourelly) Komárek					+										
Diplosphaera chodatii Bialosuknia em. Vischer	+		+						+	+	+		+		
Diplosphaera sp.	+	+	+	+						+	+	+	+		
Elliptochloris subsphaerica (Reisigl) Ettl & Gärtner					+						+				
Excentrochloris fraunhoferiana Hofbauer, Gärtner, Rennebarth, Sedlbauer, Mayer & Breuer										+					
Stichococcus bacillaris Nägeli	+	+		+						+	+	+			
Stichococcus minutus Grintzesco & Péterfi	+	+			+	+			+		+		+	+	+
Trentepohlia iolithus (Linnaeus) Wallroth											+				
Protonemata													+		
Acremonium strictum W. Gams				+								+			
Alternaria alternata (Fr.) Keissler	+	+	+	+	+	+			+	+	+	+	+		+
Alternaria tenuissima (Kunze ex Pers.) Wiltshire	+	+		+							+				
Aureobasidium pullulans (De Bary) Arnaud	+	+	+							+	+				+
Botrytis cinerea Pers.												+			
Cladosporium cladosporioides (Fresen.) de Vries	+	+	+	+	+	+	+	+		+	+	+	+	+	
Cladosporium herbarum (Pers.) Link ex S. F. Gray	+	+	+	+	+	+	+		+	+	+	+	+		+
Cladosporium sp.													+		
Drechslera poae (Baudys) Shoemaker		+		+					+		+	+		+	
Epicoccum nigrum Link	+	+	+	+	+	+		+	+	+	+	+	+		+
Penicillium brevicompactum Dierckx													+		
Penicillium corylophilum Dierckx														+	
Phoma eupyrena Sacc.	+										+				+

	1	2	3	4	5	6	7	8	9	10	11	12	13	14	15
Phoma exigua Desm.				+						+					
Phoma pinodella (L. K. Jones) Morgan-Jones et K. B. Burch	+	+	+			+				+	+				+
Pithomyces chartarum (Berk. & Curt) M. B. Ellis						+									
Trichoderma viride Pers. ex Gray		+													
Ulocladium chartarum (Preuss) Simmons	+	+	+	+		+	+		+	+	+	+	+		+
Sterile mycelia	+	+	+							+	+				
Candelariella aurella (Hoffm.) Zahlbr.						+									
Physcia sp.													+	+	+
Xanthoria parietina (L.) Th. Fr.													+	+	+
Bacteria	+	+	+	+	+	+	+	+	+	+	+	+	+	+	+
Streptomycetes									+		+	+			

+ Taxon identified as component of initial surface growth (primary growth)

Table A.6 Overview of organisms identified on specimens of "new" variants (NV) after 2 and 3 years' outdoor exposure (series of variants in Finnentrop/Heggen)

Taxon	Variant																							
	NV1		NV2		NV3		NV4		NV5		NV6		NV10		NV11		NV12		NV13		NV14		NV15	
	I	II	I	II	I	II	I	II	I	II	I	II	I	II	I	II	I	II	I	II	I	II	I	II
Apatococcus lobatus (Chodat) J. P. Petersen														+				+						
Calothrix parietina Thuret ex Bornet & Flahault																	+							
Chlorella angusto-ellipsoidea Hanagata & Chihara															+	+								
Chlorella luteoviridis Chodat		+																		+				
Chlorella trebouxioides Punčochářová			+	+	+	+							+	+		+	+	+		+				
Chlorococcus sp.																	+							
Choricystis minor (Skuja) Fott var. *gallica* (Bourelly) Komárek		+	+	+	+	+	+	+					+	+				+	+	+		+		+
Diplosphaera chodatii Bialosuknia em. Vischer																			+	+		+		
Diplosphaera sp.					+	+								+		+	+	+		+				
Elliptochloris reniformis (S. Watanabe) Ettl & Gärtner																				+				
Leptosira terricola (Bristol) Printz																	+	+						
Nostoc sp.																	+	+						
Stichococcus bacillaris Nägeli		+																						+
Stichococcus chlorelloides Grintzesco & Péterfi														+										
Stichococcus minutus Grintzesco & Péterfi																+					+	+		
Mossprotonemata (*Hypnum cupressiforme* Hedw.)																		+						
Mossprotonemata (*Orthotrichum anomalum* Hedw.)		+		+				+												+				
Brachythecium salebrosum (Web. & Mohr) B. S. G.																		+						
Orthotrichum anomalum Hedw. (plantlets)				+																				

Schistidium apocarpum (Hedw.) B. S. G.																			+							
Tortula muralis Hedw.																			+							
Acremonium strictum W. Gams			+																							
Alternaria alternata (Fr.) Keissler	+	+	+	+	+	+	+	+	+	+	+	+	+	+	+	+	+	+	+	+	+	+	+	+	+	+
Alternaria tenuissima (Kunze ex Pers.) Wiltshire	+	+	+	+										+	+		+				+					
Arthrinium phaeospermum (Corda) M. B. Ellis														+												
Aspergillus fumigatus Fres.																			+							
Aureobasidium pullulans (De Bary) Arnaud				+	+	+	+	+			+			+	+	+	+	+				+	+	+		+
Botrytis cinerea Pers.				+											+											
Chaetomium perlucidum Sergejeva															+											
Cladosporium cladosporioides (Fresen.) de Vries		+	+	+	+	+	+	+	+	+		+	+	+	+	+	+	+	+	+	+	+	+	+	+	+
Cladosporium herbarum (Pers.) Link ex S. F. Gray				+	+	+							+	+	+	+			+							+
Coniothyrium sp.															+	+										
Epicoccum nigrum Link	+	+	+	+		+		+				+	+	+	+	+	+	+	+	+	+	+	+	+	+	+
Phoma exigua Desm.		+		+			+	+			+	+	+	+	+	+	+	+	+	+						+
Phoma pinodella (L. K. Jones) Morgan-Jones et K. B. Burch					+	+			+	+			+			+					+					
Phoma pomorum Thüm.	+																									
Phoma sp.								+																		
Pleospora herbarum (Pers. ex Fr.) Rabenh.	+																									
Preussia sp.			+													+										
Stachybotrys chartarum (Ehrenb. ex Link) Hughes															+		+									
Ulocladium chartarum (Preuss) Simmons	+	+	+	+	+	+	+	+			+		+	+	+	+			+	+	+	+	+	+	+	+
Ulocladium oudemannsii Simmons	+	+																		+	+					
Sterile mycelia		+		+		+		+	+	+		+		+		+			+	+	+			+		+

Caloplaca holocarpa (Hoffm. ex Ach.) Wade				+		+								+				+		+				
Candelariella aurella (Hoffm.) Zahlbr.						+														+				
Physcia sp.		+	+	+		+		+				+		+		+		+		+	+	+		
Sarcogyne regularis Körber														+										
Xanthoria parietina (L.) Th. Fr.																				+				
Bacteria	+	+	+	+	+	+	+	+	+	+	+		+	+	+	+	+	+	+	+	+	+	+	+
Streptomycetes															+			+				+		

I Findings according to random samples after 2 years' weathering exposure; II findings according to random samples after 3 years' weathering exposure; + taxon identified as component of initial surface growth (primary growth)

Table A.7 Overview of organisms identified on specimens of "new" variants (NV) after 2 and 3 years' outdoor exposure (series of variants in Ernsthofen/Oberrramstadt)

Taxon	NV1 I	NV1 II	NV2 I	NV2 II	NV3 I	NV3 II	NV4 I	NV4 II	NV5 I	NV5 II	NV6 I	NV6 II	NV10 I	NV10 II	NV11 I	NV11 II	NV12 I	NV12 II	NV13 I	NV13 II	NV14 I	NV14 II	NV15 I	NV15 II
Apatococcus lobatus (Chodat) J. P. Petersen																				+				
Chlorella luteoviridis Chodat														+				+						
Chlorella saccharophila (Krüger) Migula																					+	+	+	+
Chlorella trebouxioides Punčochářová	+	+	+	+		+		+					+	+	+	+				+		+		+
Chlorella vulgaris Beijerinck								+								+		+						
Chlorococcum infusionum (Schrank) Meneghini																	+	+						
Choricystis minor (Skuja) Fott var. *gallica* (Bourelly) Komárek					+	+										+		+		+				
Diplosphaera sp.	+	+		+		+		+			+	+		+	+	+	+	+	+	+	+			+
Gloeocapsopsis pleurocapsoides (Nováček) Komárek et Anagnostidis																	+							
Nostoc commune Vaucher ex Bornet & Flahault																	+	+						
Stichococcus bacillaris Nägeli	+															+				+				
Stichococcus minutus Grintzesco & Péterfi				+				+	+							+	+	+						+
Alternaria alternata (Fr.) Keissler	+	+	+	+			+	+		+	+	+	+	+	+	+	+	+	+	+	+	+	+	+
Alternaria tenuissima (Kunze ex Pers.) Wiltshire	+				+								+	+	+	+	+							+
Aspergillus fumigatus Fres.					+			+										+			+		+	
Aspergillus versicolor (Vuill.) Tiraboschi																								+
Aureobasidium pullulans (De Bary) Arnaud				+		+	+	+								+	+	+		+	+	+		+
Botrytis cinerea Pers.				+				+																

Cladosporium cladosporioides (Fresen.) de Vries		+	+	+	+	+	+	+					+	+	+	+	+	+	+	+	+	+	+	+
Cladosporium herbarum (Pers.) Link ex S. F. Gray													+		+		+		+			+		
Dematiaceae indet.															+			+		+				
Epicoccum nigrum Link	+	+	+	+	+	+	+	+			+		+	+	+	+	+	+	+	+	+	+	+	+
Paecilomyces sp.																		+						
Penicillium brevicompactum Dierckx																	+	+					+	
Penicillium solitum Westling																		+						
Phoma eupyrena Sacc.			+																					
Phoma exigua Desm.			+				+	+				+	+						+		+	+	+	+
Phoma pinodella (L. K. Jones) Morgan-Jones et K. B. Burch				+			+	+							+	+		+		+				
Pleospora herbarum (Pers. ex Fr.) Rabenh.								+												+				
Preussia sp.																						+	+	
Stachybotrys chartarum (Ehrenb. ex Link) Hughes																+								
Trichoderma viride Pers. ex Gray																	+	+		+				
Ulocladium chartarum (Preuss) Simmons	+	+	+	+			+	+				+	+	+	+	+	+	+	+	+	+	+		+
Sterile mycelia			+				+		+		+	+	+			+	+	+	+	+		+	+	+
Caloplaca holocarpa (Hoffm. ex Ach.) Wade																		+		+				
Physcia sp.						+						+								+				+
Bacteria	+	+	+	+	+	+	+	+	+	+	+	+	+	+	+	+	+	+	+	+	+	+	+	+
Streptomycetes																		+		+	+	+		

I Findings according to random samples after 2 years' weathering exposure; II findings according to random samples after 3 years' weathering exposure; + taxon identified as component of initial surface growth (primary growth)

Table A.8 part 1 Overview of organisms identified on specimens of "additional" variants (AV) and surface investigations

Taxon	Additional variants and surface investigations												
	AV 1	AV 2	AV 3	AV 4	AV 5	AV 6	AV 7	AV 8	AV 9	AV1 0	AV1 1	AV1 2	AV1 3
Actinotaenium cucurbita (Brébisson) Teiling													
Apatococcus lobatus (Chodat) J. P. Petersen		+	+	+		+					+		
Bracteacoccus sp.			+	+			+						
Chlamydomona s sp.													+
Chlorella ellipsoidea Gerneck agg.		+++	+++	+++	+				+				
Chlorella luteoviridis Chodat										+			
Chlorella minutissima Fott & Nováková													
Chlorella saccharophila (Krüger) Migula													
Chlorella sp.					+					+	+		
Chlorella trebouxioides Punčochářová						+		+		++	+		
Chlorella vulgaris Beijerinck					++		+		+				+
Chlorellidium astigmatum Schwarz							++						
Chlorococcum infusionum (Schrank) Meneghini													
Chroococcus pallidus Nägeli						+							
Coenochloris sp.							+						
Cosmarium sp.													
Cyanothece aeruginosa (Nägeli) Komárek							++						
Cylindrocystis brebissonii (Ralfs) De Bary													
Desmococcus olivaceus (Pers.) Laundon							+			+			
Desmonostoc muscorum						+							

(Agardh) Hrouzek & Ventura														
Diplosphaera chodatii Bialosuknia em. Vischer									+					
Diplosphaera sp.	+	+	+	+	+					+	+	++		
Eustigmatos magnus (J. B. Petersen) Hibberd					++									
Fragillaria sp.										+				
Gloeocapsa atrata Kützing														
Gloeocapsa sanguinea (Agardh) Kützing							++							
Gloeocapsa sp.							+							
Gloeocystis polydermatica (Kützing) Hindák		++	++	+										
Graesiella sp.														
Haematococcus pluvialis Flotow em. Wille								+++						+++
Hantzschia amphioxis (Ehrenberg) Grunow														
Heterococcus sp.						+								
Klebsormidium flaccidum (Kützing) Silva, Mattox & Blackwell	+	+++	+	++		++	++	+++	+	+				+
Klebsormidium nitens (Meneghini) Lokhorst					+			+						
Klebsormidium sterile (Deason & Bold) Silva, Mattox & Blackwell						++	+	++	+					
Leptolyngbya foveolarum (Rabenhorst) Anagnostidis & Komárek						+	+			+				
Leptolyngbya notata (Schmidle) Anagnostidis & Komárek					+	+	+	+	+	+				

Leptosira terricola (Bristol) Printz	+++		+	+												
Navicula sp.										+						
Nostoc commune Vaucher ex Bornet & Flahault						+				+						
Nostoc ellipsosporum Rabenh.							+									
Nostoc sp.					+	+										
Oscillatoria sp.					+	+		+								
Phormidium autumnale (Agardh) Trevisan					++	+		+	+	+						
Phormidium schroeteri (Hansgirg) Anagnostidis									+							
Phormidium sp.			+	+	+	+		+		+						+
Pinnularia sp.										+						
Planktolyngbya limnetica (Lemmermann) Komárková-Legnerová & Cronberg					+											
Porphyridium purpureum (Bory) K. M. Drew & R. Ross										+++						
Prasiola crispa (Lightfoot) Kützing																
Prasiolopsis ramosa Vischer											+++					
Scenedesmus acutus Meyen																++
Scytonema stuposum Bornet ex Bornet & Flahault																
Stichococcus bacillaris Nägeli		+		+			+		+	+	+	++	++			
Stichococcus exiguus Gerneck				+												
Stichococcus minutus Grintzesco & Péterfi						+				+		+	++			
Stichococcus sp.																
Symploca cf. *elegans* Kützing						+										
Trebouxia																

arboricola Puymaly													
Trebouxia sp.		+	+		+								
Trentepohlia aurea (Linnaeus) Martius													
Protonemata			+	+		+				+	+		
Bryum argenteum Hedw.					++								
Pohlia annotina (Hedw.) Lindb. agg.													
Acremonium strictum W. Gams										+			
Alternaria alternata (Fr.) Keissler	+	+	++	++		++	+	+	+	+	+	++	+
Alternaria sp.													
Alternaria tenuissima (Kunze ex Pers.) Wiltshire												+	
Aspergillus fumigatus Fres.													
Aspergillus sp.										+			
Aureobasidium pullulans (De Bary) Arnaud	++		+	+	+				++	+	+	+++	
Cladosporium cladosporioides (Fresen.) de Vries	++		++	++				++	++	+	++	+	+
Cladosporium herbarum (Pers.) Link ex S. F. Gray							+	++	+		+	+	
Coniothyrium sp.			+	+									
Dematiaceae indet.													
Epicoccum nigrum Link	+	+	+	+	++	+	+		++	+	+	+	
Fusarium sp.										+	+		++
Hormiactis sp.													
Mucor plumbeus Bonord.													
Mucor racemosus Fres.													
Paecilomyces sp.													
Penicillium sp.			+	+	+					+			
Phoma eupyrena Sacc.	+									+			
Phoma exigua Desm.									+				

Phoma glomerata (Corda) Wollenw. & Hochapfel								+					
Phoma pinodella (L. K. Jones) Morgan-Jones et K. B. Burch													
Phoma sp.	+	+	+	+	++	++		+			+		
Pleospora herbarum (Pers. ex Fr.) Rabenh.	+												
Rhodotorula mucillaginosa (Jörgensen) Harrison													
Sordaria sp.				+						+			
Stachybotrys chartarum (Ehrenb. ex Link) Hughes													
Trichoderma viride Pers. ex Gray							+			+			
Ulocladium chartarum (Preuss) Simmons				+							+	++	
Ulocladium oudemannsii Simmons													
Sterile mycelia	+	+	+	+	++	+	++	+	+	+			+
Caloplaca holocarpa (Hoffm. ex Ach.) Wade		+++					+						
Soredia		+++	+++	+++	+		+			+			
Bacteria	+	+	+	+	+	+	+	+	+	+	+		++
Streptomycetes						+				+			
Amoeba													
Chironomida													+++
Ciliata		+											+
Nematoda										+			
Rotatoria			+	+						+			+
Tardigrada							++						

+++ Taxon identified as dominant component of surface growth; ++ taxon identified as subdominant (abundant) component of surface growth; + taxon identified as sporadic component of surface growth

Table A.8 part 2 Overview of organisms identified on specimens of "additional" variants (AV) and surface investigations

Taxon	Additional variants and surface investigations											
	AV14	AV15	AV16	AV17	AV18	AV19	AV20	AV21	AV22	AV23	AV24	AV25
Actinotaenium cucurbita (Brébisson) Teiling						+						
Apatococcus lobatus (Chodat) J. P. Petersen		+		+++	+				+			+
Bracteacoccus sp.												
Chlamydomonas sp.												
Chlorella ellipsoidea Gerneck agg.	+			+								
Chlorella luteoviridis Chodat		+			++	++					++	+
Chlorella minutissima Fott & Nováková											+	
Chlorella saccharophila (Krüger) Migula					+			+++	+++			+
Chlorella sp.					+							
Chlorella trebouxioides Punčochářová		+	+		+++			++		+	+	+
Chlorella vulgaris Beijerinck						+				+	+	+++
Chlorellidium astigmatum Schwarz												
Chlorococcum infusionum (Schrank) Meneghini										+		
Chroococcus pallidus Nägeli						+						
Coenochloris sp.												
Cosmarium sp.						+						
Cyanothece aeruginosa (Nägeli) Komárek						++						
Cylindrocystis brebissonii (Ralfs) De Bary						++						
Desmococcus olivaceus (Pers.) Laundon		+										
Desmonostoc												

muscorum (Agardh) Hrouzek & Ventura													
Diplosphaera chodatii Bialosuknia em. Vischer	+	+									+		
Diplosphaera sp.	+	++	+		+				+				++
Eustigmatos magnus (J. B. Petersen) Hibberd													
Fragillaria sp.						+	+				++		
Gloeocapsa atrata Kützing	+												
Gloeocapsa sanguinea (Agardh) Kützing													
Gloeocapsa sp.	+						+						
Gloeocystis polydermatica (Kützing) Hindák											+		
Graesiella sp.	+												
Haematococcus pluvialis Flotow em. Wille						+++	+++			+++			
Hantzschia amphioxys (Ehrenberg) Grunow						++							
Heterococcus sp.	+					+					+		
Klebsormidium flaccidum (Kützing) Silva, Mattox & Blackwell		+				++	+			+			
Klebsormidium nitens (Meneghini) Lokhorst							+			++			
Klebsormidium sterile (Deason & Bold) Silva, Mattox & Blackwell	+		+			+							
Leptolyngbya foveolarum (Rabenhorst) Anagnostidis & Komárek					+	+					+		
Leptolyngbya notata (Schmidle)						+	+				+	+	

Anagnostidis & Komárek												
Leptosira terricola (Bristol) Printz												++
Navicula sp.											+	
Nostoc commune Vaucher ex Bornet & Flahault	+					++					+	
Nostoc ellipsosporum Rabenh.	+					+						
Nostoc sp.	+					+					+	
Oscillatoria sp.	+		+			+	+				+	
Phormidium autumnale (Agardh) Trevisan	+					+	+				+	
Phormidium schroeteri (Hansgirg) Anagnostidis												
Phormidium sp.	+		+		+		+				+	
Pinnularia sp.						+	+				+	
Planktolyngbya limnetica (Lemmermann) Komárková-Legnerová & Cronberg												
Porphyridium purpureum (Bory) K.M. Drew & R. Ross											+++	
Prasiola crispa (Lightfoot) Kützing		+++										
Prasiolopsis ramosa Vischer												
Scenedesmus acutus Meyen												
Scytonema stuposum Bornet ex Bornet & Flahault	+											
Stichococcus bacillaris Nägeli	+				+	++	++	+		+	+	+
Stichococcus exiguus Gerneck												
Stichococcus minutus Grintzesco &		+	+		+					++	++	+

Péterfi												
Stichococcus sp.						+						
Symploca cf. *elegans* Kützing												
Trebouxia arboricola Puymaly			+++									
Trebouxia sp.					++		+					
Trentepohlia aurea (Linnaeus) Martius	+++											
Protonemata	+					+	+			+	+	
Bryum argenteum Hedw.												
Pohlia annotina (Hedw.) Lindb. agg.						+						
Acremonium strictum W. Gams					+							
Alternaria alternata (Fr.) Keissler	+	+	++		+++	+	++	++	+	++	+	++
Alternaria sp.						+						
Alternaria tenuissima (Kunze ex Pers.) Wiltshire					+		+	+				
Aspergillus fumigatus Fres.	+							+				+
Aspergillus sp.			+								+	
Aureobasidium pullulans (De Bary) Arnaud		+			+++	+		+		++	++	+
Cladosporium cladosporioides (Fresen.) de Vries	++		++	+	+++	+	++	++	+	++	+	++
Cladosporium herbarum (Pers.) Link ex S. F. Gray	+	++	+	++	+++	+	+	+	++	+	+	+
Coniothyrium sp.										+		+
Dematiaceae indet.					+							
Epicoccum nigrum Link		+	+		+	+	+	+	++	++	+	+
Fusarium sp.	++	++				+	+			+	++	+
Hormiactis sp.					++							
Mucor plumbeus Bonord.					+	+	+					
Mucor						+					+	

racemosus Fres.												
Paecilomyces sp.					+							
Penicillium sp.	+	+			+						+	
Phoma eupyrena Sacc.							+					+
Phoma exigua Desm.	+	++	+			+	+		+	+		+
Phoma glomerata (Corda) Wollenw. & Hochapfel												
Phoma pinodella (L. K. Jones) Morgan-Jones et K. B. Burch								+		+	+	
Phoma sp.	+	+	+	+		+					+	+
Pleospora herbarum (Pers. ex Fr.) Rabenh.							+				+	
Rhodotorula mucillaginosa (Jörgensen) Harrison					++							
Sordaria sp.						+	+					
Stachybotrys chartarum (Ehrenb. ex Link) Hughes					+	++						
Trichoderma viride Pers. ex Gray	+	+					+	+		+	+	+
Ulocladium chartarum (Preuss) Simmons					++	++	+	+		++		+++
Ulocladium oudemannsii Simmons					+						+	+
Sterile mycelia	+++	++	++	+	+	++	+		+		++	+
Caloplaca holocarpa (Hoffm. ex Ach.) Wade												
Soredia	++	+					++					
Bacteria	+	+	+	+	++	++	++	+	+	++	++	+
Streptomycetes	+	+		+	+	+	+			+	+	+
Amoeba	+	+									+	
Chironomida												
Ciliata												
Nematoda	+					+				+	+	
Rotatoria		+				+						
Tardigrada												

+++ Taxon identified as dominant component of surface growth; ++ taxon identified as subdominant (abundant) component of surface growth; + taxon identified as sporadic component of surface growth

Table A.9 Identified organisms in air germ measurements of ambient air at the IBP in Holzkirchen

Taxon	Date of air germ collection						
	Nr. 1: 18.09. 06	Nr. 2: 27.08. 06	Nr. 3: 09.10. 06	Nr. 4: 21.06. 07	Nr. 5: 04.07. 07	Nr. 6: 12.07. 07	Nr. 7: 17.07. 07
Acremonium sp.					+	+	
Acremonium strictum W. Gams				+			
Alternaria alternata (Fr.) Keissler	+		+	+	+	+	+
Alternaria tenuissima (Kunze ex Pers.) Wiltshire			+				+
Arthrinium phaeospermum (Corda) M. B. Ellis			+				
Aspergillus candidus Link ex Link				+	+		
Aspergillus fumigatus Fres.	+	+	+	+	+		+
Aspergillus sp.			+				
Aspergillus terreus Thom	+						
Aureobasidium pullulans (De Bary) Arnaud		+	+	+	+	+	+
Botrytis cinerea Pers. ex Pers.	+		+	+		+	+
Cladosporium cladosporioides (Fresen.) de Vries	+++	++	+++	+++	+++	+++	+++
Cladosporium herbarum (Pers.) Link ex S. F. Gray	+	+	+	+	++	+	
Cladosporium sphaerospermum Penz.	+++	+	+	+	+		+
Dematiaceae indet.		+	+	+	+	+	+
Drechslera poae (Baudys) Shoemaker							+
Epicoccum nigrum Link	+	+	+		+	+	
Eurotium amstellodamii Mangin						+	
Eurotium sp.	+		+	+	+		+
Fusarium sp.				+	+	+	
Humicola sp.			+				
Mucor sp.					+		
Paecilomyces sp.		+					
Penicillium brevicompactum Dierckx	+	+	+	+	+	+	+
Penicillium chrysogenum Thom		+	+		+		
Penicillium citrinum Thom					+		
Penicillium frequentans Westling				+		+	+
Penicillium solitum Westling					+		
Penicillium sp.		+	+		+		
Phoma pinodella (L. K. Jones) Morgan-Jones et K. B. Burch				+			
Phoma pomorum Thüm.					+		
Phoma sp.					+		
Pleospora herbarum (Pers. ex Fr.) Rabenh.				+	+		
Rhodotorula mucillaginosa (Jörgensen) Harrison						+	+
Trichoderma viride Pers. ex Gray							+
Ulocladium chartarum (Preuss) Simmons	+		+	+	+	+	+
Sterile mycelia	++	+++	++	++	++	++	++
Yeast			+				
Bacteria	+	+	+	+	+	+	+
Streptomycetes			+		+		

+++ Taxon identified as dominant component of air germs; ++ taxon identified as subdominant (abundant) component of air germs; + taxon identified as sporadic component of air germs

Table A.10 Identified organisms in sedimentation samples of ambient air at the IBP in Holzkirchen

Taxon	Date of sample										
	1: 27. 09. 06	**2:** 17. 10. 06	**3:** 25. 06. 07	**4:** 25. 06. 07	**5:** 25. 06. 07	**6:** 03. 07. 07	**7:** 03. 07. 07	**8:** 13. 07. 07	**9:** 13. 07. 07	**10:** 17. 07. 07	**11:** 17. 07. 07
	w	w	w	s	s	w	s	w	s	w	s
	Duration of measurement (h)										
	1	1	1.3	1.3	4	2	2	2.3	2.3	2	2
Apatococcus lobatus (Chodat) J. P. Petersen											+
Chlamydomonas sp.										+	
Chlorella trebouxioides Punčochářová			+		+					+	
Chlorella vulgaris Beijerinck					+++					+	
Chlorococcum acidum Archibald et Bold										+	
Choricystis minor (Skuja) Fott var. *gallica* (Bourelly) Komárek					+						
Diplosphaera chodatii Bialosuknia em. Vischer			+		+				+		
Diplosphaera sp.	+				+						
Elliptochloris reniformis (S. Watanabe) Ettl & Gärtner					+			+		+	
Geminella terricola J. B. Petersen					+						
Leptolyngbya foveolarum (Rabenhorst ex Gomont) Anagnostidis et Komárek								+			
Nostoc commune Vaucher ex Bornet & Flahault								+			
Dictyochloropsis cf. *splendida* Geitler					+						
Stichococcus bacillaris Nägeli					+		+	+		+	+
Symploca cf. *elegans* Kützing ex Gomont					+						
Moosprotonemata			+	+	+		+	+	+	+	+
Fernprothallia			+	+	++	+	+	+	+	+	+
Acremonium strictum W. Gams										+	
Alternaria alternata (Fr.) Keissler	+		+	+	+	+	+	++	+	+	+
Alternaria sp.			+		+						
Alternaria tenuissima (Kunze ex Pers.) Wiltshire			+		+	+	+	+		+	+
Arthrinium phaeospermum (Corda) M. B. Ellis								+			+
Aspergillus fumigatus Fres.	+		+		+			+			
Aspergillus terreus Thom	+				+						
Aureobasidium pullulans (De Bary) Arnaud										+	+
Botrytis cinerea Pers.		+			+						
Cladosporium	+	+++	+++	+++	+++	+++	+++	+++	+++	++	+++

cladosporioides (Fresen.) de Vries											
Cladosporium herbarum (Pers.) Link ex S. F. Gray	++	++	++	++	++	++	++	++	++	+	+
Cladosporium macrocarpum Preuss			+		+					+	
Cladosporium sphaerospermum Penz.		+									
Dematiaceae indet.	++		+	++	+	+	++	+	+	+	+
Drechslera poae (Baudys) Shoemaker		+				++				+	+
Epicoccum nigrum Link	+++	++	+	+	+		+	++	+	+	+
Fusarium sp.			+		+					+	
Harzia acremonioides (Harz) Cost.										+	
Mycotypha microspora Fenner								+			
Oidiodendron sp.	+		+			+					
Penicillium brevicompactum Dierckx		+									
Penicillium sp.								+		+	
Phoma exigua Desm.	+	+		+				+			
Phoma herbarum Westend.				++	+						
Phoma pinodella (L.K. Jones) Morgan-Jones et K.B. Burch					+					+	
Phoma sp.				+	+	+	+				
Pleospora herbarum (Pers. ex Fr.) Rabenh.				+	+			+	+		
Torula herbarum (Pers.) Link ex F. S. Gray					+						
Trichoderma viride Pers. ex Gray		+	+								
Ulocladium chartarum (Preuss) Simmons				++	+		+	+		+	+
Ulocladium oudemannsii Simmons			+		+						
Sterile mycelia	+	+++	+	+	+	+	+	++	+		
Fungi indet.					+						+
Bacteria	++	+++	+	+	+	++	+	+	+	+++	+
Streptomycetes			+	+++	+++			+	+	+	+

+++ dominant; ++ subdominant; + sporadic; w sedimentation plate vertical; s sedimentation plate horizontal

Table A.11 Identified organisms in driving rain at the IBP in Holzkirchen

Taxon	Date of sample												
	1: 24.10.06	**2:** 02.11.06	**3:** 14.11.06	**4:** 21.11.06	**5:** 04.12.06	**6:** 21.12.06	**7:** 08.01.07	**8:** 01.02.07	**9:** 12.02.07	**10:** 26.02.07	**11:** 13.03.07	**12:** 26.03.07	**13:** 16.04.07
Chlorella saccharophila (Krüger) Migula						+++							
Chlorella sp.		++	++						+				
Chlorella vulgaris Beijerinck	+++		+	+			+++			+			
Chlorococcum cf. *acidum* Archibald et Bold				+++			++						
Choricystis minor (Skuja) Fott var. *gallica* (Bourelly) Komárek		+	++		++								
Diplosphaera chodatii Bialosuknia em. Vischer		+											
Diplosphaera sp.					+								
Elliptochloris reniformis (S. Watanabe) Ettl & Gärtner		++			+++	++	+	+++	+++			+++	
Elliptochloris subsphaerica (Reisigl) Ettl & Gärtner									+				
Radiococcus papuanus (S. Watanabe) Kostikov et al.					+								
Geminella terricola J. B. Petersen													+
Klebsormidium flaccidum (Kützing) Silva et al.			+	+			+						
Scenedesmus acutus Meyen								+					
Stichococcus bacillaris Nägeli			+										
Stichococcus exiguus Gerneck								+					
Stichococcus minutus Grintzesco & Péterfi								+					
Xanthonema solidum (Vischer) Silva			++	++									
Mossprotonemata										+		+	
Acremonium strictum W. Gams	+												
Alternaria alternata (Fr.) Keissler	+											+	+
Aureobasidium pullulans (De Bary) Arnaud	+	+++						++	+	+++		+++	+++
Aspergillus versicolor (Vuill.) Tiraboschi								+					
Chaetomium globosum Kunze ex Steud.	+				+	+	+						
Cladosporium cladosporioides (Fresen.) de Vries	++	+								++		+++	
Cladosporium herbarum (Pers.) Link ex S. F. Gray	+												
Dematiaceae indet.		+											
Epicoccum nigrum Link	+	+										+	

Fusarium sp.		+											+
Oidiodendron sp.												+	
Penicillium brevicompactum Dierckx									+++				
Penicillium sp.		+										+	+
Phoma eupyrena Sacc.	+												
Phoma exigua Desm.		+										+	+
Preussia sp.	+										+		
Stachybotrys chartarum (Ehrenb. ex Link) Hughes										+			
Ulocladium oudemannsii Simmons				+									
Sterile mycelia	++	+										+	
Bacteria	x	x	x	x	x	x	x	x	x	x	x	x	x
Streptomycetes		+											
+++ dominant; ++ subdominant; + sporadic; x assessed as whole group only, therefore no estimate of dominance													

Glossary

Acidophilic Living in or preferring a pH range below pH 7

Acrocarpic Spore capsule on top of the main axis of the bryophyte plant

Aerophytic "Living in the air," expression used to characterize microorganisms living in places which are exposed to air in contrast to terrestrial, limnic and marine habitats

Aeroterrestric Forms which live in and on soil, on rock surfaces and in comparable habitats

Airborne germ measurement Procedure to count and identify airborne germs

Akinete Thick-walled resting cell

Alkaliphilic Organisms which grow well at pH between 10 and 11, but grow less at neutral pH

Alkalitolerant Species which tolerate pH values above 9, but have their optimum at neutral pH or below

Anamorph Development stage of fungi which reproduce asexually, often by conidia

Anisogamy Fusion of unequal gametes during sexual reproduction

Aplanospore Non-motile asexual reproduction spore

Apophysis Basal sterile part of the capsule of bryophytes (musci) or of a sporangium of the zygomycetes

Apothecium Fruiting body of ascomycetes (respectively, their lichens), cup- or plate-like

Arthroconidium Vegetative reproduction cell in fungi consists of hyphae which divide into spores

Ascoma Fruiting body of ascomycetes

Ascospores Meiospores produced in a specialized sporangial cell called an ascus, usually eight per ascus

Ascus Tube-like cells in ascomycetes where meiospores develop (sporangium), usually eight ascospores (endospores) are developed in an ascus

Association Technical term in our context used to address typical combinations/communities of different organisms in a habitat, an association of, e.g., plants forms a central unit in a hierarchy, like a species in taxonomy

© Springer-Verlag GmbH Deutschland, ein Teil von Springer Nature 2020

W. K. Hofbauer and G. Gärtner, *Microbial life on Façades,*

https://doi.org/10.1007/978-3-662-54833-2

Autospore Asexual reproduction spore with the same shape as the parental cell

Background concentration A "normal" not elevated concentration of an usually very abundant agent/component, e.g., air spora usually are found in changing but low concentrations all the time in our natural environment, but may be elevated in case of a water damage indoors

Basal cell Foot cell that facilitates attachment to the substrate in algae

Basidiospores Exogenous meiospores produced on a basidium at the tip of sterigmata, usually four per basidium

Basidium Typical sporangium of basidiomycetes which produces meiospores

Binding agent/binder Material component of a coating recipe which binds together all ingredients and therefore is responsible for its physical strength

Biocorrosion Damage in structure and stability or loss of material as a result of the activities or secretion of microorganisms

Biodeterioration Worsening, soiling or destruction of materials, surfaces, etc., by microorganisms, in our context of building parts and surfaces

Biofouling Development of a microbial layer (biofilm) on material surfaces which causes changes in the physicochemical properties eventually resulting in damage or even disintegration

Bitunicate Ascus with two wall layers

Blastoconidia Conidiospores formed by budding

Building component Part of a building which can be made of a single material or is in itself a compound structure

Capsal Algal layer organised in a common mucilage, cells show still monadoid character

Cauloid Stem-like thallus of lower plants (mosses and macroalgae)

Caulonema The secondary development stage of a moss protonema, often brown colored, with less chloroplasts and oblique cross-walls

CFU Colony-forming units

Chlamydospore Thick-walled resting spore, developed within or at the end of vegetative hyphae, may consist of one or several cells

Chloroplast DNA containing organelle of cells of eukaryotes contains photosynthetic pigments

Chromatoplasm External colored part of protoplasm in cyanoprokaryotes

Cleistothecium Closed fruiting body of ascomycetes, spores are released by rupture of wall

Coating system Whole composition of a compound coating with different layers and components

Coccal Algal cells without characters of a monadoid organization and exogenous cell wall, with vegetative propagation

Coenobia Colony of algal cells in specific number and arrangement

Coenoblast Multinucleate cell

Columella Sterile structure in the center of the capsule of musci or in the sporangia of zygomycetes

Conidium Spore of vegetative reproduction in fungi, non-motile fungal mitospore not formed inside a sporangium

Contact angle Defined geometrically as the angle formed by a drop of liquid at the three-phase boundary where liquid, gas and solid intersect, it is an indicator of the wettability of a surface

Contractile vacuoles Organelles of osmoregulation (also pulsating vacuoles)

Cyst Resting cell of microorganisms which is protected against adverse environmental conditions by an exogenous or endogenous produced thick cell wall

DBU German Federal Environmental Foundation ("Deutsche Bundesstiftung Umwelt")

Dew point The temperature to which air must be cooled to become saturated with water vapor

DGGE Denaturing gradient gel electrophoresis is a technique to separate short- to medium-length DNA fragments based on their melting characteristics. It has been used as a molecular fingerprinting method without the need for DNA sequencing

DIN German National Organization for Standardization ("Deutsches Institut für Normung")

DNA Deoxyribonucleic acid

Driving rain Rain propelled by driving wind and thus arriving on a vertical surface

EN European standard ("Europäische Norm")

Endoplasmatic reticulum System of biomembranes in the cell, site of protein synthesis

ETICS External thermal insulation compound system, building coating system with external thermal insulation involved

Eukaryont "Nucleated entity": Cells with a true nucleus and usually complex organelles like mitochondria and plastids

Eutrophic With a high load of nutrients

Eyespot (stigma) Orange-red spot containing carotenoids in the apical area of flagellated unicells, often associated with a chloroplast, only rarely a completely independent organelle

False branching In filamentous Cyanoprokaryota, lateral branches are formed by rupture of a loop, thus giving the impression of a second filament growing sideways

Flagellate Motile cell equipped with one or more flagella

Flagellum Thread-like organelle for movement of cells

Frustule Silica enclosure of diatom cells, linked together in two halves (epi- and hypotheca)

Gametangium Cells whose content divides or transforms into gametes

Gamete Haploid (possessing a single set of chromosomes) generative (sexual) cell

Hematochrome Extraplastidial red color matter contains astaxanthin

Heterocyte Thick-walled cells in Cyanoprokaryota (site of nitrogen fixation)

Heterokont Two different flagella

Heterotrich Thallus consisting of different kinds of trichomes, often prostrate and erect trichomes

Hormogonium Few celled fragment of a filament in Cyanoprokaryota (functions in asexual reproduction) often shows autonomous movement

Hyalodermis External layer of inflated colorless cells on moss stemlets

Hypertrophic With extreme load of nutrients

Hypnoblast Synonymous to akinete

Hypnospore Resting cell that develops a secondary thick wall for protection

IBP Fraunhofer Institute for Building Physics

IPI Iodine/potassium iodide ("Jod-Kaliumjodid," JKJ) = Lugol's solution, dye used for the proof of starch that turns red to violet to black

ISO International Standardization Organization

Isogamy Sexual reproduction by gametes with identic morphology

Isokont Two or more flagella of the same length and structure

Isopleth Line of identical condition, here understood as line of the same germination time or the same daily growth rate

Isopleth system Array of curves of different germination times or growth rates in dependence of relative humidity and temperature

Isthmus Connecting part of the two semicells of desmids

LC-MS Liquid chromatography–mass spectrometry

Leaf-ear cells Cells in the basal corners of moss leaves, often showing a typical form or color

MAA Mycosporine like amino acid, derivate of the amino acid mycosporine, regarded as additional protection against UV irradiation

Material indices In our context indices which define important physical and chemical properties of the used materials

Metula Several phialides connect on one metula (cell), regularly shaped cells in a complex conidiophore, which support the conidiogenous cells, as in many species of *Penicillium*

Mitochondria DNA containing organelles in eukaryotic cells, the site of cellular respiration

Mixotrophic Organisms that can thrive by a mix of autotrophic and saprotrophic alimentation

Monadoid Equipped with flagella

MPA Official material testing institute in Germany ("Materialprüfungsanstalt")

Mycobiont Fungal partner in lichen symbiosis

Nanocyte Very small reproductive cells in Cyanoprokaryota formed after several cell divisions

Oligotrophic Minor content of nutrients

Oogamy Fusion of a motile small male gamete and a larger non-motile female egg cell

Outdoor weathering Material tests where the specimens are exposed to environmental weather conditions

Palmella-stage Non-motile algal cells embedded in mucilage

Perithecium Globular fruiting body of ascomycetes (lichens), inserted in a fruiting thallus or attached to a hyphal layer, opens with a pore at maturity

Phialide Bottle-like cell which forms conidiospores in basipetal succession from an open end without any change of the length of the cell

Photobiont Algal (or cyanoprokaryote) partner in lichens

Phragmoplast Microtubules in certain green algae arranged perpendicularly to the plane of cell division

Phycobiont Algal partner in lichen symbiosis

Phylloid Leaf-like organ

Plasmodesmata Thin protoplasmatic connections between adjacent cells

Plaster A mixture of cement and/or lime or gypsum or another binder with sand, perhaps further ingredients and water which is applied as a coating to building surfaces, similar to stucco or mortar

Plastid Cell organelle with DNA and a double membrane contains photosynthetic pigments or stores carbohydrates (e.g., starch)

Pleurocarpic Sporophyte (capsule) inserted at lateral branches in moss plants

Prokaryota "Prior to nucleated cells," organisms with a cell organization without nucleus

Protonema First development stage of mosses, sometimes like a filamentous green alga

Pseudofilament A row of algal cells distantly spaced in mucilage

Pseudothecium Ascostroma similar to a perithecium, type of fruiting body in ascomycetes

Pt-100 Platinum resistance temperature detector with a nominal resistance of 100 Ω at a temperature of 0 °C

Pyknidium Bottle-like structure in which conidia are formed (ascomycetes, ascolichenes)

Pyrenoid Protein body in plastids of many algae, associated with storage of polysaccharides

RNA Ribonucleic acid

Raphe Longitudinal canal in one or both valves of pennate diatoms

Relative humidity (RH) Ratio of the partial pressure of water vapor to the equilibrium vapor pressure of water at a given temperature

Rhizoid Root-like filament, involved in attachment

Ribosome Granular cell structures, involved in protein synthesis

Rpm Rotation per minute

Sarcinoid Three-dimensional packet-like arrangement of cells

Sclerotium A firm mass of hyphae, resting stage in fungi, often surrounded by dark and tight wall layer

Semicell A half-cell of desmids

Seta Stalk which bears the moss capsule

Slope angle Angle between horizontal and vertical plane when droplets descend, in our context used to characterize hydrophobicity

Sporangium Cell whose content divides into spores

Sporodochium A cushion-shaped multihyphal asexual fructification producing conidia

Sporopollenin Very durable biopolymer in cell walls of different organisms

SQL Structured query language, a database language

SRP Soluble reactive phosphate, water soluble salts of phosphoric acid

Stigma Eye spot, red pigmented cell organelle which is involved in light reception

Stolon Runner of hyphae for vegetative reproduction

Subaerophytic Living in habitats, which are aerophytic and are regularly provided with liquid water

Surface growth In our context growth of organisms on the surface and in the surface near pores and cracks of a coating

Teleomorph Sexual manifestation of a fungus in ascomycetes, basidiomycetes, zygomycetes and oomycetes

Thallus Rather undifferentiated vegetation body without real leafs, stems, vessels and roots

Thermophilic Organisms that can grow at temperatures above 40 °C

Thylakoid Flat, membranous structure in cyanoprokaryotic cells and plastids of eukaryotes

Trichal Filamentous organization

Trichome Filament of cells in Cyanoprokaryota and algae without any mucilage sheath

TWEEN 80 Mild detergent

UV Ultraviolet irradiation

Vacuole Cavity within the cytoplasm, surrounded by a membrane and filled with fluid

Valve Flat upper part with down-turned edges of every halve of the frustule in diatoms

Vapor pressure Defined as the pressure exerted by a vapor in thermodynamic equilibrium with its condensed phases at a given temperature in a closed system

VdL Association of the German paint and varnish industry ("Verband der deutschen Lack- und Druckfarbenindustrie")

VdL-RL Directive (Richtlinie) of the association of the German paint industry

VKI Association for consumer information ("Verein für Konsumenteninformation")

Water activity (a$_w$) Partial vapor pressure of water in a substance divided by the standard state partial vapor pressure of water

Wet storage function Functional context between ambient relative humidity and water content that equilibrates in a porous material

WUFI® Warmth and moisture transient ("Wärme und Feuchte instationär"), calculation tool for the assessment of hygrothermal behavior of building materials, constructions and rooms, developed at the IBP

Xerothermic Dry and warm habitats

Zoospore Asexual reproductive spore with flagella

Zygospore Thick-walled resting spore or generative spore, produced through fusion of two gametes or two cells of adjacent filaments

Zygote Fusion product of gametes, diploid (with a double set of chromosomes)

References

van der Aa HA, van Kesteren HA (1971) The identity of Phyllosticta destructiva Desm. and similar Phoma-like Fungi described from Malvaceae and Lycium halimifolium. Acta botanica neerlandica 20(5): 552–563

Abarca ML, Accensi F, Cano J, Cabañes FJ (2004) Taxonomy and significance of black aspergilli. Antonie van Leeuwenhoek 86:33–49

Abdelahad N (1985) Observationi su alcune Cianoficee cavernicole rare e interessanti. Giornale Botanico Italiano 119:45–46

Abe A, Asano K, Sone T (2010) A Molecular Phylogeny-Based Taxonomy of the Genus *Rhizopus*. Biosci Biotechnol Biochem 74(7):1325–1331

Abe K, Mihara H, Hirano M (1998) Characteristics of growth and carotenoid accumulation of the aerial microalga *Trentepohlia aurea* in liquid culture. J Mar Biotechnol 6:53–58

Abe K, Nishimura N, Hirano M (1999) Simultaneous production of β-carotene, vitamin E and vitamin C by the aerial microalga *Trentepohlia aurea*. J Appl Phycol 11:331–336

Abe K, Takizawa H, Kimura S, Hirano M (2004) Characteristics of chlorophyll formation of the aerial microalga *Coelastrella striolata* var. *multistriata* and its application for environmental biomonitoring. J Biosci Bioeng 98(1):34–39

Abler, S.W. (2003): Ecology and taxonomy of *Leptosphaerulina* spp. associated with turfgrasses in the United States. Thesis. Virginia Polytechnic Institute and State University, 56 pp

Aboal M, Egidos AI, Marin JP, Asencio AD (2002) *Trentepohlia iolithus* (L.) Wallroth 1833 (Chlorophyta, Ulvophyceae) in subaerial habitats from southeastern Spain. Algol Stud 107:153–162

Ahmed SI, Cain RF (1972) Revision of the genera *Sporormia* and *Sporormiella*. Can J Bot 50:419–477

Ahrens (2001) 7. *Bryum* Hedw. Birnmoos. In: Nebel M, Philippi G (eds) Die Moose Baden-Württembergs. Band 2: Spezieller Teil (Bryophytina II, Schistostegales bis Hypnobryales). Ulmer, Stuttgart, pp 47–103

Aichele D, Schwegler H-W (1984) Unsere Moos- und Farnpflanzen. Eine Einführung in die Lebensweise, den Bau und das Erkennen heimischer Moose, Farne Bärlappe und Schachtelhalme. 9. Aufl. Franckh'sche Verlagshandlung, Stuttgart

Ajlani G, Kirilovsky D, Picaud M, Astier C (1989a) Molecular analysis of *psbA* mutations responsible for various herbicide resistance phenotypes in *Synechocystis* 6714. Plant Mol Biol 13:469–479

Ajlani G, Meyer I, Vernotte C, Astier C (1989b) Mutation in phenoltype herbicide resistance maps within the *psbA* gene in *Synechocystis* 6714. FEBS Lett 246:207–210

© Springer-Verlag GmbH Deutschland, ein Teil von Springer Nature 2020
W. K. Hofbauer and G. Gärtner, *Microbial life on Façades*,
https://doi.org/10.1007/978-3-662-54833-2

Albertano P (1990) The role of photosynthetic microorganisms on ancient monuments. In: Baer NS, Sabbioni C, Sors A (eds) Science, tecnology and Europaean cultural heritage: proceedings of the European symposium, Bologna, Italy, 13.–16. June 1989, pp 151–159

Albertano P, Pollio A, Taddei R (1991) *Viridiella friedericiana* (Chlorococcales, Chlorophyta), a new genus and species isolated from extremely acid environments. Phycologia 30:346–354

Alexander M (1982) Most probable number method for microbial populations. In: Page AL, Miller RH, Keeney DR (eds) Methods of soil analysis. Part 2. American Society of Agronomy Inc. & Soil Science Society America Inc., Madison, USA, pp 815–820

Alexopoulos CJ, Mins CW, Blackwell M (1996) Introductory mycology, 4th edn. John Wiley & Sons Inc., New York

Almgren K (1966) Ecology and distribution of algae belonging to *Haematococcaceae*. I. Notes on nomenclature and history. Svensk Bot Tidskr 60(1):49–73

Al-Mousawi AHA, Whitton BA (1983) Influence of environmental factors on algae in rice-field soil from the Iraqi marshes. Arab Gulf J Sci Res 1:237–253

Alum A, Mobasher B, Rashid A, Abbaszadegan M (2009) Image Analyses-Based Nondisruptive Method to Quantify Algal Growth on Concrete Surfaces. Journal of Environmental Engineeering 135(3):185–190

Anagnostidis K, Komárek J (1985):Modern approach to the classification system of cyanophytes, 1 – Introduction. Arch Hydrobiol/Algol Stud 38/39:291–302

Anagnostidis K, Komárek J (1988) Modern approach to the classification system of cyanophytes, 3 – Oscillatoriales. Arch Hydrobiol/Algol Stud 50/51: 327–472

Anagnostidis K, Komárek J (1990) Modern approach to the classification system of cyanophytes, 5 – Stigonematales. Arch Hydrobiol/Algol Stud 59:1–73

Anagnostidis K, Economou-Amili A, Roussomoustakaki M (1983) Epilithic and chasmolithic microflora (Cyanophyta, Bacillariophyta) from marbles of the Parthenon (Acropolos-Athens, Greece). Nova Hedwigia 38:227–287

Anandi V, John TJ, Walter A, Shastry JCM, Lalitha MK, Padhye AA, Ajello L, Chandler FW (1989) Cerebral Phaeohyphomycosis Caused by *Chaetomium globosum* in a Renal Transplant Recipient. J Clin Microbiol 27(10):2226–2229

Andersen B, Nielsen KF, Thrane U, Szaro T, Taylor J, Jarvis BB (2003) *Stachybotrys chlorohalonata*, a new species from water-damaged buildings. Mycologia 95:1225–1236

Andersen RA (ed) (2005) Algal culturing techniques. Amsterdam, Elsevier, p 578

Andersen RA, Brett RW, Potter D, Sexton JP (1998) Phylogeny of the Eustigmatophyceae based upon 18S rRNA, with emphasis on *Nannochloropsis*. Protist 149:61–74

Angelini E, Grassini S, Mombello D, Neri A, Parvis M (2010) An imaging approach for a contact-less monitoring of the conservation state of metallic works of art. Appl Phys A 100:919–925

Apinis AE (1963) Occurrence of thermophilous microfungi in certain alluvial soils near Nottingham. Nova Hedwigia 5:57–78

Apinis AE (1964) On fungi isolated from soils and *Ammophila* debris. Kew Bull 19:127–131

Ariño X, Saiz-Jimenez C (1996) Colonization and deterioration processes in Roman mortars by cyanobacteria, algae and lichens. Aerobiologia 12:9–18

Ariyadej C, Tansakul R, Tansakul P, Angsupanich S (2004) Phytoplankton diversity and its relationship to the physico-chemical environment in the Banglang Reservoir, Yala Province. Songklanakarin J Sci Technol 26(5):595–607

Armitage AD, Barbara DJ, Harrison RJ, Lane CR, Sreenivasaprasad S, Woodhall JW, Clarkson JP (2015) Discrete lineages within *Alternaria alternata* species group: Identification using new highly variable loci and support from morphological characters. Fungal Biol 119:994–1006

Årsvoll K (1975) Fungi causing winter damage on cultivated grasses in Norway. Meld Norg Landbrhoeisk 54:49

Asada K (1994) Production and action of active oxygen species in photosynthetic tissues. In: Foyer CH, Mulineaux PM (eds) Cases of photooxidative stress and amelioration of defense systems in plants. CRC Press, Boca Raton, pp 77–104

Asencio AD, Aboal M (1996) Cyaonphytes from Andragulla abrigo (Murcia, Spain) and their environmental conditions. Arch Hydrobiol/Algol Stud 83:55–72

ASTM D 3274-09 (Reapproved 2013) Standard test method for evaluating degree of surface disfigurement of paint films by fungal or algal growth, or soil and dirt accumulation. 2013.

ASTM 3456-86 (Reapproved 2012) Standard practice for determining by exterior exposure tests the susceptibility of paint films to microbiological attack. 2012.

ASTM D 3719-00 (2000) Standard test method for quantifying dirt collection on coated exterior panels. 2000.

ASTM D 5589-09 (reapproved 2013) Standard test method for determining the resistance of paint films and related coatings to algal defacement

Avenot HF, Michailides TJ (2007) Resistance to boscalid fungicide in Alternaria alternata isolates from pistachio in California. Plant Dis 91:1345–1350

Aveskamp MM, de Gruyter J, Woudenberg JHC, Verkley GJN, Crous PW (2010) Highlights of the Didymellaceae: A polyphasic approach to characterise Phoma and related pleosporalean genera. Stud Mycol 65:1–60

Ayerst G (1969) The effects of moisture and temperature on growth and spore germination in some fungi. J Stored Prod Res 5:127–141

Aytoun RSC (1953) The genus Trichoderma. Its relationship with Armillaria mellea and Polyporus schweinitzii, together with preliminary observations on its ecology in woodland soils. Trans Proc Bot Soc Edin 36:99–114

Bachmann G, Gärtner G (1992) Observations on some free-living and lichenized Trentepohlia species in culture. The 2nd Intern. Lichenological symposium IAL 2 1992, Hemmeslöv, Bastad, Sweden, Abstracts, p 28

Bachmann G (1992) Zur Kultur und Taxonomie freilebender und lichenisierter Trentepohliaarten (Grünalgen – Chlorophyta). Diplomarbeit, Leopold Franzens Universität Innsbruck

Baddley JW, Pappas PG, Smith AC, Moser SA (2003) Epidemiology of Aspergillus terreus at a University Hospital. J Clin Microbiol 41(12):5525–5529

Bagda E, Wühl-Couturier G, Lindner W (1999) Algen und Pilzen auf Fassadenbeschichtungen auf der Spur. Farbe & Lack 105(3):106–115

Baker NR, Bowyer JR (eds) (1994) Photoinhibition of photosynthesis from molecular mechanisms to the field. BIOS Scientific Publishers, Oxford

Bakerspigel A, Lowe D, Rostas A (1981) The isolation of Phoma eupyrena from a human lesion. Arch Derm 117:362–363

Balajee SA, Marr KA (2006) Phenotypic and genotypic identification of human pathogenic aspergilli. Future Microbiol 1(4):435–445

Balajee SA, Gribskov JL, Hanley E, Nickle D, Marr KA (2005) Aspergillus lentulus sp. nov., a new sibling species of A. fumigatus. Eukaryot Cell 4:625–632

Balajee SA, Nickle D, Varga J, Marr KA (2006) Molecular Studies Reveal Frequent Misidentification of Aspergillus fumigatus by Morphotyping. Eucaryot Cell 5(10):1705–1712

Barberousse H, Tell G, Yéprémian C, Couté A (2006) Diversity of algae and cyanobacteria growing on building façades in France. Algol Stud 120:81–105

Barkman JJ (1969) Phytosociology and ecology of cryptogamic epiphytes. Including a taxonomic survey and description of their vegetation units in Europe. Van Gorcum, Assen, Netherlands

Barron MA, Sutton DA, Veve R, Guarro J, Rinaldi M, Thompson E, Cagnoni PJ, Moultney K, Madinger NE (2003) Invasive Mycotic Infections Caused by Chaetomium perlucidum, a New Agent of Cerebral Phaeohyphomycosis. J Clin Microbiol 41(11):5302–5307

Bates JW, Bell JNB, Farmer AM (1990) Epiphyte recolonization of oaks along a gradient of air Pollution in south-east England, 1979-1990. Environ Pollut 68:81–99

Bates JW, Bell JNB, Massara AC (2001) Loss of *Lecanora conizaeoides* and other fluctuations of epiphytes on oak in S.E. England over 21 years with declining SO_2 concentrations. Atmos Environ 25:2557–2568

Bates JW, McNee PJ, McLeod AR (1996) Effects of sulphur dioxide and ozone on lichen colonization of conifers in the Liphook Forest Fumigation Project. New Phytol 132:653–660

Becker EW (1982) Physiological studies on antarctic *Prasiola crispa* and *Nostoc commune* at low temperatures. Polar Biol 1:99–104

Bellinger EG, Sigee DC (2015) Freshwater algae: identification, enumeration and use as bioindicators, 2nd edn. Wiley Blackwell, USA, p 275

Benny GL (2001) The zygomycota: zygomycetes. In: McLaughlin DJ, McLaughlin EG, Lemke PA (eds) The Mycota, vol 7A. Systematics and evolution. Springer, Berlin, pp 147–160

Benny GL, Benjamin RK (1976) Observations on Thamnidiaceae (Mucorales). II. *Chaetocladium, Cokeromyces, Mycotypha,* and *Phascolomyces.* Aliso 8:391–424

Benny GL, Humber RA, Morton JB (2001) The zygomycota: zygomycetes. In: McLaughlin DJ, McLaughlin EG, Lemke PA (eds) The Mycota, vol 7A. Systematics and evolution. Springer, Berlin, pp 113–146

Benny GL, Kirk PM, Samson RA (1985) Observations on Thamnidiaceae (Mucorales). III. Mycotyphaceae fam. nov. and a re-evaluation of *Mycotypha* sensu Benny & Benjamin illustrated by two new species. Mycotaxon 22:119–148

Bensch K, Braun U, Groenewald JZ, Crous PW (2012) The genus *Cladosporium.* Stud Mycol 72:1–401

Bergman B, Gallon JR, Rai AN, Stal LJ (1997) N_2 fixation by non-heterocystous cyanobacteria [review]. FEMS Microbiol Rev 19:139–185

Berner M, Wanner G, Lubitz W (1997) A comparative study of the fungal flora present in medieval wall paintings in the chapel of the Castle Herberstein and in the Parish Church of St. Georgen in Styria, Austria. Int Biodeterior Biodegradation 40(1):53–61

Bertsch A (1966) CO_2-Gaswechsel und Wasserhaushalt der aerophilen Grünalge *Apatococcus lobatus.* Planta 70:46–72

Bertsch K (1959) Moosflora von Südwestdeutschland, 2nd edn. Eugen Ulmer, Stuttgart

Bertsch K (1964) Flechtenflora von Südwestdeutschland, 2nd edn. Eugen Ulmer, Stuttgart

Beuf L, Kurano N, Miyachi S (2000) Effect of external pH on inorganic carbon assimilation in unicellular marine green algae. Phycol Res 48(1):47–54

Bhati HS, Gaur RD (1979) Studies on aerobiology – atmospheric fungal spores. New Phytol 82:519–527

Bhattacharya D, Friedl T, Damberger S (1996) Nuclear-encoded rDNA group I introns: origin and phylogenetic relationships of insertion sites to lineages in the green algae. Mol Biol Evol 13:978–989

Bischoff WW, Bold HC (1963) Phycological studies. IV. Some soil algae from Enchanted Rock and related algal species. Univ Texas Publ 6318:1–95

Björkman O, Holmgren P (1963) Adaptability of the photosynthetic apparatus to light intensity in ecotypes from exposed and shaded habitats. Physiol Plant 16:889–914

Blaser P (1975) Systematische Untersuchungen über die *Aspergillus glaucus*-Gruppe. Mykosen 18:87–89

Blom HH (1998) 96. *Schistidium* Bruch & Schimp. in B.S.G., Bryol. Eur., 1845 nom. cons. In: Nyholm, E. (ed.): Illustrated Flora of Nordic Mosses. Fasc. **4**. Aulacomniaceae – Meesiaceae – Catascopiaceae – Bartramiaceae – Timmiaceae – Encalyptaceae – Grimmiaceae – Ptychomitraceae – Hedwigiaceae – Orthotrichaceae. Nord. Bryol. Soc., Copenhagen, Lund, pp 287–330

Bock C, Jacob A, Kirst GO, Leibfritz D, Mayer A (1996) Metabolic changes of the Antarctic green alga *Prasiola crispa* subjected to water stress investigated by *in vivo* ^{31}P NMR. J Exp Bot 47(295):241–249

Bock C, Krienitz L, Pröschold T (2011) Taxonomic reassessment of the genus *Chlorella* (Trebouxiophyceae) using molecular signatures (barcodes), including description of seven new species. Fottea 11:293–312

Bock E, Koops H-P, Möller UC, Rudert M (1990) A new facultatively nitrite oxidizing bacterium, Nitrobacter vulgaris sp. nov. Arch Microbiol 153:105–110

Bock E, Sand W (1993) The microbiology of masonry biodeterioration. J Appl Bacteriol 74:503–514

Boerema GH, Dorenbosch MMJ (1973) The *Phoma* and *Ascochyta* species described by Wollenweber and Hochapfel in their study on fruit-rotting. Stud Mycol Baarn 3:50

Boerema GH (1976) The *Phoma* species studied in culture by Dr. R.W.G. Dennis. Trans Br Mycol Soc 67:289–319

Bold HC (1931) The life history and cell structure of *Chlorococcum infusionum*. Bull Torrey Bot Club 57:577–604

Bold HC (1951) Cytology of algae. In: Smith GM (ed) Manual of phycology. The Chronica Botanica Co., Waltham, Mass., pp 203–227

Boltansky H, Kwon-Chung KJ, Macher AM, Gallin JI (1984) *Acremonium strictum*-related pulmonary infection in a patient with granulomatous disease. J Infect Dis 149:653

Boneh Sh (1953) Contribution to the knowledge of some textile destroying fungi in Israel. Palest J Bot Ser J 6:107–113

Boone DR, Castenholz RW, Garrity GM (eds) (2001) Bergey's Manual® of systematic bacteriology, 2nd edn. The archaea and the deeply branching phototrophic bacteria. Springer, New York, Berlin, Heidelberg, p 721

Booth C (1971a) Fungal culture media. In: Booth C (ed) Methods in microbiology. Academic Press, London, pp 49–94

Booth C (1971b) The genus *Fusarium*. Commonwealth Mycological Institute, Kew, p 237

Booth C (1977) *Fusarium* laboratory guide to the identification of the major species. Commonwealth Mycological Institute, Kew, p 58

Booton GC, Floyd GL, Fuerst PA (1998) Polyphyly of tetrasporalean green algae inferred from nuclear small subunit ribosomal DNA. J Phycol 34:306–311

Born A, Ermuth J (1999) Copyright by nature – Neue Micro-Silikonharzfarbe mit Lotuseffekt für trockene und saubere Oberflächen. Farbe Lack 3:96–104

Bourrelly P (1966) Les algues d'eau douce. Initiation a la systematique. Tome 1: Les algues vertes. Boubée & Cie, Paris

Bourrelly P (1970) Les algues d'eau douce. Tome III: Les algues bleues et rouges, les Eugléniens, Peridiniens et Cryptomonadines. Boubée & Cie, Paris, p 512

Boussiba S, Vonshak A (1991) Astaxanthin accumulation in the green alga *Haematococcus pluvialis*. Plant Cell Physiol 32(7):1077–1082

Braams J (1992) Ecological studies on the fungal microflora inhabiting historical sandstone monuments. Dissertation, Universität Oldenburg

Brahamanage RS, Hyde KD, Li XH, Zhang W, Liu M, Jayawardena RS, McKenzie EHC, Yan JY (2018) Are pathogenic isolates of *Stemphylium* host specific and cosmopolitan? Plant Pathol Quarantine 8(2):153–164

Brain APR, Young TWK (1979) Ultrastructure of the asexual apparatus in *Mycotypha*. Microbios 25:93–106

Brand F, Stockmayer S (1925) Analyse der aerophilen Grünalgenanflüge, insbesondere der proto-pleurococcoiden Formen. Arch Protistenk 52:265–354

Brenner DJ, Staley JT, Krieg NR (2001) Classification of procaryotic organisms and the concept of bacterial speciation. In: Boone DR, Castenholz RW, Garrity GM (eds) Bergey's Manual® of systematic bacteriology, vol One. The Archaea and the Deeply Branching Phototrophic Bacteria. Springer, New York, Berlin, Heidelberg, pp 27–31

Breuer K, Hofbauer W, Krueger N, Mayer F, Scherer C, Schwerd R, Sedlbauer K (2012) Wirksamkeit und Dauerhaftigkeit von Bioziden in Bautenbeschichtungen. Bauphysik 34(4), pp 170–182.

Breuer K, Hofbauer W, Krus M; Scherer C, Schwerd R, Krueger N, Mayer F, Sedlbauer K (2011) Bedeutung des bioziden Wirkstoffeinsatzes bezüglich der Dauerhaftigkeit von Fassadenbeschichtungen. In: Venzmer H (ed.): Fassadensanierung: Praxisbeispiele, Produkteigenschaften, Schutzfunktionen. Deutsches Institut für Normung e.V., Rudolf Müller Verlagsgesellschaft, Köln, pp 53–77.

Briand J-F, Jacquet S, Bernard C, Humbert J-F (2003) Health hazards for terrestrial vertebrates from toxic cyanobacteria in surface water ecosystems. Vet Res 34:361–377

Bristol BM (1920) On the algal flora of some desiccated English soils, an important factor in soil biology. Ann Bot 34:35–80

Broady PA (1979) The terrestrial algae of Signy Island, South Orkney Islands. Br. Antarct Surv Sci Rep 98:1–117

Broady PA (1982) New records of chlorophycean micro-algae cultured from Antarctic terrestrial habitats. Nova Hedwigia 36:445–484

Broady PA (1989a) Survey of algae and other terrestrial biota at Edward VII Peninsula, Marie Byrd Land. Antarct Sci 1(3):215–224

Broady PA (1989b) The distribution of *Prasiola calophylla* (Carmich.) Menegh. (Chlorophyta) in Antarctic freshwater and terrestrial habitats. Antarct Sci 1(2):109–118

Broady PA, Ingerfeld M (1993) Three new species and a new record of chaetophoracean (Chlorophyta) algae from terrestrial habitats in Antarctica. Eur J Phycol 28:25–31

Brook PJ (1963) Ecology of the fungus *Pithomyces chartarum* (Berk. & Curt.) M.B. Ellis in pasture in relation to facial eczema disease of sheep. NZ J Agri Res 6:147–228

Brooks FT, Hansford CG (1923) Mould growth upon cold-stored meat. Trans Br Mycol Soc 8:113–142

Brown DE, Halsted DJ (1975) The effect of acid pH on the growth kinetics of *Trichoderma viride*. Biotechnol Bioeng 17:1199–1210

Brown JC (1958) Soil fungi of some British sand dunes in relation to soil type and succession. J Ecol 46:641–664

Brown RM Jr, McLean RJ (1969) New taxonomic criteria in classification of *Chlorococcum* species II. Pyrenoid fine structure. J Phycol 5:114–118

Brown RM Jr, Larson DA, Bold HC (1964) Airborne algae: their abundance and heterogeneity. Science 143:583–585

Brown RM, Arnott HJ, Bisalputra T, Hoffman LR (1967) The pyrenoid: its structure, distribution and function. J Phycol 3:5–7

Bubrick P, Galun M, Frensdorff A (1984) Observation in free-living *Trebouxia* and *Pseudotrebouxia* and evidence that both symbionts from *Xanthoria parietina* can be found free-living in nature. New Phytol 97:455–462

Buchecker R, Eugster CH, Kjøsen H, Liaanen-Jensen S (1973) Absolute Konfiguration von β, ε-Carotin-2-ol, β, β-Carotin-2-ol und β, β-Carotin-2,2'-diol. Helv Chim Acta 56(8):2899–2901

Buchheim MA, Chapman RL (1991) Phylogeny of the colonial green flagellates: a study of 18S and 26S RNA sequence data. BioSystems 25:85–100

Buchheim MA, Lemieux CC, Otis C, Gutell R, Chapman RL, Turmel M (1996) Phylogeny of the Chlamydomonadales (Chlorophyceae): a comparison of ribosomal RNA gene sequences from the nucleus and the chloroplast. Mol Phylogenet Evol 5:391–402

Buchheim MA, Michalopulos EA, Buchheim JA (2001) Phylogeny of the Chlorophyceae with spezial reference to the Spaeropleales: a study of 18S and 26s rDNA data. J Phycol 37:819–835

Buchheim MA, Turmel M, Zimmer EA, Chapman RL (1990) Phylogenetic systematics of *Chlamydomonas* based on cladistic analysis of nuclear 18S rRNA sequence data. J Phycol 26:689–699

Büdel B, Henssen A (1988) *Trebouxia aggragata* und *Gloeocapsa sanguinea*, Phycobionten in *Euopsis granatina* (Lichinaceae). Plant Syst Evol 158(2–4):235–241

Burge HP, Boise JR, Rutherford JA, Solomon WR (1977) Comparative recoveries of airborne fungus spores by viable and non-viable modes of volumetric collection. Mycopathologia 61(1):27–33

Burzlaff A (2007) Keiner klebt so stark wie *Caulobacter*. Mikrokosmos 96(6):378–380

Butt ZL, Ghaffar A (1974) Effect of certain physico-chemical factors on growth and antifungal property of *Stachybotrys atra*. Z Pflphysiol 71:463–466

Byrne PJ, Jones EBG (1975) Effect of salinity on spore germination of terrestrial and marine fungi. Trans Br Mycol Soc 64:497–503

Cain RF (1961) Studies on coprophilous ascomycetes. 7 Preussia. Can J Bot 39:1633–1666

Cambra J, Hernandez-Mariné MC (1998) Observaciones sobre algas corticolas del nordeste y sudeste de espana. Anales del jardin botánoco de Madrid 46(1):115–126

Campbell ME (1938) An investigation of the Mucorales in the soil. Trans R Soc Edinb 59:411–436

Caneva G (1993) Ecological approach to the genesis of calcium oxalate patinas on stone monuments. Aerobiologia 9:149–156

Caneva G, Salvadori O (1987) Biodeterioration of stone. In: UNESCO (ed) The deterioration and conservation of stone. Studies and documents on the cultural heritage. UNESCO, Paris, pp 182–234

Carmichael JW, Kendrick WB, Conners IL, Sigler L (1980) Genera of hyphomycetes. Univ Alberta Press, Utrecht, p 386

Castenholz RW (1969) Thermophilic blue-green algae and the thermal environment. Bacteriol Rev 33(4):476–504

Castenholz RW (2001a) Phylum BX. Cyanobacteria. Oxygenic photosynthetic bacteria. In: Boone DR, Castenholz RW, Garrity GM (eds) Bergey's Manual® of systematic bacteriology, vol One. The archaea and the deeply branching phototrophic bacteria. Springer, New York, Berlin, Heidelberg, p 474

Castenholz RW (2001b) General characteristics of the cyanobacteria. In: Boone DR, Castenholz RW, Garrity GM (eds) Bergey's Manual® of systematic bacteriology, vol One. The archaea and the deeply branching phototrophic bacteria. Springer, New York, Berlin, Heidelberg, pp 474–487

Castenholz RW (2004) Phototrophic bacteria under UV stress. In Seckbach J (ed) Cellular origins, life in extreme habitats and astrobiology. Origins, Kluyver Academic Publishers, Netherlands, pp 445–461

Castenholz RW, Herdman M, Rippka R (2001a) Form-genus IX. *Scytonema* Agardh 1824. In: Boone DR, Castenholz RW, Garrity GM (eds) Bergey's Manual® of systematic bacteriology, vol One. The archaea and the deeply branching phototrophic bacteria. Springer, New York, Berlin, Heidelberg, pp 580–582

Castenholz RW, Herdman M, Rippka R (2001b) Form-genus XV. *Symploca* Kützing 1843. In: Boone DR, Castenholz RW, Garrity GM (eds) Bergey's Manual® of systematic bacteriology, vol One. The archaea and the deeply branching phototrophic bacteria. Springer, New York, Berlin, Heidelberg, pp 474–487

Castenholz RW, Rippka R, Herdman M (2001c) Form-genus IX. *Oscillatoria* Vaucher 1803 (sensu Rippka, Deruelles, Waterbury, Herdman and Stanier). In: Boone DR, Castenholz RW, Garrity GM (eds) Bergey's Manual® of systematic bacteriology, vol One. The archaea and the deeply branching phototrophic bacteria. Springer, New York, Berlin, Heidelberg, pp 550–553

Castenholz RW, Rippka R, Herdman M, Wilmotte A (2001d) Subsection III. (Formerly Oscillatoriales Elenkin 1934). In: Boone DR, Castenholz RW, Garrity GM (eds) Bergey's Manual® of systematic bacteriology, vol One. The archaea and the deeply branching phototrophic bacteria. Springer, New York, Berlin, Heidelberg, pp 539–542

Castenholz RW, Rippka R, Herdman M, Wilmotte A (2001e) Form-genus V. *Leptolyngbya* Anagnostidis and Komarek 1988. In: Boone DR, Castenholz RW, Garrity GM (eds) Bergey's Manual® of systematic bacteriology, vol One. The archaea and the deeply branching phototrophic bacteria. Springer, New York, Berlin, Heidelberg, pp 544–546

Catani SC, Peterson JL (1967) Antagonistic relationships between *Verticillium dahliae* and fungi isolated from the rhizosphere of *Acer platanoides*. Phytopathology 57:363–366

CDC (1994) Acute pulmonary hemorrhage/hemosiderosis among infants – Cleveland, January 1993 – November 1994. MMWR 43:881–883

CDC (1995) Acute pulmonary hemorrhage among infants – Chicago, April 1992 – November 1994. MMWR 44:73–74

CDC (2000) Update: Acute pulmonary hemorrhage/hemosiderosis among infants – Cleveland, Ohio, 1993–1996. MMWR 49:180–184

Çelen E, Kiliç MA (2004) Isolation and characterization of aerobic denitrifies from agricultural soil. Turk J Biol 28:9–14

Cerman Z, Barthlott W, Neinhuis C (2003) Der Lotus-Effekt: Selbstreinigende biologische Oberflächen und Möglichkeiten ihrer technischen Nutzung. In Venzmer (Hrsg) Altbauinstandsetzung **5/6**. Algen an Fassadenbaustoffen II. 4. Dahlberg-Kolloquium Mai 2003. Sonderheft Dahlberg Kolloquium, Verlag Bauwesen, S 191–198

Cerman Z, Stosch AK, Barthlott W (2004) Der Lotus-Effekt®. Selbstreinigende Oberflächen und ihre Übertragung in die Technik. Biologie in unserer Zeit 5:290–296

Cerniglia CE (1993) Biodegradation of polycyclic aromatic hydrocarbons. Curr Opin Biotechnol 4:331–338

Chadha A, Pandey DC (1993) Certain observations on *Scytonema stuposum* (Kuetz.) Born. (Cyanophyta, Nostocales). Hydrobiologia 62(2):105–106

Chapman ES, Fergus CL (1979) Germination of ascospores of *Chaetomium globosum*. Mycologia 67:1048–1052

Chapman RL, Waters DA (2004) Lichenization of the Trentepohliales. Complex algae and odd relationships. In: Seckbach J (ed) Symbiosis. Mechanisms and model systems book series: cellular origin, life in extreme habitats and astrobiology. Springer, Netherlands, pp 359–371

Chapman RL, Borkhsenious RC, Brown RC, Henk MC, Waters DA (2001) Phragmoplast-mediated cytokinesis in *Trentepohlia*: results of TEM and immunofluorescence cytochemistry. Int J Syst Evol Microbiol 51:759–765

Chapman RL, Buchheim MA, Delwiche CF, Friedl T, Huss VAR, Karol KG, Lewis LA, Manhart J, McCourt RM, Olsen JL, Waters DA (1998) Molecular systematics of the green algae. In: Soltis PS, Soltis DE, Doyle JJ (eds) The molecular systematics of plants II. Kluwer Academic Publishers, Norwell, Massachusets, USA, pp 508–540

Chen AW (1964) Soil fungi with high salt tolerance. Trans Kans Acad Sci 67:36–40

Chen AW (1966) Soil physical factors and the ecology of fungi. 5. Further studies in relatively dry soils. Trans Br Mycol Soc 49:419–426

Chen AW, Griffin DM (1966) Soil physical factors and the ecology of fungi. 6. Interaction between temperature and soil moisture. Trans Br Mycol Soc 49:551–561

Chen Q, Hou LW, Duan WJ, Crous PW, Cai L (2017) Didymellaceae revisited. Stud Mycol 87:105–159

Cherepanova NP (1975) Effects of environmental conditions on the production of fruit bodies by species of *Chaetomium* Fr. Mikol Fitopat 9:391–396

Chodat R (1913) Monographie d'algues en culture pure. Beitr Kryptogamenfl Schweiz 4(2):1–266

Chodat R (1928) Les clones chez les algues inférieures. Zeitschr. Indukt. Abst.-Vererb. Suppl. (Verhandl V. Internat. Kongr. Vererbungswiss., Berlin 1927) 1, Verl. Borntraeger, Leipzig, S 522–530

Chodat R, Grintzesco J (1900) Sur les méthodes de culture pure les algues vertes. Congrés International de Botanique, Paris. Extrait du Compte-rendu. Imprimerie Lucien Declume, Lons-le-Saunier, S 157–162

Christensen M (1969) Soil microfungi of dry to mesic conifer-hardwood forests in northern Wisconsin. Ecology 50:9–27

Chung YL, Chang S-N, Hann S-K, Cho H-J, Suh S-B, Lee KH (1999) Spontaneously healed primary cutaneous alternariosis: reports of 2 cases. Kor J Med Mycol 4:137–142

Cifuentes AS, González MA, Vargas S, Hoeneisen M, González N (2003) Optimization of biomass, total carotenoids and astaxanthin production in *Haematococcus pluvialis* Flotow strain Steptoe (Nevada, USA) under laboratory conditions. Biol Res 36:343–357

Ciubuc N (2005) Algele edaphice din solurile municipiului Chişinău. Dissertation Universitatea de Stat din Moldova, Chişinău, 147 pp

Clare NT (1944) Photosensitivity diseases in New Zealand. 3. The photosensitizing agent in facial eczema. NZ Jl Sci Technol 25:202–220

Claus G (1955) Algae and their Mode of Life in the Baradla Cave at Aggtelek. Acta Bot. Acad. Sci. Hung. 2:1–26

Claus G (1958) Algae and their Mode of Life in the Baradla Cave at Aggtelek II. Speleology 1:13–17

Cohn F (1850) Nachträge zur Naturgeschichte des *Protococcus pluvialis* Kützing (*Haematococcus pluvialis* Flotow). Nov Act Leop Carol 22(2):605–764

Collin RG, Towers NR (1995a) First reported isolation from New Zealand pasture of *Pithomyces chartarum* unable to produce sporidesmin. Mycopythologia 130:37–40

Collin RG, Towers NR (1995b) Competition of a sporidesmin-producing *Pithomyces* strain with a non-toxigenic *Pithomyces* strain. NZ Vet J 43:149–152

Collin RG, Smith BL, Towers NR (1996) Lack of toxicity of a nonsporidesmin-producing strain of *Pithomyces chartarum* in cell culture and when dosed to lambs. NZ Vet J 44:131–134

Colón-López MS, Sherman DM, Sherman LA (1997) Transcriptional and transitional regulation of nitrogenase in light-dark- and continuous-light-grown cultures of the unicellular cyanobacterium, *Cyanothece* sp. ATCC 51142. J Bacteriol 179:4319–4372

Contet-Audonneau N, Barbaud A, Guérin V, Basile AM, Guiguen C, Rassemusse C, Percebois G (1991) Alternariose cutanée et syndrome de Cushing: nouvelle observation. J Mycol Méd 1:82–83

Cooke WB (1970) Fungi in burned and unburned chapparal soils. Sydowia 24:164–168

Corry JEL (1973) The water relations and heat resistance of microorganisms. Prog Indust Microbiol 12:73–108

Couderchet M, Rumbolz J, Kring F, Böger P (1995) Characteristics of a metazachlor-resistant *Scenedesmus acutus* cell line. Pest Biochem Physiol 52:222–233

Coutinho MAGLD (2015) Biological colonization on majolica glazed tiles: biodeterioration, bioreceptivity and mitigation atrategies. Doctoral tesis, Facultade de Ciencias e Technologia Universitade Nova de Lisboa, Lisboa

Cowell BC (1960) A quantitative study of the Winter Plankton of Urschel's Quarry. Ohio J Sci 60(3):183–191

Craig WM, Broderick GA, Ricker DB (1987) Quantitation of microorganisms associated with the particulate phase of ruminal ingesta. J Nutr 117:56–62

Crisam EV (1973) Current concepts of thermophilism and the thermophilic fungi. Mycologia 65:1171–1189

Crispim CA, Gaylarde CC (2005) Cyanobacteria and biodeterioration of cultural heritage: a review. Microb Ecol 49:1–9

Crispim CA, Gaylarde PM, Gaylarde CC, Neilan BA (2006) Deteriogenic Cyanobacteria on historic buildings in Brazil detected by culture and molecular techniques. Int Biodeterior Biodegradation 57(4):239–243

Crous PW, Braun U, Groenewald JZ (2007) *Mycosphaerella* is polyphyletic. Stud Mycol 58:1–32

Culberson WL (1970) Chemosystematics and ecology of lichen-forming fungi. Ann Rev Ecol Syst 1:153–170

Curran PMT (1971) Sporulation in some members of the *Aspergillus glaucus* group in response to osmotic pressure, illumination and temperature. Trans Br Mycol Soc 57:201–211

Cutler NA, Viles HA, Ahmad S, McCabe S, Smith BJ (2013) Algal 'greening' and the conservation of stone heritage structures. Sci Total Environ 442:152–164

Czurda V (1928) Morphologie und Physiologie des Algenstärkekornes. Beih Bot Centralbl 45(1):97–270

Czygan F-C (1970) Blutregen und Blutschnee: Stickstoffmangel-Zellen von *Haematococcus pluvialis* und *Chlamydomonas nivalis*. Arch. Mikrobiol. 74:69–76

Dalla-Chiesa M, Friso G, Deák Z, Vass I, Barber J, Nixon PJ (1997) Reduced turnover of the D_1 polypeptide and photoactivation of electron transfer in novel herbicide mutants of *Synechocystis* sp. PCC 6803. Eur J Biochem 248:731–740

Dalla Torre KW, von Sarnthein LG (1901) Flora der gefürsteten Grafschaft Tirol, des Landes Vorarlberg und des Fürstenthumes Liechtenstein II. Wagner, Innsbruck, p 210

Darienko T, Gruber M, Pröschold T, Schagerl M (2013) Terrestrial microalgae on Viennese buildings. Final report of project H-2081/2010 funded by Hochschuljubiläumsstiftung der Stadt Wien. Universität Wien, 1–20 + 20 pp appendices

Darienko T, Gustavs L, Mudimu O, Rad Menendez C, Schumann R, Karsten U, Friedl T, Pröschold T (2010) *Chloroidium*, a common terrestrial coccoid green alga previously assigned to *Chlorella* (Trebouxiophyceae, Chlorophyta). Eur J Phycol 45:79–95

Darienko T, Hoffmann L (2003) Algal growth on cultural monuments in Ukraine. Biol Bratisl 58(4):575–587

Darienko T, Pröschold T (2019) The genus *Jaagichlorella* Reisigl (Trebouxiophyceae, Chlorophyta) and its close relatives: an evolutionary puzzle. Phytotaxa 388(1):47–68

Davey MC, Clarke KJ (1991) The spatial distribution of microalgae on Antarctic fjellfield soils. Antar Sci 3(3):257–263

Davison S, Marbrook J (1965) The effect of temperature on the toxicity of spores of *Pithomyces chartarum* (Berk. & Curt.) M.B. Ellis. NZ Jl Agric Res 8:126–130

Deason TR, Bold HC (1960) Phycological studies. I. Explanatory studies of texas soil algae. Univ Texas Publ 6022:1–72

Deason TR, Silva PC, Watanabe S, Floyd GL (1991) Taxonomic status of the green algal genus *Neochloris*. Pl Syst Evol 177:213–219

Degen J (2002) Entwicklung eines Photobioreaktors mit verbesserter Lichtnutzung für Mikroalgen. Dissertation Universität Hohenheim, 166 pp

Degen J, Uebele A, Retze A, Schmid-Staiger U, Trösch W (2001) A novel photobioreactor with baffles for improved light utilization through the flashing light effect. J Biotechnol 92:89–94

Deichfelder K (1985) Geschichte der Medizin. Englisch Verlag, Wiesbaden

Del Monte M, Sabbioni C, Zappia G (1987) The origin of calcium oxalates on monuments, historical buildings and natural outcrops. Sci Total Environ 67:17–39

Di Bonaventura M-P, Del Gallo M, Cacchio P, Ercole C, Lepidi A (1999) Microbial formation of oxalate films on monument surfaces: bioprotection or biodeterioration? Geomicrobiol J 16(1):55–64

di Menna ME, Campbell J, Mortimer PH (1970) Sporidesmin production and sporulation in *Pithomyces chartarum*. J Gen Microbiol 61:87–96

Dickson JH (2000) Bryology and the Iceman: Chorology, ecology and ethnobotany of the mosses *Neckera complanata* Hedw. and *N. crispa* Hedw. In: Bortenschlager S, Oeggl K (eds) The man in the ice, Volume **4**: The Iceman and his natural environment: palaeobotanical results. Springer, Wien, New York, pp 77–88

Dickson JH (2003) Low to moderate altitude mosses at the iceman site and their significance. In: Fleckinger A (Hrsg) Die Gletschermumie aus der Kupferzeit 2. Neue Forschungsergebnisse zum Mann aus dem Eis. Schriften des Südtiroler Achaeologiemuseums, **3**, 27–34

Dickson JH, Bortenschlager S, Oeggl K, Porley R, McMullen A (1996) Mosses and the Tyrolean Iceman's southern provenance. Proc R Soc Lond B 263:567–571

Dickson, J.H. & Hofbauer, W. (2005): Bryophytes and the iceman. XVII International Botanical Congress. Abstracts (www.ibc2005.ac.at), p 75

Dickson JH, Oeggl KD, Kofler W, Hofbauer WK, Porley R, Rothero GP, Schmidl A, Heiss AG (2019) Seventy-five mosses and liverworts found frozen with the late Neolithic Tyrolean Iceman: Origins, taphonomy and the Iceman's last journey. PLoS ONE 14(10):e0223752. https://doi.org/10.1371/journal.pone.0223752

Dillon JG, Castenholz RW (1999) Scytonemin, a cyanobacterial sheath pigment, protects against UVC radiation: implications for early photosynthetic life. J Phycol 35:673–681

Dillon JG, Tatsumi CM, Tandigan PG, Castenholz RW (2002) Effect of environmental factors on the synthesis of scytonemin, a UV-screening pigment, in a cyanobacterium (*Chroococcidiopsis* sp.). Arch Microbiol 177:322–331

DIN EN 60751: Industrielle Platin-Widerstandsthermometer und Platin-Meßwiderstände. Deutsche Fassung EN 60751, Ausgabe 07-1996, Beuth, Berlin, Wien, Zürich.

Dingley JM (1962) *Pithomyces chartarum*, its occurrence, morphology and taxonomy. NZ Jl Agric Res 5:49–61

Dodds WK, Gudder DA, Mollenhauer D (1995) The ecology of nostoc. J Phycol 31:2–18

Domsch KH (1960) Das Pilzspektrum einer Bodenprobe. 3. Nachweis der Einzelpilze. Arch Mikrobiol 35:310–339

Domsch KH, Gams W, Anderson T-H (1980) Compendium of soil fungi. Reprint der Ausgabe von 1980, 1993, IHW-Verlag, Eching

Domsch KH, Gams W, Anderson T-H (2007) Compendium of soil fungi. Second edition, taxonomically revised by Walter Gams. IHW-Verlag, Eching

Done J, Mortimer PH, Taylor A (1961) The production of sporidesmin and sporidesmolides by *Pithomyces chartarum*. J Gen Microbiol 26:207–222

D´Orazio M, Cursio G, Graziani L, Aquilanti L, Osimani A, Clementi F, Yéprémian C, Lariccia V, Amoroso S (2014) Effects of water absorption and surface roughness on the bioreceptivity of ETICS compared to clay bricks. Build Environ 77:20–28

Douglas AE, Huss VAR (1986) On the characteristica and taxonomic position of symbiontic *Chlorella*. Arch Microbiol 145:80–84

Dreyfuss M (1976) Taxonomische Untersuchungen innerhalb der Gattung *Chaetomium*. Sydowia 28:50–132

Dřímalová D, Poulíčková A (2003) Filament fragmentation of *Klebsormidium flaccidum*. Biol Bratisl 58(4):525–527

Drouhet E, Luciani J, Frantz P, Chomette G, Ravisse P, Dupont B (1991) Alternariose cutanée et sarcome de Kaposi chez un greffe rénal. J Mycol Méd 1:84–87

Duran MT, Del Pozo J, Yebra MT, Crespo MG, Paniagua MJ, Cabezon MA, Guarro J (2003) Cutaneous infection caused by *Ulocladium chartarum* in a heart transplant recipient: case report and review. Acta Der Venereol 83(3):218–221

Durrell LW, Shields LM (1960) Fungi isolated in culture from soils of the Nevada test site. Mycologia 52:636–641

Eckhardt FEW (1996) Microbial diversity and airborne contamination. In: Heitz E, Flemming H-C, Sand E (eds) Microbially influenced corrosion of materials. Springer, Berlin, Heidelberg, pp 75–95

Edelmann RE, Klomparens KL (1995) Low temperature scanning electron microscopy of the ultrastructural development of zygospores and sporangiospores in *Mycotypha africana*, and the effects of cultural conditions on sexual versus asexual reproduction. Mycol Res 99:539–548

Edlich F (1936) Einwirkung von Temperatur und Wasser auf aerophile Algen. Arch Mibrobiol 7:62–109

Ehling-Schulz M, Bilger W, Scherer S (1997) UV-B-induced synthesis of photoprotective pigments and extracellular polysaccharides in the terrestrial cyanobacterium *Nostoc commune*. J Bacteriol 179(6):1940–1945

Ehrlich HL (1998) Geomicrobiology: its significance for geology. Earth-Sci Rev 45:45–60

Ehrlich HL (2002) Geomicrobiology, 4th edn. Marcel Decker, New York

Eliáš M, Amaral R, Fawley KP, Fawley MW, Němcová Y, Neustupa J, Přibyl P, Santos LMA, Ševčiková T (2017) Eustigmatophyceae. In: Archibald JM, Simpson AGB, Slamovits CH (eds) Handbook of the protists, 2nd edn. Springer Intern, Publishing AG, Cham, Switzerland, pp 367–406

Ellenberg H (1986) Vegetation Mitteleuropas mit den Alpen in ökologischer Sicht. Ulmer, Stuttgart, p 989

Elliot AM (1934) Morphology and life history of *Haematococcus pluvialis*. Arch Protistenk 82:250–272

Ellis JJ (1985) Species and varieties in the *Rhizopus arrhizus – Rhizopus oryzae* group as indicated by their DNA complemantarity. Mycologia 77:243–247

Ellis M (1931) Some experimental studies on *Pleospora herbarum* (Pers.) Rabenh. Trans Br Mycol Soc 16:102–114

Ellis MB (1960) Dematiaceous Hyphomycetes. 1. Mycol Pap 76:1–36

Ellis MB, Ellis JP (1998) Microfungi on miscellaneous substrates. An identification handbook, New enlarged edn. The Richmond Publishing Co., Ltd., Slough, p 246

Ellis MB (1971) Dematiaceous hyphomycetes. C A B International, Kew, London, p 601

Ellis MB (1976) More dematiaceous hyphomycetes. C A B International, Kew, London, p 507

Engelhart S, Loock A, Skutlarek D, Sagunski H, Lommel A, Färber H, Exner M (2002) Occurrence of Toxigenic *Aspergillus versicolor* Isolates and Sterigmatocystin in Carpet dust from damp indoor environments. Appl Environ Microbiol 68(8):3886–3890

Entwistle AR (1972) Study of *Phoma exigua* populations in the field. Trans Br Mycol Soc 58:217–223

Escadeillas G, Bertron A, Blanc P, Dubosc A (2006) Accelerated testing of biological stain growth on external concrete walls. Part 1: Development of the growth test. Materials and Structures. online first. https://doi.org/10.1617/s11527-006-9205-x

Esser K (2000) Kryptogamen 1. Cyanobakterien, Algen, Pilze, Flechten: Praktikum und Lehrbuch. Springer-Verlag, Berlin, Heidelberg, New York

Ettl H (1976) Die Gattung *Chlamydomonas* Ehrenberg. Beih Nova Hedwigia 49:1–1122

Ettl H (1978) Band 3: Xanthophyceae. 1. Teil. In: Ettl H, Gerloff J, Heynig H (Hrsg) (1978) Süßwasserflora von Mitteleuropa. 3. Gustav Fischer, Stuttgart, New York, 530 pp

Ettl H (1980) Grundriß der allgemeinen Algologie. G. Fischer, Jena, p 549

Ettl H, Komárek J (1982) Was versteht man unter dem Begriff „coccale Grünalgen"? Algological Studies 29: 345–374

Ettl H (1983) Chlorophyta I. Phytomonadina. In Ettl H, Gerloff J, Heynig H, Mollenauer D (Hrsg) Süßwasserflora von Mitteleuropa. 9. Gustav Fischer, Stuttgart, New York, 807 pp

Ettl H, Gärtner G (1988a) Chlorophyta II. Tetrasporales, Chlorococcales, Gloeodendrales. In Ettl H, Gerloff J, Heynig H, Mollenauer D (Hrsg) Süßwasserflora von Mitteleuropa, 10. G. Fischer, Stuttgart, 436 pp

Ettl H, Gärtner G (1988b) Eine einfache Methode zur Darstellung der Struktur der Stärkehüllen von Pyrenoiden bei Grünalgen (Chlorophyta). Arch Protistenkd 135:179–181

Ettl H, Gärtner G (1995) Syllabus der Boden-Luft- und Flechtenalgen. Gustav Fischer, Stuttgart, Jena, New York, p 721

Ettl H, Gärtner G (2014) Syllabus der Boden-, Luft- und Flechtenalgen, 2nd edn. Springer Spektrum, Berlin, Heidelberg, p 773

Evans RD, Johansen JR (1999) Microbiotic crusts and ecosystem processes. Crit Rev Plant Sci 18:183–225

Eveleigh DE (1961a) The growth requirements of *Phoma violacea*, with reference to its disfiguration of painted surfaces. Ann Appl Biol 49:412–423

Eveleigh DE (1961b) *Phoma* species associated with painted surfaces. Trans Br Mycol Soc 44:573–585

Eveleigh DE (1970) Fungal disfigurement of paper, and soft rot of cedar shingles. Appl Microbiol 19:872–874

Fahrthofer K (2004) Beitrag zur Kultur und Systematik aero-terrestrischer Grünalgen. Die Gattung *Stichococcus* Nägeli. Dipl.-Arb. Leopold-Franzens-Universität Innsbruck, 130 pp

Famintzin A (1871) Die anorganischen Salze als ausgezeichnetes Hilfsmittel zum Studium der Entwicklung niederer chlorophyllhaltiger Organismen. Bull Acad Sci St Petersb 17:31–70

Fan L, Vonshak A, Boussiba S (1994) Effect of temperature and irradiance on growth of *Haematococcus pluvialis*. J Phycol 30:829–833

Farmer AM, Bates JW, Bell JNB (1991) Seasonal variations in acidic pollutant inputs and their effects on the chemistry of stemflow, bark and epiphyte tissues in three oak woodlands in N.W. Britain. New Phytol 118:441–451

Farmer AM, Bates JW, Bell JNB (1992) Ecophysiological effects of acid rain on bryophytes and lichens. In: Bates JW, Farmer AM (eds) Bryophytes and lichens in a changing environment. Clarendon Press, Oxford, pp 284–313

Farr DF, Bills GF, Chamuris GP, Rossman AY (1989) Fungi on plants and plant products in the United States. APS Press, St. Paul, Minnesota, p 1252

Fassier M, Faugeron C, Gloaguen V, Ducroquetz C, Dupont O (2013) Behaviour of roofing materials facing to micro-organisms. Green Sustain Chem 3:8–14

Fehér D (1936) Untersuchungen über die regionale Verbreitung der Bodenalgen. Archiv Mikrobiol 7:439–476

Fehér D, Besenyei Z (1933) Untersuchungen über die mikroskopischen Pilze des Waldbodens. Forstl Versuche 35:55–77

Ferry AP, Abedi S (1983) Diagnosis and management of rhino-orbitocerebral mucormycosis (phycomycosis). Ophthalmology 90:1096–1104

Fife AJ (2000) A synopsis of the New Zealand species of *Schistidium* (Grimmiaceae; Musci), with observations on a little known species of *Racomitrium*. NZ J Bot 38:191–204

Fitzgerald JM, Collin RG, Towers NR (1998) Biological control of sporidesmin-producing strains of *Pithomyces chartarum* by biocompetitive exclusion. Lett Appl Microbiol 26:17–21

Fleming ED, Castenholz RW (2007) Effects of periodic desiccation on the synthesis of the UV-screening compound, scytonemin, in cyanobacteria. Environ Microbiol 9(6):1448–1455

Foged N (1953) Diatomeer indslæbt med mellemskarv, *Phalacrocorax carbo sinensis*. Bot Tidsskr 50:63–74

Fonseca AJ, Pina F, Macedo MF, Leal N, Romanowska-Deskins A, Laiz L, Gómez-Bolea A, Saiz-Jimenez C (2010) Anatase as an alternative application for preventing biodeterioration of mortars: Evaluation and comparison with other biocides. Int Biodeter Biodegradation 64:388–396

Fontaine KM, Stocker-Wörgötter E, Booth T, Piercey-Normore MD (2013) Genetic diversity of lichen-forming alga, *Diplosphaera chodatii,* in North America and Europe. Lichenologist 45(6):799–813

Forgacs J (1972) Stachybotryotoxicosis. In: Kadis S, Ciegler A, Ajl SJ (eds) Microbial toxins, vol 8. Academic Press, New York, pp 95–128

Fott B, Nováková M (1969) A monograph of the genus *Chlorella*. The fresh water species. In: Fott B (ed) Studies in phycology. Academia, Praha, pp 10–59

Frahm J-P (2008) Nitrophile Moose und Flechten nehmen zu. Überdüngung und Versalzung durch Katalysatoren? Biol Unserer Zeit 38(2):92–102

Frahm J-P, Solga A (1999) Der Einfluß von Stickstoffemissionen auf Moose und Flechten. Bryologische Rundbriefe 28(1):4–10

Frändberg E, Olsen M (1999) Performance of DG18 media, a collaborative study. J Food Mycol 2:239–249

Franz G (1975) (1975): Temperaturansprüche mikroskopischer Bodenpilze aus klimatisch und geographisch verschiedenen Standorten. Z Pflanzenernähr Bodenk 1:73–87

Frey W, Frahm JP, Fischer E, Lobin W (2006) The liverworts, mosses and ferns of Europe. Harley Books, Colchester

Friedl T (1995) Inferring taxonomic positions and testing genus level assignments in coccoid green lichen alga: a phylogenetic analysis of 18S ribosomal RNA sequences from *Dictyochloropsis reticulata* and from members of the genus *Myrmecia* (Chlorophyta, Trebouxiophyceae cl. nov.). J Phycol 31:632–639

Friedl T (1997) The evolution of the green algae. In: Bhattacharya D (ed) Origin of algae and their plastids. Springer, Wien, pp 87–101

Friedl, T. (2002): Internationale Tagung: Algen als biologische Ressourcen erforschen und nutzen. Uniprotokolle (2002), Universität Göttingen

Friedl T, O´Kelly CJ (2002) Phylogenetic relationships of green algae assigned to the genus *Planophila* (Chlorophyta) evidence from 18S rDNA sequence data and ultrastructure. Eur J Phycol 37:373–384

Friedmann EI (1961a) *Chroococcidiopsis kashayii* sp. n. and the genus *Chroococcidiopsis*. Österr Bot Z 108:354–367

Friedmann EI (1961b) The ecology of the atmophytic nitrate-alga *Chroococcidiopsis kashayii* Friedmann. IV. Studies on cave algae from Israel. Archiv für Mikrobiologie 42:42–45

Frisvad JC (1989) The connection between the penicillia and aspergilli and mycotoxins with special emphasis on misidentified isolates. Arch Environ Contam Toxicol 18:452–467

Frisvad JC, Thrane U (2002) Mycotoxin production by common filamentous fungi. In: Samson RA, Hoekstra ES, Frisvad JC, Filtenborg O (eds) Introduction to food- and airborne fungi, 6th edn. Centraalbureau voor Schimmelcultures, Utrecht, pp 321–331

Fučíková K, Lewis LA (2012) Intersection of *Chlorella, Muriella* and *Bracteacoccus*: Resurrecting the genus *Chromochloris* Kol & Chodat (Chlorophyceae, Chlorophyta). Fottea 12:83–93

Furtado JS, Olive LS (1971) Ascospore discharge and ultrastructure of the ascus in *Leptosphaerulina australis*. Nova Hedwigia 19(3/4):799–823

Gärtner G (1974) Beitrag zur Ökologie und Systematik von Rindenalgen. Diss. Phil. Fak. Universität Innsbruck, 107 pp

Gärtner G (1985e) Flechten. In Tarmann G (Hrsg) Kein Leben ohne Pilze. Beiheft zur Ausstellung im Tiroler Landeskundlichen Museum Zeughaus 1985, S 106–116

Gärtner G (1986) Über Aufgabe und Bedeutung von Kulturen in der Algentaxonomie (am Beispiel von Boden-, Luft- und Flechtenalgen). Sauteria 1:149–157

Gärtner G (1994) Zur Taxonomie aerophiler grüner Algenanflüge an Baumrinden. Berichte des Naturwissenschaftlich-medizinischen Vereins Innsbruck **81**, 51–59

Gärtner G (1996) Soil Algae. In: Schinner F, Öhlinger R, Kandeler E, Margesin R (eds) Methods in soil biology. Springer, Berlin, Heidelberg, New York, pp 295–305

Gärtner G (2004) ASIB: The culture collection of algae at the Botanical Institute, Innsbruck. Nova Hedwigia 79:71–76

Gärtner G, Ettl H (1988) Neugliederung der Gattung *Chlorococcum* Meneghini (Chlorophyta, Chlamydophyceae, Chlorococcales). Nova Hedwigia 47:271–278

Gärtner G, Ingolić E (1987) Additional observations on cytology and reproduction processes of the lichen alga *Trebouxia*. Bibliotheca Lichenologica 25:73–76

Gärtner G, Ingolić E (1989) Ein Beitrag zur Kenntnis von *Apatococcus lobatus* (Chlorophyta, Chaetophorales, Leptosiroideae). Pl Syst Evol 164:133–143

Gärtner G, Ingolić E (1995) Fortschritte und Probleme in der Taxonomie aeroterrestrischer Grünalgen der Gattungen *Chlorococcum* und *Tetracystis* (Chlorococcales, Chlamydophyceae). Carinthia II, 53. Sonderh., S 35–36

Gärtner G, Ingolić E (1997) Additional observations on the morphology of zygotes and hypnoblasts of *Chlorococcum*- and *Tetracystis*-species. Intern. symposium biology and taxonomy of green Algae III, Smolenice, Slovakia, 6.–10. Oct. 1997, abstracts S 27

Gärtner G, Ingolić E (2000) Zur Taxonomie häufiger Rinden- und Bodenalgen: die Gattung *Desmococcus* BRAND emend. VISCHER. 9. Österr. Botanikertreffen Illmitz, 28.9.-1.10.2000, Kurzfass. D. Vorträge u. Poster, Linzer biol. Beitr. 32(2):634–635

Gärtner G, Ingolić E (2003) Further studies on *Desmococcus* Brand emend. Vischer (Chlorophyta, Trebouxiophyceae) and an new species *Desmococcus spinocystis* sp. nov. from soil. Biol Bratisl **58**(4): 517–523

Gärtner G, Stoyneva MP (2003) First study of aerophytic cryptogams in Bulgaria. Ber. nat.-med. Verein Innsbruck 90:73–82

Galbraith JC, Smith JE (1969) Filamentous growth of *Aspergillus niger* in submerged shake cultures. Trans Br Mycol Soc 52:236–246

Galun M (ed) (1988) CRC handbook of lichenology. Volume I–III, CRC Press, Boca Raton, Florida.

Gams H (1927) Von den Follatères zur Dent de Morcles. Vegetationsmonographie aus dem Wallis. Beitr geobot Landesaufn Schweiz 15:1–760

Gams, H (1950) Die Rotalgen Tirols. Der Schlern 1950, Bozen, pp 170–172

Gams W (1971) *Cephalosporium*-artige Schimmelpilze (Hyphomycetes). G. Fischer, Stuttgart

Garcia-Pichel F, Castenholz RW (1991) Characterization and biological implications of scytonemin, a cyanobacterial sheath pigment. J Phycol 27:395–409

Garcia-Pichel F, Castenholz RW (1993) Occurrence of UV-absorbing, mycosporine-like compounds among cyanobacterial isolates and an estimate of their screening capacity. Appl Environ Microbiol 59(1):163–169

Garcia-Pichel F, Sherry ND, Castenholz RW (1992) Evidence for an ultraviolet sunscreen role of the extracellular pigment scytonemin in the terrestrial cyanobacterium *Chlorogloeopsis* sp. Photochem Photobiol 56:17–23

Garcia-Pichel F, Sherry ND, Castenholz RW (1992b) Evidence for an ultraviolet sunscreen role of the extracellular pigment scytonemin in the terrestrial cyanobacterium *Chlorogloeopsis* sp. Photochem. Photobiol 56:17–23

Garcia-Pichel F, Wingard CE, Castenholz RW (1993) Evidence regarding the UV sunscreen role of a mycosporine-like compound in the cyanobacterium *Gloeocapsa* sp. Appl Environ Microbiol 59(1):170–179

Garrity GM, Holt JG (2001) The road map to the manual. In: Boone DR, Castenholz RW, Garrity GM (eds) Bergey's Manual® of systematic bacteriology, vol One. The archaea and the deeply branching phototrophic bacteria. Springer, New York, Berlin, Heidelberg, pp 119–166

Gaylarde PM, Englert G, Ortega-Morales O, Gaylarde C (2006) Lichen-like colonies of pure *Trentepohlia* on limestone monuments. Int Biodeter Biodegradation 58(3–4):119–123

Gaylarde PM, Gaylarde CC (2000) Algae and cyanobacteria on painted buildings in Latin America. Int Biodeter Biodegradation 46:93–97

Gehl KA, Colman B (1985) Effect of external pH on the internal pH of *Chlorella saccharophila*. Plant Physiol 77:917–921

Geiser DM, Frisvad JC, Taylor JW (1998) Evolutionary relationships in *Aspergillus* section *Fumigati* inferred from partial b-tubulin and hydrophobin DNA sequences. Mycologia 90(5):831–845

Geissler U, Gerloff J (1965) Das Vorkommen von Diatomeen in menschlichen Organen und in der Luft. Nova Hedwigia 10:565–577

Geitler L (1932) Cyanophyceae von Europa unter Berücksichtigung der anderen Kontinente. In Kolkwitz R (Hrsg) Rabenhorst's Ktyptogamen-Flora von Deutschland, Österreich und der Schweiz. 2.Aufl., Bd. 14, S 1–1196, Leipzig

Geitler L (1942) Morphologie, Entwicklungsgeschichte und Systematik neuer bemerkenswerter atmophytischer Algen aus Wien. Flora 136:1–29

Geitler L (1944a) Die angebliche stammesgeschichtliche Verwandtschaft von Cyanophyceen und Rhodophyceen. Wiener Botanische Zeitschrift 93(3/4):123–126

Geitler L (1944b) Furchungsteilung, simultane Mehrfachteilung, Lokomotion, Plasmoptyse und Ökologie der Bangiacee *Porphyridium cruentum*. Flora 37:300–333

Geitler L (1966) Die Chlorococcalen *Dictyochloris* und *Dictyochloropsis* nov. gen. Österr Bot Z 113:155–164

Geitler L (1980) Zellteilung und Bildung von Innenschalen bei *Hantzschia amphioxys* und *Achnanthes coarctata*. Pl Syst Evol 136:275–286

Gerlach W (1970) Suggestions to an acceptable modern *Fusarium* system. Ann Acad Sci Fenn A IV Biol 168:37–49

Gerlach W, Nirenberg H (1982) The genus *Fusarium*, a pictoral atlas. Mitt Biol Bundesanst Land-Forstw Berlin-Dahlem 209:1–406

Gibbs SP (1962a) The ultrastructure of the chloroplasts of algae. J Ultrastr Res 7:41–84

Gibbs SP (1962b) The ultrastructure of the pyrenoids of algae exclusive of the green algae. J Ultrastr Res 7:247–261

Gibson IAS (1953) Crown rot, a seedling disease of groundnuts caused by *Apergillus niger*. Trans Br Mycol Soc 36:198–209

Gilbert OL (1992) Lichen reinvasion with declining air pollution. In: Bates JW, Farmer AM (eds) Bryophytes and lichens in a changing environment. Clarendon Press, Oxford, pp 159–177

Gimmler H (2001) Acidophilic and acidotolerant algae. In: Rai LC, Gaur JP, Hindu B (eds) Algal adaption to environmental stresses. Springer, Berlin, pp 259–290

Gimmler H, Degenhardt B (2001) Alkaliphilic and alkali tolerant algae. In: Rai LC, Gaur JP (eds) Algal adaptation to environmental stresses: physiological, biochemical and molecular mechanisms. Springer, Berlin, pp 291–321

Gingrich JC, Buzby JS, Stirewalt VI, Bryant DA (1988) Genetic analysis of two new mutations resulting in herbicide resistance in the cyanobacterium *Synechococcus* sp. PCC 7002. Photosynth Res 16:83–99

Giovagnoli A, Paradisi A, Polidori C, Santonico G (2009) Il videomicroscopio per la valutazione e il monitoraggio in situ dei trattamenti conservative. VII. Congresso Nazionale IGIIC, Lo Stato dell'Arte, Napoli, 8.–10. October 2009

Girgin S, Kazanci N, Dügel M (2004) On the limnology of deep and saline lake burdur in Turkey. Acta Hydrochim Hydrobiol 32(3):189–200

Golden SS, Haselkorn R (1985) Mutation to herbicide resistance maps within the psbA gene of *Anacystis nidulans* R2. Science 229:1104–1107

Golden SS, Sherman LA (1984) Biochemical and biophysical characterization of herbicide-resistant mutants of the unicellular cyanobacterium, *Anacystis nidulans* R2. Biochim Biophys Acta 764:239–246

Gontcharov AA, Birger M, Melkonian M (2004) Are combined analyses better than single gene phylogenies? A case study using SSU rDNA and *rbc*L sequence comparison in the zygnematophyceae (Streptophyta). Mol Biol Evol 21(3):612–624

Gonzalez JM, Saiz-Jimenez C (2004) Microbial diversity in biodeteriorated monuments as studied by denaturing gradient gel electrophoresis. J Sep Sci 27:174–180

Gorbushina AA, Heyrman J, Dornieden T, Gonzalez-Delvalle M, Krumbein WE, Laiz L, Petersen K, Saiz-Jimenez C, Swings J (2003) Bacterial and fungal diversity and biodeterioration problems in mural painting environments of St. Martins church (Greene-Kreiensen, Germany). Int Biodegradation Biodeter 53:13–24

Gorbushina AA, Lyalikova NN, Vlasov DY, Khizhnyak TV (2002) Microbial communities on the monuments of Moscow and St. Petersburg: biodiversity and trophic relations. Microbiology 71(3):409–417

Gostinčar C, Ohm RA, Kogej T, Sonjak S, Turk M, Zajc J, Zalar P, Grube M, Sun H, Han J, Sharma A, Chiniquy J, Ngan CY, Lipzen A, Barry K, Grigoriev IV, Gunde-Cimerman N (2014) Genome sequencing of four *Aureobasidium pullulans* varieties: biotechnological potential, stress tolerance, and description of new species. BMC Genom 15(549):1–28

Grabherr W (1941) Beiträge zur Flora der Umgebung von Innsbruck. Mit besonderer Berücksichtigung des Voldertales bei Hall in Tirol. Pl Syst Evol 90(1):53–62

Graham JH, Luttrell ES (1961) Species of *Leptosphaerulina* on forage plants. Phytopathology 51:680–693

Graham LE, Graham JM, Wilcox LW (2009) Algae, 2nd edn. Pearson Benjamin Cummings, San Francisco, p 616

Gravesen S (1979) Fungi as cause of allergic disease. Allergy 34:28–30

Greenwood JR, Steenbergen JF (1976) A unique thermal aquatic environment: a microbial characterization. Hydrobiologia 51:265–273

Gregory PH, Lacey ME (1964) The discovery of *Pithomyces chartarum* in Britain. Trans Br Mycol Soc 47:25–30

Griffin DM (1963a) Soil moisture and the ecology of soil fungi. Biol Rev 38:141–166

Griffin DM (1963b) Soil physical factors and the ecology of fungi. 3. Activity of fungi in relatively dry soil. Trans Br Mycol Soc 46:373–377

Gronda I, Monte E, Rives V, Vicente MA (1997) Lichenized association between Septonema tormes sp. nov., a coccoid cyanobacterium, and a green alga with an unforeseen biopreservation effect of Villamayor sandstone at Casa Lis of Salamanca, Spain. Mycol Res 101:1489–1495

Grübl P, Weigler H, Karl S (2001) Beton, Arten. Herstellung und Eigenschaften. Ernst & Sohn, Berlin

Grung M, Souza FML, Borowitzka M, Liaanen-Jensen S (1992) Algal carotenoids 51. Secondary carotenoids 2. *Haematococcus pluvialis* aplanospores as a source of (3S,3′S)-astaxanthin esters. J Appl Phycol 4:165–172

de Gruyter J, Woudenberg JHC, Aveskamp MM, Verkley GJM, Groenewald JZ, Crous PW (2012) Redisposition of phoma-like anamorphs in Pleosporales. Stud Mycol 75:1–36

Guerin M, Huntley ME, Olaizola M (2003) *Haematococcus* astaxanthin: applications for human health and nutrition. TRENDS Biotechnol 21(5):210–216

Guillitte O (1995) Bioreceptivity: a new concept for building ecology studies. Sci Total Environ 167:215–220

Guiry MD, Guiry GM (2019) Algae base. World-wide electronic publication, National University of Ireland, Galway. https://www.algaebase.org; searched on 19 December 2019

Gustavs L, Schumann R, Karsten U (2016) Mixotrophy in the terrestrial green alga *Apatococcus lobatus* (Trebouxiophyceae, Chlorophyta). J Phycol 52:311–314

Häder DP, Nultsch W (1971) Untersuchungen zur Abgrenzung der Arten *Phormidium uncinatum* und *Phormidium autumnale*. Aquat Sci 33(2):566–577

Hakulinen R (1954) Die Flechtengattung *Candelariella* Müller Argoviensis, mit besonderer Berücksichtigung ihres Auftretens und ihrer Verbreitung in Fennoskandien. Ann Bot Soc Vanamo 27(3):1–127

Halder CA, Taber RA, Camp BJ (1979) High-performance liquid chromatography of the mycotoxin, sporidesmin, from *Pithomyces chartarum* (Berk. & Curt.) M.B. Ellis. J Chromatogr 175:356–361

Halder CA, Taber RA, Camp BJ (1981) Absence of Sporidesmin Production by Twelve Texas Isolates of *Pithomyces* spp. Appl Environ Microbiol 41(1):212–215

Hamilton WD, Lenton TM (1998) Spora and Gaia: how microbes fly with their clouds. Ethol Ecol Evol 10:1–16

Hanagata N (1998) Phylogeny of the subfamily Scotiellocystoideae (Chlorophyceae, Chlorophyta) and related taxa inferred from 18S ribosomal RNA gene sequence data. J Phycol 34:1049–1054

Hanagata N, Karube I, Chihara M (1997) Bark-inhabiting Green Algae in Japan (3). *Chlorella trebouxioides* and *Ch. angusto ellipsoidea*, sp. nov, (Chlorelloideae, Chlorellaceae, Chlorococcales). J Jpn Bot 72(1):36–43

Hanagata N, Karube I, Chihara M, Silva PC (1998) Resonsideration of the taxonomy of ellipsoidal species of *Chlorella* (Trebouxiophyceae, Chlorophyta), with establishment of *Watanabea* gen. nov. Phycol Res 46:221–229

Hansgirg A (1892) Beiträge zur Kenntnis der Süsswasseralgen- und Bacterienflora von Tirol und Böhmen. Sitzungsber. Böhm. Ges. Wiss, S 105–156

Harrach B, Bata A, Sandor G, Vanyi A, Vezer F (1988) Stachybotryotoxicosis in sport horses verified by chromatographic analysis. Magyar Allatorvosok Lapja 43:681

Harvey R (1967) Air-spora studies at Cardiff. 1. *Cladosporium*. Trans Br Mycol Soc 50:479–495

Hasija SK (1970) Physiological studies of *Alternaria citri* and *A. tenuis*. Mycologia 62:289–295

Häubner N, Schumann R, Karsten U (2006) Aeroterrestrial microalgae growing in biofilms on facades – response to temperature and water stress. Microb Ecol 51:285–293

Hauck M (2003) Epiphytic lichen diversity and forest dieback: the role of chemical site factors. Bryologist 106:257–269

Hauck M (2005) Epiphytic lichen diversity on dead and dying conifers under different levels of atmospheric pollution. Environ Pollut 135:111–119

Hauck M, Hesse V, Runge M (2002) The significance of stemflow chemistry for epiphytic lichen diversity in a dieback-affected spruce forest on Mt. Brocken, northern Germany. Lichenologist 34:415–427

Hauck M, Jung R, Runge M (2001) Relevance of element content of bark for the distribution of epiphytic lichen in a montane spruce forest affected by forest dieback. Environ Pollut 112:221–227

Hawkes CV, Flechtner VR (2002) Biological soil crusts in a xeric florida shrubland: composition, abundance, and spatial heterogeneity of crusts with different disturbance histories. Microb Ecol 43:1–12

Hawksworth DL (1976) The natural history of Slapton Ley Natural Reserve. 10. Fungi. Fld Stud 4:391–439

Hawksworth DL, Hill DJ (1984) The Lichen-forming Fungi. Blackie, Glasgow

Hawksworth DL, Kirk BC, Sutton BC, Pegler DN (1995) Ainsworth & Bisby's dictionary of fungi, 8th edn. CAB International, Wallingford

Hawksworth DL, Rose F, Coppins BJ (1973) Changes in the lichen flora of England and Wales attributable to pollution of the air by sulfur dioxide. In: Ferry BW, Baddeley MS, Hawksworth DL (eds) Air pollution and lichens. Athlone Press, London, pp 330–367

Hazen TE (1899) The life history of *Sphaerella lacustris* (*Haematococcus pluvialis*). Mem Torrey Bot Cl 6(3):211–247

Hegewald E, Hanagata N (2000) Phylogenetic studies on Scenedesmaceae (Chlorophyta). Algol Stud 100:29–49

Hegewald E, Hanagata N (2002) Validation of the new combinations of *Coelastrella* and *Neodesmus* and the description of the new subfamily Desmodesmoideae of the Scenedesmaceae (Chlorophyta). Algol Stud 105:7–9

Hegewald E, Silva PC (1988) Annotated catalogue of *Scenedesmus* and nomenclaturally related genera, including original descriptions and figures. Bibliotheca Phycologica Band 80. J Cramer, Berlin, Stuttgart, 587 pp

Heintzeler I (1939) Das Wachstum der Schimmelpilze in Abhängigkeit von den Hydraturverhältnissen unter verschiedenen Außenbedingungen. Arch Mikrobiol 10:92–132

Henley WJ, Hironaka JL, Guillou L, Buchheim MA, Buchheim JA, Fawley MW, Fawley KP (2004) Phylogenetic analysis of the „*Nannochloris*-like" algae and diagnoses of *Picochlorum oklahomensis* gen. et spec. nov. (Trebouxiophyceae, Chlorophyta). Phycologia 43(6):641–652

Hennebert GL, Gilles GL (1958) Epidemiologie de *Botrytis cinerea* Pers. sur les fraisiers. Meded Landbhogesch Opzoekstns Gent 23:864–888

Herdman M, Castenholz RW, Iteman I, Rippka R (2001a) Form-genus VIII. *Gloeocapsa* Kützing 1843. In: Boone DR, Castenholz RW, Garrity GM (eds) Bergey's Manual® of systematic bacteriology, vol One. The archaea and the deeply branching phototrophic bacteria. Springer, New York, Berlin, Heidelberg, pp 503–504

Herdman M, Castenholz RW, Iteman I, Waterbury JB, Rippka R (2001b) Subsection I. (Formerly Chroococcales Wettstein 1924, emend. Rippka, Deruelles, Waterbury, Herdman and Stanier 1979). In: Boone DR, Castenholz RW, Garrity GM (eds) Bergey's Manual® of systematic bacteriology, vol One. The archaea and the deeply branching phototrophic bacteria. Springer, New York, Berlin, Heidelberg, pp 493–495

Herdman M, Castenholz RW, Rippka R (2001c) Form-genus VIII. *Nostoc* Vaucher 1803. In: Boone DR, Castenholz RW, Garrity GM (eds) Bergey's Manual® of systematic bacteriology, vol One. The archaea and the deeply branching phototrophic bacteria. Springer, New York, Berlin, Heidelberg, pp 575–580

Hershkovitz N, Oren A, Cohen Y (1991) Accumulation of trehalose and sucrose in cyanobacteria exposed to matric water stress. Appl Environ Microbiol 57(3):645–648

Hesseltine CW (1965) A millenium of fungi, food, and fermentation. Mycologia 57:149–197

Hesseltine CW, Ellis JJ (1973) Mucorales. In: Ainsworth GG, Sparrow FK, Sussman AF (eds) The Fungi IVb. Academic, New York, pp 187–217

Hibberd DJ, Leedale GF (1970) Eustigmatophyceae: a new algal class with unique organization of the motile cell. Nature 225:758–760

Hibberd DJ, Leedale GF (1971) A new algal class: the Eustigmatophyceae. Taxon 20:523–525

Hibberd DJ, Leedale GF (1972) Observations on the cytology and ultrastructure of the new algal class Eustigmatophyceae. Ann Bot 36:49–71

Hill DR, Hladun SL, Scherer S, Potts M (1994) Water stress proteins of *Nostoc commune* (Cyanobacteria) are secreted with UV-A/B-absorbing pigments and associate with 1,2-β-D-Xylanxylanohdrolase activity. J Biol Chem 269(10):7726–7734

Hindák F (1978) The genus *Gloeocystis* (Chlorococcales, Chlorophyceae). Preslia, Praha 50:3–11

Hindák F (2008) Colour atlas of cyanobacteria. Veda, Bratislava, p 253

Hirschberg J, Ohad N, Pecker I, Rahad A (1987) Isolation and characterization of herbicide resistant mutants in the cyanobacterium *Synechococcus* R2. Z Naturforsch 42C:758–761

Hirst JM, Stedman OJ, Hogg WH (1967) Long distance spore transport: methods of measurement, vertical spore profiles and the detection of imigrant spores. J Gen Microbiol 48:329–355

Hladik M (2003) Was trocken bleibt, bleibt algenfrei! In Venzmer (Hrsg) Altbauinstandsetzung 5/6. Algen an Fassadenbaustoffen II. 4. Dahlberg-Kolloquium Mai 2003. Sonderheft Dahlberg Kolloquium, Verlag Bauwesen, S 95–106

Ho KK, Tan KH, Wee YC (1983) Growth conditions of *Trentepohlia odorata* (Chlorophyta, Ulotrichales). Phycologia 22(3):303–308

Hocking AD, Pitt JI (1980) Dichloran-glycerol medium for enumeration of xerophilic fungi from low moisture foods. Appl Environ Microbiol 39:488–492

Hodges P, Ronaldson JW, Taylor A, White EP (1963) Sporidesmin and sporidesmin-B. Chem Ind 1963:42–43

van den Hoek C, Mann DG, Jahns HM (1995) Algae. An introduction to phycology. Cambridge University Press, Cambridge

Hoekstra ES, Samson RA, Summerbell RC (2002) Methods for the detection and isolation of fungi in the indoor environments. In: Samson RA, Hoekstra ES, Frisvad JC, Filtenborg O (eds) Introduction to food- and airborne fungi, 6th edn. Centraalbureau voor Schimmelcultures, Utrecht, pp 298–305

Hofbauer W (2005) Biologische Analyse von Aufwuchs an Fassaden. Bioskop 8(3):12–15

Hofbauer W (2007) Aerophytische Organismen an Bauteiloberflächen. Thesis. Leopold-Franzens Universität, Innsbruck, 436 pp

Hofbauer WK (2015) 9. Class Eustigmatophyceae D.J. Hibberd & Leedale. In: Frey W (ed.) Syllabus of Plant Families. Adolf Engler's Syllabus der Pflanzenfamilien. 13th Edition. Part 2/1. Photoautotrophic eukaryotic Algae. Glaucocystophyta, Cryptophyta, Dinophyta/Dinozoa, Haptophyta, Heterokontophyta/Ochrophyta, Chlorarachniophyta/Cercozoa, Euglenophyta/Euglenozoa, Chlorophyta, Streptophyta p.p.. Borntraeger Verlagsbuchhandlung, Stuttgart: pp 109–117

Hofbauer W, Breuer K, Fitz C, Sedlbauer K (2006) Mikrobielles Wachstum auf Fassaden – Biologische Zusammenhänge und Grundlagen. In Venzmer H (Hrsg) Fassadenbiofilme. Sonderheft Dahlberg-Kolloquium, S 143–154

Hofbauer W, Breuer K, Gärtner G (2005a) Pioneer aerophytic microorganisms on outer building surfaces – a comprehensive approach. XVII International Botanical Congress. Vienna, Austria, Europe. Austria Center Vienna. 17–23 July 2005. Abstracts: 118

Hofbauer W, Breuer K, Krueger N, Sedlbauer K (2005b) Toxic mould versus façade-jungle – a comparison of undesirable biological growth on indoor surfaces and outer building coatings. Proceedings of the 10th international conference on indoor air quality and climate. Volume II(2). Indoor Air 2005 September 4–9 2005, Beijing, China, S 2450–2454

Hofbauer W, Breuer K, Sedlbauer K (2003) Algen, Flechten, Moose und Farne auf Fassaden. Bauphysik 25(6):383–396

Hofbauer W, Breuer K, Tschaikner A, Krus M, Sedlbauer K, Schoch T (2005b) VergleichendeUntersuchungen zum Schimmelpilzwachstum auf verschiedenen Baustoffoberfl. chen – materialspezifische Wachstumsisoplethen. IBP-Mitteilung 32:457

Hofbauer W, Fitz C, Krus M, Sedlbauer K, Breuer K (2006) Prognoseverfahren zum biologischen Befall durch Algen, Pilze und Flechten an Bauteiloberflächen auf der Basis bauphysikalischer und mikrobieller Untersuchungen. Bauforschung für die Praxis Band 77. Fraunhofer IRB Verlag, Stuttgart, 304 p

Hofbauer W, Gärtner G, Breuer K (2006c) Zwei neue Nachweise von *Porphyridium purpureum* (Bory) Ross in Drews & Ross 1965 (Rhodophyta, Porphyridiaceae) in Nordtirol und Bayern und Bemerkungen zu ihrer Kultur. Ber. nat.-med. Verein Innsbruck 93:31–38

Hofbauer W, Forrest LL, Hollingsworth ML, Rennebarth T, Breuer K (2014) Unexpected diversity in mosses on walls of modern buildings. IBP-Mitteilung **41**: 532. (Englische Ausgabe)

Hofbauer W, Gärtner G, Rennebarth T, Sedlbauer K, Mayer F, Breuer K (2011) Excentrochloris fraunhoferiana sp. nov. (Botrydiopsidaceae, Xanthophyceae), a new aerophytic species from the surfaces of modern buildings. Fottea 11(2): 279–291

Hofbauer WK, Forrest LL, Hollingsworth PM, Hart ML (2016) Preliminary insights from DNA barcoding into the diversity of mosses colonising modern building surfaces. Bry Div Evo 38(1):1–22

Hoffmann K, Pawłowska J, Walther G, Wrzosek M, de Hoog GS et al (2013) The family structure of the Mucorales: a synoptic revision based on comprehensive multigene-genealogies. Persoonia 30:57–76

Hoffmann L (1986) Cyanophycees aeriennes et subaeriennes du Grand-Duche de luxembourg. Bull du Jardin botanique national de Belgique 56(1/2):77–127

Hoffmann L, Ector L, Kostikov I (2007) Algal flora from limed and unlimed forest soils in the Ardenne (Belgium). Syst Geogr Pl 77:15–90

Hogg B (1966) Micro-fungi on leafes of *Fagus sylvatica*. 2. Duration of survival, spore viability and cellulolytic activity. Trans Br Mycol Soc 49:193–204

Holmes PE (1986) Bacterial enhancement of vinyl fouling by algae. Appl Environ Microbiol 52(6):1391–1393

Holz I (2000) *Schistidium* Bruch & Schimp. Spalthütchen. In: Nebel M, Philippi G (eds) Die Moose Baden-Württembergs. Band 1: Allgemeiner Teil. Spezieller Teil (Bryophytina I, Andreaeales bis Funariales). Ulmer, Stuttgart, pp 379–398

Hollingsworth PM, Forrest LL, Spouge JL, Hajibabaei M, Ratnasingham S, van der Bank M, Chase MW, Cowan RS, Erickson DL, Fazekas AJ, Graham SW, James KE, Kim K-J, Kress WJ, Schneider H, van AlphenStahl J, Barrett SCH, van den Berg C, Bogarin D, Burgess KS, Cameron KM, Carine M, Chacón J, Clark A, Clarkson JJ, Conrad F, Devey DS, Ford CS, Hedderson TAJ, Hollingsworth ML, Husband BC, Kelly LJ, Kesanakurti PR, Kim JS, Kim Y-D, Lahaye R, Lee H-L, Long DG, Madriñán S, Maurin O, Meusnier I, Newmaster SG, Park C-W, Percy DM, Petersen G, Richardson JE, Salazar GA, Savolainen V, Seberg O, Wilkinson MJ, Yi D-K, Little DP (2009) A DNA barcode for land plants. PNAS 106(31):12794–12797

Holzinger A, Roleda MY, Lütz C (2009) The vegetative arctic freshwater green alga Zygnema is insensitive to experimental UV exposure. Micron 40:831–838

Hong S-B, Go S-J, Shin H-D, Frisvad JC, Samson RA (2005) Polyphasic taxonomy of *Aspergillus fumigatus* and related species. Mycologia 97(6):1316–1329

Hong S-B, Shin H-D, Hong J, Frisvad JC, Nielsen PV, Varga J, Samson RA (2007) New taxa of *Neosartorya* and *Aspergillus* section *Fumigati*. Antonie van Leeuwenhoek 93(1–2):87–98. https://doi.org/10.1007/s10482-007-9183-1

de Hoog GS, Guarro J, Gené J, Figueras HJ (2000) Atlas of clinical fungi. 2nd ed. Centraalbureau voor Schimmelcultures, Utrecht/Universitat Rovira I Virgili, Reus, 1126 pp

Houbraken J, Samson RA (2011) Phylogeny of *Penicillium* and the segregation of Trichocomaceae into three families. Stud Mycol 70:1–51

Hoppert M, König S, Hegermann J (2005) Mikroalgen auf Oberflächen von Baumaterialien. Z dt Ges Geowiss 156(1):93–101

Howland LJ (1929) The moisture relations of terrestrial algae. IV. Periodic observations of *Trentepohlia aurea* Martius. Ann Bot 43:173–202

Hrouzek P, Ventura S, Lukešová A, Mugnai MA, Turicchia S, Komárek J (2005) Diversity of soil *Nostoc* strains: phylogenetic and phenotypic variability. Algol Stud 117:251–264

Huber R (1993) Hyperthermophile Lebensgemeinschaften in Vulkangebieten. In: Hausmann K, Kremer BP (eds) Extremophile: Mikroorganismen in ausgefallenen Lebensräumen. VCH, Weinheim, New York, Basel, Cambridge, Tokyo, pp 49–67

Hudson HJ (1969) Aspergilli in the air-spora at Cambridge. Trans Br Mycol Soc 52:153–159

Hughes SJ, Dickinson CH (1968) New Zealand fungi. 11. Gliomastix Guéguen. NZ J Bot 6:106–114

Huss VAR, Sogin ML (1990) Phylogenetic position of some *Chlorella* species within the Chlorococcales based upon complete small-subunite ribosomal RNA sequences. J Mol Evol 31:432–442

Inderbitzin P, Gareth-Jones EB, Vrijmoed LP (2000) A new species of *Leptosphaerulina* from decaying mangrove wood from Hong Kong. Mycoscience 41:233–237

Infante F, Castro A, Dominguez E, Guárdia A, Méndez J, Sabariego S, Vega A (1999) A comparative study of the incidence of *Cladosporium* conidia in the atmosphere of five spanish cities. Polen 10:15–23

Ingold CT (1971) Fungal spores: Their liberation and dispersal. Clarendon Press, Oxford, p 302

Ingolić E, Gärtner G (1989) Zur Ultrastruktur und Taxonomie der Grünalge *Diplosphaera* Bialosuknia (Chlorophyta, Chaetophorales, Leptosiroideae). Dreiländer-Tagung Elektronenmikrosk. Salzburg 1989, Abstracts, p 59

Inui T, Takeda Y, Iizuka H (1965) Taxonomical studies on the genus *Rhizopus*. J Gen Appl Microbiol 11(Supp):121

Ionita I (1973) Contributions to the study of the biodeterioration of the works of art and historical monuments. 4. Fungi involved in the deterioration of mural painting from the monasteries of Moldavia. Revue Roum Biol Ser Bot 18:179–189

Irwin JAG, Davis RD (1985) Taxonomy of some *Leptosphaerulina* spp. on legumes in Eastern Australia. Australian Journal of Botany 33:233–237

ISO 2810 (2004): Paints and varnishes – Natural weathering of coatings – Exposure and assessment.

Izaguirre I, Pizarro H (1998) Epilithic algae in a glacial stream at Hope Bay (Antarctica). Polar Biol 19:24–31

Jaag O (1945) Untersuchungen über die Vegetation und Biologie der Algen des nackten Gesteins in den Alpen, im Jura und im schweizerischen Mittelland. Beiträge zur Kryptogamenflora der Schweiz. Band **9**(3): 560 pp

Jacob A, Wiencke C, Lehmann H, Kirst GO (1992) Physiology and ultrastructure of desiccation in the green alga *Prasiola crispa* from Antarctica. Bot mar 34(4):297–303

Jahnke J (2001) Algenbiofilme auf Bodenoberflächen: Strukturanalyse an Paraffinschnitten. Mikrokosmos 90(3):149–156

James EJ (1935) An investigation of the algal growth in some naturally occuring soils. Beih Bot Centralbl 53/A:519–553

Jansson C, Debus RJ, Osiewaez HD, Gurevitz M, McIntosh L (1987) Construction of an obligate photoheterotrophic mutant of the cyanobacterium *Synechocystis* 6803. Plant Physiol 85:1021–1025

Jarvis WR (1977) *Botryotinia* and *Botrytis* species. Taxonomy, physiology, and pathogenicity. Res Stn Can Dep Agric Harrow Monogr **15**: 195 pp

Jensen PA, Schafer MP (1998) Sampling and characterization of bioaerosols. In: National Institute for Occupational Safety and Health (ed) NIOSH manual of analytical methods 4th ed., 2nd Supplement Publication 98–119: 82–112

Joffe AZ (1962) Biological properties of some toxic fungi isolated from overwintered cereals. Mycopath Mycol Appl 16:201–221

Joffe AZ (1974) A modern system of *Fusarium* taxonomy. Mycopath Mycol Appl 53:201–228

Johansen DA (1940) Plant microtechnique. McGraw-Hill Book Company, New York, Toronto, London

John DM, Whitton BA, Brook AJ (2011) The freshwater algal flora of the British isles. An identification guide to the freshwater and terrestrial algae, 2nd edn. Cambridge University Press, Cambridge, p 896

John RP (1942) An ecological and taxonomic study of the algae of British soils. I. The distribution of the surface-growing algae. Ann Bot 6(22):323–349

Johnson A (1962) A short note on some soil algae from New Guinea. Garden Bull Singapore 19:375–377

Johnson LF, Osborne TS (1963) Survival of fungi in soil exposed to gamma radiation. Can J Bot 42:105–113

Johnson LR (2002) Phylum Xanthophyta (Yellow-Green Algae). In: John DM, Whitton BA, Brook AJ (eds) The freshwater algal flora of the british isles. An indentification guide to freshwater and terrestrial algae. Cambridge University Press, Cambridge, pp 245–261

Johnston PR (1981) *Phoma* on New Zealand grasses and pasture legumes. NZ J Bot 19:173–186

Jones D (1988) Lichens and pedogenesis. In: Galun M (ed) CRC handbook of lichenology, vol III. CRC Press. Boca Raton, Florida, pp 109–124

Jong SC, Davis EE (1976) Contributons to the knowledge of *Stachybotrys* and *Memnoniella* in culture. Mycotaxon 3:409–485

Jozsa PG, Gehrke T, Sand W, Zapel K (1994) Mikrobielle Werkstoffzerstörung – Simulation, Schadensfälle und Gegenmaßnahmen für anorganische nichtmetallische Werkstoffe: Mikrobiologisch beeinflußte Schädigung von Ziegelmauerwerk in Abwasserleitungen. Werkst Korros 45:119–121

Jurjevic Z, Peterson SW, Horn BW (2012) *Aspergillus* section *Versicolores*: nine new species and multilocus DNA sequence based phylogeny. IMA Fungus 3(1):59–79

Kärnefelt I (1989) Morphology and phylogeny in the Teloschistales. Crypt Bot 1:147–203

Kakizono T, Kobayashi M, Nagai S (1992) Effect of carbon/nitrogen ratio on encystment accompanied with astaxanthin formation in a green alga, *Haematococcus pluvialis*. J Ferm Bioeng 74:403–405

Kalina T, Punčochárová M (1987) Taxonomy of the subfamily Scotiellocystoideae Fott 1976 (Chlorellaceae, Chlorophyceae). Arch Hydrobiol Suppl **73**(4) (Algological Studies **45**): 473–521

Kalina T (1996) *Glaphyrella*, a new generic name, proposed for *Chlorella saccharophila* group (Chlorophyceae). 1st Europ Phycol Congr Cologne 1996 Abstr, Köln Abstr Nr 246

Kanevskaya IG (1966) Decomposition of methylcellulose by soil fungi (in Russian). Mikrobiologiya 35:868–870

Kantz TS, Theriot EC, Zimmer EA, Chapman RL (1990) The Pleurastrophyceae and Micromonadophyceae: a cladistic analysis of nuclear rRNA sequence data. J Phycol 26:711–721

Kappen L (1993) Terrestrische Mikroalgen und Flechten in der Antarktis. In: Hausmann K, Kremer BP (eds) Extremophile: Mikroorganismen in ausgefallenen Lebensräumen. VCH, Weinheim, New York, Basel, Cambridge, Tokyo, pp 3–25

Karlson B, Potter D, Kuylenstierna M, Andersen RA (1996) Ultrastructure, pigment composition, and 18S rRNA gene sequence for *Nannochloropsis granulata* sp. nov. (Monodopsidaceae, Eustigmatophyceae), a marine ultraplankter isolated from Skagerrak, northeast Atlantic Ocean. Phycologia 35:253–260

Karsten U, Friedl T, Schuhmann R, Hoyer K, Lembke S (2005a) Mycosporine-like amino acids and phylogenies in green algae: *Prasiola* and its relatives from the Trebouxiophyceae (Chlorophyta). J Phycol 41:557–566

Karsten U, Herburger K, Holzinger A (2017) Photosynthetic plasticity in the green algal species *Klebsormidium flaccidum* (Streptophyta) from a terrestrial and a freshwater habitat. Phycologia 56:213–220

Karsten U, Klimant I, Holst G (1996) A new in vivo fluorimetric technique to measure growth of adhering phototrophic microorganisms. Appl Environ Microbiol 62(1):237–243

Karsten U, Schuhmann R, Häubner N, Friedl T (2005b) Aeroterrestrische Mikroalgen – Lebensraum Fassade. Biologie in unserer Zeit 35:20–30

Karsten U, Schumann R, Mostaert AS (2007) Aeroterrestrial algae growing on man-made surfaces: what are the secrets of their ecological success? In: Seckbach J (ed) Algae and cyanobacteria in extreme environments, vol 11. Cellular origin, life in extreme habitats and astrobiology. Springer, Dordrecht, pp 585–597

Katz ME, Dougall AM, Weeks K, Cheetham BF (2005) Multiple genetically distinct groups revealed among clinical isolates identified as atypical *Aspergillus fumigatus*. J Clin Microbiol 43(2):551–555

Kauffmann J (1953) Rôle des bactéries nitrificantes dans l'altération des pierres calcaires des monuments. Corros Anticorros 1:33–41

Kaufman DD, Blake J (1970) Degradation of atrazine by soil fungi. Soil Biol Biochem 2:297–308

Kaufman DD, Kearney PC, Sheets TJ (1963) Simazine degradation by soil microorganisms. Science NY 142:405–406

Kearny PC, Kaufman DD, Sheets TJ (1965) Metabolites of simazine by *Aspergillus fumigatus*. J Agric Fd Chem 13:369–372

Kendrick WB, Carmichael JW (1973) 10. Hyphomycetes. In: Ainsworth GC, Sparrow Fk, Sussman AS (eds) The fungi. An advanced treatise, vol IVA. A taxonomic review with keys: ascomycetes and fungi imperfecti. Academic Press, New York, London, pp 323–512

Kessler E (1965) Physiologische und biochemische Beiträge zur Taxonomie der Gattung *Chlorella*. I. Säureresistenz als taxonomisches Merkmal. Arch Mikrobiol 52:291–296

Kessler E (1967) Physiologische und biochemische Beiträge zur Taxonomie der Gattung *Chlorella*. III: Merkmale von 8 autotrophen Arten. Arch Mikrobiol 55:346–357

Kessler E (1972) Physiologische und biochemische Beiträge zur Taxonomie der Gattung *Chlorella*. VII. Die Thermophilie von *Chlorella vulgaris* f. *tertia* Fott et Nováková. Arch Mikrobiol 87:243–248

Kessler E (1974) Physiologische und biochemische Beiträge zur Taxonomie der Gattung *Chlorella*. IX: Salzresistenz als taxonomisches Merkmal. Arch Mikrobiol 100:51–56

Kessler E (1985) Upper limits of temperature for growth in *Chlorella* (Chlorophyceae). Pl Syst Evol 151:67–71

Kessler E (1992) *Chlorella*. Biochemische Taxonomie einer für Forschung und Biotechnologie wichtigen Gattung einzelliger Grünalgen. Naturwissenschaften 79:260–265

Kessler E, Huss VAR, Rahat M (1988) Species-specific ability of *Chlorella* strains (Chlorophyceae) to form stable symbioses with *Hydra viridis*. Pl Syst Evol 160:241–246

Kessler E, Kramer H (1960) Physiologische Untersuchungen an einer ungewöhnlich säureresistenten *Chlorella*. Arch Mikrobiol 37:245–255

Khan SR, Talbot PHB (1975) Monosporous sporangiola in *Mycotypha* and *Cunninghamella*. Trans Br Mycol Soc 65:29–39

Khaybullina LS, Gaysina LA, Johansen JR, KrautováÁ M (2010) Examination of the terrestrial algae of the Great Smoky Mountains National Park, USA. Fottea 10:201–215

Khobragade CN, Rao RS, Borkar PS, Yangade RS (2006) Microbially induced impact on physico-chemical properties of porous lime stones: a case from Kandahar fort. Curr Sci 91(19):1318–1320

Kiel G, Gaylarde C (2006) Bacterial diversity in biofilms on external surfaces of historic buildings in Porto Allegre. World J Microbiol Biotechnol 22(3):293–297

King AD, Hocking AD, Pitt JI (1979) Dichloran-rose Bengal medium for enumeration of moulds from foods. Appl Environ Microbiol 37:959–964

King AD, Pitt JI, Beuchat LR, Corry JEL (1986) Methods for the mycological examination of Food. Plenum Press, New York, London

Kirk PM, Cannon PF, David JC, Stalpers JA (eds) (2001) Ainsworth and Bisby's dictionary of the fungi, 9th edn. CABI Bioscience, Egham, p 624

Klich MA (2002) Identification of Common *Aspergillus* Species. Centraalbureau voor Schimmelcultures, Utrecht, p 116

Klich MA, Pitt JI (1988) A laboratory guide to the common *Aspergillus* species and their teleomorphs. Commonwealth Scientific and Industrial Research organisation, Division Food Processing, 116 pp

Klinkhammer F (1997) Vom Irrationalismus des "Pilzwahns". Deutsches Ärtzteblatt 94(3):22–23

Kłyszejko A, Kubus Ż, Żakowska Z (2005) Mycological analysis of cereal samples and screening of *Fusarium* strains' ability to form Deoxynivalenole (DON) and Zearalenone (ZEA) Mycotoxins – a pilot study. Pol J Microbiol 54:21–25

Knebel G (1935) Monographie der Algenreihe der Prasiolales, insbesondere von *Prasiola crispa*. Inaugural-Dissertation, Friedrich-Wilhelms-Universität, Berlin, 120 pp

Koenig F (2001) Eucaryotic algae, cyanobacteria and pesticides. In: Rai LC, Gaur JP, Hindu B (eds) Algal adaption to environmental stresses. Springer, Berlin, pp 389–406

Kogej T, Gostinčar C, Volkmann M, Gorbushina AA, Gunde-Cimerman N (2006) Mycosporines in Extremophilic Fungi – Novel Complementary Osmolytes? Environ Chem 3:105–110

Koike ST, Henderson DM, Butler EE (2001) Leaf spot disease of spinach in California caused by *Stemphylium botryosum*. Plant Dis 85(2):126–130

Kok B (1956) On the inhibition of photosynthesis by intense light. Biochim Biophys Acta 21:234–244

Komárek J (1976) Taxonomic reviews of the genera *Synechocystis* Sauv. 1892; *Synechococcus* Näg., 1949, and *Cyanothece* gen. nov. (Cyanophyceae). Arch Protistenkd 118:119–179

Komárek J (1994) Current trends and species delimination in the cyanoprocaryote taxonomy. Arch Hydrobiol/Algol Stud 75:11–29

Komárek J (2013) Cyanoprokaryota. In: Büdel B, Gärtner G, Krienitz L, Schagerl M (eds) Süßwasserflora v. Mitteleuropa, 19/3. Springer Spektrum, Heidelberg, p 1130

Komárek J, Anagnostidis K (1986) Modern approach to the classification system of cyanophytes. 2 – Chroococcales. Arch Hydrobiol/Algol Stud 43:157–226

Komárek J, Anagnostidis K (1989) Modern approach to the classification system of cyanophytes. 4 – Nostocales. Arch Hydrobiol/Algol Stud 56:247–345

Komárek J, Anagnostidis K (1995) Nomenclatural novelties in chlorococcalean cyanoprocaryotes. Preslia 67:15–23

Komárek J, Anagnostidis K (1998) Cyanoprokaryota 1. Teil: Chroococcales. In: Ettl H, Gärtner G, Heynig H, Mollenhauer D (eds) Süsswasserflora von Mitteleuropa 19/1. Gustav Fischer, Jena, Stuttgart, Lübeck, Ulm, p 548

Komárek J, Anagnostidis K (2005) Cyanoprokaryota 2. Teil: Oscillatoriales. In: Büdel B, Gärtner G, Krienitz L, Schagerl M (eds) Süsswasserflora von Mitteleuropa 19/2. Elsevier, München, p 759

Komárek J, Cepák V (1998) Cytomorphological characters supporting the taxonomic validity of *Cyanothece* (*Cyanoprokaryota*). Pl Syst Evol 210:25–39

Komárek J, Fott B (1983) Chlorophyceae (Grünalgen). Ordnung Chlorococcales. In: Thienemann A, Huber-Pestalozzi G (ed) Das Phytoplankton des Süßwassers, die Binnengewässer XVI, Bd. 7(1). E. Schweizerbart: Stuttgart, pp 1–1044

Kommission Reinhaltung der Luft (KRdL) im VDI und DIN - Normenausschuß (Hrsg) (2005) VDI 3957 Blatt 13: 2005-12. Biologische Messverfahren zur Ermittlung und Beurteilung der Wirkung von Luftverunreinigungen mit Flechten (Bioindikation) – Kartierung der Diversität epiphytischer Flechten als Indikator für Luftgüte. Verband Deutscher Ingenieure, 27 pp

Koops HP, Möller UC (1992) The lithotrophic ammonia-oxidizing bacteria. In: Balows A, Trüper HG, Dworkin M, Harder W, Schleifer KH (eds) The prokaryotes, Vollume III, Chapter 137, 2nd edn. Springer, New York, pp 2625–2637

Kornmann P, Sahling P-H (1974) Prasiolales (Chlorophyta) von Helgoland. Helgoländer wiss Meeresunters 26:99–133

Kostikov IY, Romanenko GO, Demchenko EM, Darienko TM, Mikhailyuk TI, Ribchinskii OV, and Solonenko AM (2001) Algae of Ukrainean Soils (History and Methods of Investigation, System, Compendium of the Flora), Kiev: Fitosotsiotsentr

Kostikov IJu, Romanenko PO, Demchenko EM, Darienko TM, Mikhailyuk TI, Ribshinskij OB, Solonenko AM (2001) The algae of the ukrainan soils. Center of Phytosociobiology, Kiew, p 310

Kostikov I, Darienko T, Lukešová A, Hoffmann L (2002) Revision of the classification system of the Radiococcaceae Fott ex Komárek (except the subfamily Dictyochlorellaceae) (Chlorophyta). Algol Stud 104:23–58

Kouřil R, Ilík P, Tomek P, Nauš J, Poulíčková A (2001) Chlorophyll fluorescence temperature curve on *Klebsormidium flaccidum* cultivated at different temperature regimes. J Plant Physiol 158:1131–1136

Kouyeas V (1964) An approach to the study of moisture relations of soil fungi. Pl Soil 20:351–363

Kováčik L (2000) Cyanobacteria and algae as agents of biodeterioration of stone substrata of historical buildings and other cultural monuments. In: Choi S, Suh M (eds) Proceedings of the New Millenium International Forum on Conservation of Cultural Property, December 5–8, Daejon, Korea, pp 44–58

Kováčik L, Pereira AB (2001) Green alga *Prasiola crispa* and its lichenized form *Mastodia tessellata* in Antarctic environment: general aspects. Nova Hedwigia, Beiheft 123:465–478

Kozlovsky A, Zhelifonova V, Antipova T (2005) The fungus *Penicillium citrinum*, Isolated from permafrost sediments, as a producer of ergot alkaloids and new quinoline alkaloids quinocitrines. Appl Biochem Microbiol 41(5):499–502

Krammer K, Lange-Bertalot H (1988a) Bacillariophyceae. Teil 2: Bacillariaceae, Epithemiaceae, Surirellaceae. In: Ettl H, Gerloff J, Heynig H, Mollenhauer D (eds) Süßwasserflora von Mitteleuropa, vol 2/2. G. Fischer, Stuttgart, New York, p 596

Krammer K, Lange-Bertalot H (1988b) Bacillariophyceae. Teil 3: Centrales, Fragilariaceae, Eunotiaceae. In: Ettl H, Gerloff J, Heynig H, Mollenhauer D (eds) Süßwasserflora von Mitteleuropa, vol 2/3. G. Fischer, Stuttgart, New York, p 576

Kranz J (1962) Vergleichende Untersuchungen an *Phoma*-Isolierungen von der Kartoffel. Sydowia 16:1–40

Krause-Kupsch T (1993) Entwicklung einer Schnellmethode zur Identifizierung und Klassifizierung nitritoxidierender Bakterien. Dissertation, Universität Hamburg.

Kremer BP, Muhle H (1991) Flechten, Moose, Farne. Europäische Arten. Herausgegeben von Gunter Steinbach. Mosaik Verlag, München

Krieg NR (2001) Procaryotic domains. In: Boone DR, Castenholz RW, Garrity GM (eds) Bergey's Manual® of systematic bacteriology, vol One. The archaea and the deeply branching phototrophic bacteria. Springer, New York, Berlin, Heidelberg, pp 21–25

Krienitz L, Bock C, Dadheech PK, Pröschold T (2011) Taxnomoc reassessment of the genus Mychonastes (Chlorophyceae, Chlorophyta) including the description of eight new species. Phycologia 50(1):89–106

Krienitz L, Hegewald E, Hepperle D, Wolf A (2003) The systematics of coccoid green algae: 18S rRNA gene sequence data versus morphology. Biol Bratisl 58:437–446

Krienitz L, Hegewald EH, Hepperle D, Huss VAR, Rohr T, Wolf M (2004) Phylogenetic relationship of *Chlorella* and *Parachlorella* gen. nov. (Chlorophyta, Trebouxiophyceae). Phycologia 43:529–542

Kristiansen J (1996) Dispersal of freshwater algae – a review. In: Kristiansen J (ed) Biogeography of freshwater algae. Developments in hydrobiology **118**. Kluver Academic Publishers: Dordrecht (Reprint from Hydrobiologia **336**), pp 151–157.

Krueger N (2020) Leistungsfähigkeit von biozid ausgerüsteten Außenbeschichtungen im mehrjährigen Freilandversuch begleitet von Laboruntersuchungen. Thesis, TUM School of Life Sciences, Munich, pp 240

Krueger, N., Hofbauer, W., Krus, M., Fitz, C., Mayer, F., Melzer, A., Breuer, K. (2013): Effectiveness and durability of biocides in building coatings: Biological aspects. In: Freitas, V. Peixoto (Ed.): Hygrothermal Behavior, Building Pathology and Durability. Springer, Berlin: 45–59

Krumbein WE (1968) Zur Frage der biologischen Verwitterung: Einfluß der Mikroflora auf die Bausteinverwitterung und ihre Abhängigkeit von edaphischen Faktoren. Zeitschrift f Allg Mikrobiol 8(2):107–117

Krumbein WE, Jens K (1981) Biogenic rock varnishes of the negev desert (Israel) an ecological study of iron and manganese transformation by cyanobacteria and fungi. Oecologia 50:25–38

Krumbein WE, Villbrandt M (1993) Biofilme und Mikrobenmatten extremer Lebensräume. In: Hausmann K, Kremer BP (eds) Extremophile: Mikroorganismen in ausgefallenen Lebensräumen. VCH, Weinheim, New York, Basel, Cambridge, Tokyo, pp 89–112

Krysińska-Traczyk E, Dutkiewicz J (2000) *Aspergillus candidus*: a respiratory hazard associated with grain dust. A Agric Environ Med 7:101–109

Kuehn HH, Gunderson MF (1963) Psychrophilic and mesophilic fungi in frozen food products. Appl Microbiol 11:352–356

Kulik MM, Hanlin RT (1968) Osmophilic strains of some *Aspergillus* species. Mycclogia 60:961–964

Künzel HM (2000) Algenbewuchs an Fassaden: Eine Folge reiner Luft! Arconis 5(3):20–22

Künzel HM, Sedlbauer K (2001) Algen auf Wärmedämmverbundsystemen. IBP-Mitteilungen 28, Nr. 382

Künzel HM, Krus M, Sedlbauer K (2001) Algen auf Außenwänden – Bauphysik als Ursache? Bauphysik als Lösung! Tagungsband zum 3. Dahlberg-Kolloquium Mikroorganismen und Bauwerksinstandsetzung, 13.–14. Sep. 2001, Wismar, S 75–84

Künzel HM, Krus M, Fitz C, Hofbauer WK, Scherer C, Breuer K (2011) Accelerated test procedure to assess the microbial growth resistance of exterior finishes. XII DBMC international conference on durability of building materials and components, Porto, April 12th–15th, 2011, Proceedings, pp 1–8

Labuda R, Tancinova D, Hudec K (2003) Identification and enumeration of *Fusaria* in poultry feed mixtures from Slovakia. Ann Agric Environ Med 10:61–66

Lacey ME, Gregory PH (1962) Occurrence in Britain of the fungus causing facial eczema in sheep. Nature 193:85

Laiz L, Piñar G, Lubitz W, Saiz-Jimenez C (2003) Monitoring the colonisation of monuments by bacteria: cultivation versus molecular methods. Environ Microbiol 5(1):72–74

Lamenti G, Tiano P, Tomaselli L (2000) Biodeterioration of ornamental marble statues in the Boboli Gardens (Florence, Italy). J Appl Phycol 12:427–433

Lamenti G, Tiano P, Tomaselli L (2002) Molecular techniques applied to the taxonomic sudy of cyanobacteria living on stone monuments. Coalition 5(2):4–5

Lange OL (1965) Der CO_2-Gaswechsel von Flechten bei tiefen Temperaturen. Planta 64:1–19

Lange OL (2000) Photosynthetic performance of a gelatinous lichen under temperate habitat conditions: long-term monitoring of CO_2 exchange of *Collema cristatum*. Bibl Lichenol 75:307–332

Lange OL, Kilian E, Ziegler H (1986) Water vapour uptake and photosynthesis of lichens: performance differences in species with green and blue-green algae as phycobionts. Oecologia 71:104–110

Lange OL, Meyer A, Büdel B (1994) Net photosynthesis activation of a desiccated cyanobacterium without liquid water in high air humidity alone. Experiments with *Microcoleus sociatus* isolated from a desert soil crust. Func Ecol 8:52–57

Larcher W (2001) Ökolophysiologie der Pflanzen. Leben, Leistung und Stressbewältigung der Pflanzen in ihrer Umwelt. 6. Aufl. UTB, Stuttgart

Laskin AI, Lechevalier HA (1973) Handbook of microbiology. Vol. 3. Microbial products. CRC Press, Cleveland/Ohio

Lawrence DP, Rotondo F, Gannibal PB (2016) Biodiversity and taxonomy of the pleomorphic genus *Alternaria*. Mycol Prog 15:3

Leathy JG, Colwell RR (1990) Microbial degradation of hydrocarbons in the environment. Microbiol Rev 54:305–315

Le Bars J, Oswald E, Le Bars P, Bonnefoi M, Bezille P, Braun JP (1990) Ecotoxinogenesis of *Pithomyces chartarum*. Food Addit Contam 7(1):19–21

Lee KB, Wee YC (1982) Algae growing on walls around Singapore. Malayan Nat J 35:125–132

Lee RE (2008) Phycology, 4th edn. Cambridge University Press, Cambridge

Lefevre M (1974) La "maladie verte" de Lascaux. Stud Conserv 19:126–156

Lefevre M, Laporte GS (1969) The "maladie verte" of Lascaux.Diagnosis and Treatment. Stud Speleol 2:35–44

Leistner P, Kaufmann A, Koehler M, Würth M, Hofbauer WK, Dittrich S, Mair S, Gordt A, Jäger M (2018) Bauphysik urbaner Oberflächen. Bauphysik 40(5):358–368

Leith ID, Hicks WK, Fowler D, Woodin SJ (1999) Differential responses of UK upland plants to nitrogen deposition. New Phytol 141:277–289

Leith ID, Sheppard LJ, Pitcairn CER, Cape JN, Holl PW, Kennedy VH, Tang YS, Smith RI, Fowler D (2001) Comparison of the effects of wet N deposition (NH_4Cl) and dry N deposition (NH_3) on UK moorland species. Water Air Soil Pollut 130:1043–1048

Leliaert F, Smith DR, Moreau H, Herron MD, Verbruggen H, Delwiche CF, De Clerck O (2012) Phylogeny and molecular evolution of the green algae. Crit Rev Plant Sci 31:1–46

Leonard KJ, Suggs EG (1974) *Setosphaeria prolata*, the ascigerous state of *Exserohilum prolatum*. Mycologia 66:281–297

Lewi MP, Cowling EB (1969) Role of nitrogen in wood deterioration. 7. Physiological adaption of wood-destroying and other fungi to substrate deficient in nitrogen. Phytopythology 59:460–468

Lewin RA (2003) The name game. Phycol Newslett 39(2):8–9

Lewin RA (2006) Black algae. J Appl Phycol 18:699–702

Lewis LA, McCourt RM (2004) Green algae and the origin of land plants. Am J Bot 91(10):1535–1556

Lewis LA, Muller-Parker G (2004) Phylogenetic placement of "Zoochlorellae" (Chlorophyta), algal symbiont of the temperate sea anemone *Anthopleura elegantissima*. Biol Bull 207:87–92

Lewis LA, Wilcox LW, Fuerst PA, Floyd GL (1992) Concordance of molecular and ultrastructural data in the study of zoosporic chlorococcalean green algae. J Phycol 28:375–380

Leyon H (1954) The structure of chloroplasts. III. A study of pyrenoids. Exptl Cell Res 6:497–505

Lindner W (2000) Zur Chemie der Biozide an Fassaden. In: Bagda E (ed) Biozide in Bautenbeschichtungen. Kontakt & Studium, vol 545. Expert-Verlag, Renningen-Malmsheim, pp 53–76

Logan C, Khan AA (1969) Comparative studies of *Phoma* spp. associated with potato gangrene in Northern Ireland. Trans Br Mycol Soc 52:9–17

Loh WS (2002) Product development for green productivity – a case study of the development of algae resistant surface coating for building facades in the human tropics. Second world conference on green productivity: http://www.apo-tokyo.org/gp/manila_conf02/resource_papers/narrative/dr_loh_wah_sing.pdf, pp 1–10

Loh WS, Lee KT (1987) Factors contributing to the fouling of walls on Singapore buildings by algal growth. In: Fourth international conference "durability of building materials and components", Singapore, pp 1040–1048

Lokhorst GM (1992) Taxonomic studies in the Genus *Heterococcus*. Crypt Stud 3:1–246

Lokhorst GM (1996) Comparative taxonomic studies on the genus *Klebsormidium* (Charophyceae) in Europe. In: Jülich W (ed) Cryptogamic studies, vol 5. Gustav Fischer, Stuttgart, Jena, New York, p 132

Lokhorst GM, Segaar PJ (1989) Ultrastructure of zoosporogenesis in the alga *Botrydiopsis alpina* (Tribophyceae), as revealed by cryofixation and freeze substitution. Europ J Protistol 24:260–270

López-Bautista JM, Chapman RL (2003) Phylogenetic affinities of the Trentepohliales inferred from small-sununit rDNA. Int J Syst Evol Microbiol 53:2099–2106

López-Bautista JM, Kapraun DF, Chapman RL (1998) Karyology and nuclear genome quantifications in the order Trentepohliales (Chlorophyta). In: Proceedings of 52th annual meeting phycological society of America. J Phycol, vol 34, p 35

López-Bautista JM, Kapraun DF, Chapman RL (2006a) Nuclear DNA content estimates in the trentepohliales (Chlorophyta): phylogenetic considerations. Algol Stud 120:41–50

López-Bautista JM, Kapraun DF, Waters DA, Chapman RL (2002) The Trentepohliales revisited. Constancea **83**(1): 1–25. http://ucjeps.berkley.edu/constancea/83/lopez-etal/trentepohliales.html

López-Bautista JM, Rindi F, Guiry MD (2006b) Molecular systematics of the subaerial green algal order Trentepohliales: an assessment based on morphological and molecular data. In J Syst Evol Microbiol 56:1709–1715

López-Bautista JM, Waters DA, Chapman RL (2003) Phragmoplastin, green algae and the evolution of cytokinesis. Int J Syst Evol Microbiol 53:1715–1718

Loub W (1967) Untersuchungen zur Mikrobiologie spanischer Böden. A Edafol Agrobiol 26:975–1008

Lud D, Buma AGJ, van de Poll Willem, Moerdijk TCW, Huiskes HL (2001) DNA damage and photosynthetic performance in the antarctic terrestrial alga *Prasiola crispa* ssp. *antarctica* (Chlorphyta) under manipulated UV-B radiation. J Phycol 37(4):459–467

Lukešová A, Hoffmann L (1996) Soil algae from acid rain impacted forest areas of the Krušné hory Mts. 1. Algal communities. Vegetatio 125:123–136

Lund JWG (1947) Observations on soil algae II. Notes on groups other than Diatoms. New Phytol 46:35–60

Lürling M, Beekman W (1999) Grazer-induced defenses in *Scenedesmus* (Chlorococcales; Chlorophyceae): coenobium and spine formation. Phycologia 38:368–376

Lürling M, Van Donk E (1997) Morphological changes in *Scenedesmus* induced by infochemicals released in situ from zooplankton grazers. Limnol Oceanogr 42(4):783–788

Lürling M, Van Donk E (1999) Grazer-induced colony formation in *Scenedesmus acutus* (Chlorophyceae): Ecomorph expression at different temperatures. J Phycol 35(6):1120–1126

Luther H (1963) Botanical analysis of mute swan faeces. Acta Vertebrat 2:266–267

Machacek JE, Wallace HAH (1952) Longevity of some common fungi in cereal seed. Can J Bot 30:164–169

Maguire B (1963) The passive dispersal of small aquatic organisms and their colonization of isolated bodies of water. Ecol Monogr 33:161–185

Mallea M, Murray IG, Segretain G, Philpot CM, Charpin H, Gueho E, Charpin J (1972) Census of *Aspergillus* colonies in the air, comparison between London, Paris, Lyon, Marseilles. Acta Allergol 27:273–278

Mansch R (1994) Wechselwirkung zwischen Nitrifikanten und Schadgasen bei der Verwitterung von Naturstein. Dissertation, Universität Hamburg

Mansch R, Bartosch S, Bock E (1999) Schädigungsbeiträge auf und in Bauwerksgestein durch biogene Nitrit- und Nitratbelastung sowie nitritkatalysierte Sulfatbildung aufgrund von Umweltbelastungen durch Emission und Imission von Schwefel- und speziell Stickstoffverbindungen. Abschlußbericht zum BMBF-Forschungsvorhaben **Bau 5016E**, 75 pp

Mansch R, Bock E (1994) Mikrobielle Werkstoffzerstörung – Simulation, Schadensfälle und Gegenmaßnahmen für metallische und keramische Werkstoffe: Untersuchungen der Beständigkeit von keramischen Werkstoffen. Werkst Korros 45(2):96–104

Mansch R, Bock E (1996) Simulation of microbial attack on natural and artificial stone. In: Heitz E, Flemming H-C, Sand W (eds) Microbially induced corrosion of materials. Springer, Berlin, Heidelberg, pp 167–186

Mansch R, Bock E (1998) Biodeterioration of natural stone with special reference to nitrifying bacteria. Biodegradation 9:47–64

Marcos Laso B (2001) Biodiversidad y colonización liquénica de algunos monumentos en la ciudad de Salamanca (España). Bot Complutensis 25:93–102

Marshall WA, Chalmers MO (1997) Airborne dispersal of antarctic terrestrial algae and cyanobacteria. Ecography 20(6):585–595

Mataloni G, Tell G, Wynn-Williams DD (2000) Structure and diversity of soil algal communities from Cierva Point (Antarctic Peninsula). Polar Biol 23:205–211

Matthes B (2000) Die Wirkungsweise herbizidaler Chloracetamide. Dissertation. Universität Konstanz

Mattox KR, Bold HC (1962) Phycological studies. III. The taxonomy of certain ulotrichacean algae. The University of Texas Publication, vol 6222, pp 1–67

Mattox KR (1971) Zoosporogenesis and resistant-cell formation in *Hormidium flaccidum*. In: Parker BB, v RM (eds) Contributions in phycology. The students of Harold C. Bold. Allen Press, Lawrence, Kansas, pp 137–144

May E (2003) Microbes on building stone – for good or ill? Culture 24(2):5–8

McElhenney TR, Bold HC, Brown RM, McGovern JP (1962) Algae: a cause of inhalant allergy in children. Ann Allergy 20:739–743

McNeill J, Barrie FR, Buck WR, Demoulin V, Greuter W, Hawksworth DL, Herendeen PS, Knapp S, Marhold K, Prado J, Prud'home van Reine WF, Smith .F, Wiersema H, Turland NJ (2012) International code of nomenclature for algae, fungi, and plants (Melbourne Code). Regnum Vegetabile **154**, Koeltz Scientific, Königstein.

Mehrotra BR, Kakkar RK (1972) Ecological study of soil fungi of an agricultural field in Allahabad. Mycopath Mycol Appl 46:379–385

Meincke M, Krieg E, Bock E (1989) *Nitrosovibrio* spp., the dominant ammonia-oxidizing bacteria in building sandstone. Appl Environ Microbiol 55(8):2108–2110

Melkonian M, Surek B (1995) Phylogeny of the Chlorophyta – Congruence between ultrastructural and molecular evidence. Bull De La Soci Zool De France-Evol Et Zool 120:191–208

Meredith DS (1963) Violent spore release in some fungi imperfecti. Ann Bot 27:39–47

Meredith DS (1966) Diurnal periodicity and violent liberation of conidia in *Epicoccum*. Phytopathology 56:988–990

Mert HH, Dizbay M (1977) The effect of osmotic pressure and salinity of the medium on the growth and sporulation of *Aspergillus niger* and *Paecilomyces lilacinum* species. Mycopathologia 61:125–127

Messikommer E (1943) Untersuchungen über die passive Verbreitung der Algen. Schweiz Z Hydrol 9:310–316

Meunier PC, Colón-López MS, Sherman LA (1997) Temporal changes in state transitions and photosystem organization in the unicellular, diazotrophic cyanobacterium *Cyanothece* sp. ATCC 51142. Plant Physiol 115:991–1000

Meunier PC, Colón-López MS, Sherman LA (1998) Photosystem II Cyclic Heterogeneity and Photoactivation in the Diazotrophic, Unicellular Cyanobacterium *Cyanothece* Species ATCC 51142. Plant Physiol 116:1551–1562

Meunier PC, Watters JW, Colón-López MS, Sherman LA (1995) Regulation of the O_2-evolving mechanism during N_2 fixation in the diazotrophic cyanobacterium *Cyanothece* sp. ATCC 51142. In: Mathis P (ed) Research in photosynthesis, vol 2. Kluwer Academic Publishers, Dortrecht, pp 389–392

Migula W (1888) Die Verbreitungsweise der Algen. Biol Zentralbl 8:514–517

Mikawa T (1975) Materials for the fungus flora of Japan (18). Trans Mycol Soc Jpn 16:146–148

Mikawa T (1979) A taxonomic study on Japanese sporangiferous Mucorales (2). J Jpn Bot 54:5–14

Mikhailyuk TI (1999) Eusubaerial algae of Kaniv Nature Reserve (Ukraine). Algologia 56(5):507–513

Mikhailyuk TI, Demchenko EM, Kondratyuk SY (2003) Algae on granite outcrops from the left bank of the river Pivdennyi Bug (Ukraine). Biol Bratisl 58(4):589–601

Mikhailyuk TI, Sluiman HJ, Massalski A, Mudimu O, Demchenko EM, Kondratyuk SY, Friedl T (2008) New Streptophyte green algae from terrestrial habitats and an assessment of the genus *Interfilum* (Klebsormidiophyceae, Streptophyta). J Phycol 44:1586–1603

Milliger LE, Schlichting HE (1968) The passive dispersal of viable algae and protozoa by an aquatic beetle. Trans Am Microsc Soc 87:443–448

Millner PD (1977) Radial growth responses to temperature by 58 *Chaetomium* species, and some taxonomic relationships. Mycologia 69:492–502

Mills JT, Seifert KA, Frisvad JC, Abramson D (1995) Nephrotoxigenic *Penicillium* species occurring on farm-stored cereal grains in western Canada. Mycopythologia 130:23–28

Mirchink TG, Kashkina GB, Abaturov YuD (1972) The resistance of fungi with different pigments to gamma-radiation. Mikrobiologyia 41:83–86

Mislivec PB, Tuite J (1970) Temperature and relative humidity requirements of species of *Penicillium* isolated from yellow dent corn kernels. Mycologia 62:75–88

Mislivec PB, Dieter CT, Bruce VR (1975) Effect of temperature and relative humidity on spore germination of mycotoxic species of *Aspergillus* and *Penicillium*. Mycologia 67:1187–1189

Mitchell RJ, Sutton MA, Truscott AM, Leith ID, Cape JN, Pitcairn CER, Van Dijk N (2004) Growth and tissue N of epiphytic bryophytes following increased and decreased inputs of atmospheric N deposition. Func Ecol 18:322–329

Moniz MBJ, Rindi F, Novis PM, Broady PA, Guiry MD (2012) Molecular phylogeny of Antarctic *Prasiola* (Prasiolales, Trebouxiophyceae) reveals extensive cryptic diversity. J Phycol 48(4):940–955

Mortimer PH, Taylor A (1962) The experimental intoxication of sheep with sporidesmin, a metabolic product of *Pithomyces chartarum*. 1. Clinical observations and findings at post-mortem examination. Res Vet Sci 3:147–160

Morton FJ (1964) Species of *Alternaria* on *Brassica* hosts in New Zealand. NZ J Bot 2:19–33

Moser M (1978) Die Röhrlinge und Blätterpilze (Polyporales, Boletales, Agaricales, Russulales). Kleine Kryptogamenflora Band IIb/2 Basidiomyceten 2. Teil. 4. Aufl. Gustav Fischer, Stuttgart, New York, 532 pp

Moustafa AF, Al-Musallam AA (1975) Contribution to the fungal flora of Kuwait. Trans Br Mycol Soc 65:547–553

Mudimu O, Hallmann C, Zufall-Roth E, Karsten U, Friedl T (2007) Green algal biofilms on artificial hard substrates: a polyphasic approach to unravel hidden algal biodiversity. Botanikertagung Hamburg September 3rd–7th 2007, Abstract zu Vortrag S33-6. http://www.biologie.uni-hamburg.de/bt07/abstracts/lectures/451p.htm.

Mulder JL, Pugh GJF (1971) Fungal biological Flora. 2. *Epicoccum nigrum* Link. Int Biodeterior Bull 7:69–71

Nakano T, Isagi Y (1987) *Dictyochloropsis irregularis* sp. nov. (Chlorococcales, Chlorophyta) isolated from the surface of bark. Phycologia 26:222–227

Nakayama T, Watanabe S, Mitsui K, Uchida H, Inouye I (1996) The phylogenetic relationship between the Chlamydomonadales and Chlorococcales inferred from 18S rRNA sequence data. Phycol Res 44:47–55

Nash Th H III (2008) Lichen biology. 2nd ed. Cambridge Univ. Press, Cambridge, 486 pp

Nebel M (2000a) Grimmiaceae. Kissenmoose. In: Nebel M, Philippi G (Hrsg) Die Moose Baden-Württembergs. Band 1: Allgemeiner Teil. Spezieller Teil (Bryophytina I, Andreaeales bis Funariales). Ulmer, Stuttgart, S 377–378

Nebel, M. (2000b): *Grimmia* Hedw.. Kissenmoos. In: Nebel, M. & Philippi, G. (Hrsg.): Die Moose Baden-Württembergs. Band 1: Allgemeiner Teil. Spezieller Teil (Bryophytina I, Andreaeales bis Funariales). Ulmer, Stuttgart: 398-435.

Nebel M (2001) 4. *Pohlia* Hedw. mbergs. Band 2: Spezieller Teil (Bryophytina II, Schistostegales bis Hypnobryales). UlmPohlmoos. In: Nebel M, Philippi G (Hrsg) Die Moose Baden-Württembergs, Stuttgart, S 15–46

Nebel M, Heinrichs J (2000) *Tortula* Hedw. Drehzahnmoos. In: Nebel M, Philippi G (Hrsg) Die Moose Baden-Württembergs. Band 1: Allgemeiner Teil. Spezieller Teil (Bryophytina I, Andreaeales bis Funariales). Ulmer, Stuttgart, S 235–264

Nebel M, Philippi G (Hrsg) (2000–2005) Die Moose Baden-Württembergs.Ulmer, Stuttgart, S 1–3

Nebel M, Sauer M, Schoepe G (2001) Brachytheciaceae. Kurzbüchsenmoose, Kegelmoose. In: Nebel M, Philippi G (Hrsg) Die Moose Baden-Württembergs. Band 2: Spezieller Teil (Bryophytina II, Schistostegales bis Hypnobryales). Ulmer, Stuttgart, S 355–428

Nedelcu AM, Lee RW, Lemieux C, Gray MW, Burger G (2000) The complete mitochondrial DNA sequence of *Scenedesmus obliquus* reflects an intermediate stage in the evolution of the green algal mitochondrial genome. Genome Res 10:819–831

Neuenstein H (1914) Über den Bau des Zellkerns bei den Algen und seine Bedeutung für ihre Systematik. Arch F Zellforschung 13:1–91

Neustupa J (1998) *Prasiola crispa* (Lightfoot) Meneghini in Královská obora in Prague. Novit Bot Univ Carol Praha 12:35–39

Neustupa J, Albrechtova J (2003) Aerial algae on spruce needles in the Krušné Hory Mts., Czech Republic. Czech Phycol Olomuc 3:161–167

Neustupa J, Nemcová Y, Eliás M, Skaloud P (2009) *Kalinella bambusicola* gen. et sp. nov. (Trebouxiophyceae, Chlorophyta), a novel coccoid *Chlorella*-like subaerial alga from Southeast Asia. Phycol Res 57:159–169

Nicholls VO (1956) Fungi on chalk soils. Trans Br Mycol Soc 39:233–236

Nierman WC, Pain A, Anderson MJ, Wortman JR, Kim HS, Arroyo J, Berriman M, Abe K, Archer DB, Bermejo C, Bennett J, Bowyer P, Chen D, Collins M, Coulsen R, Davies R, Dyer PS, Farman M, Fedorova N, Fedorova N, Feldblyum TV, Fischer R, Fosker N, Fraser A, García JL, García MJ, Goble A, Goldman GH, Gomi K, Griffith-Jones S, Gwilliam R, Haas B, Haas H, Harris D, Horiuchi H, Huang J, Humphray S, Jiménez J, Keller N, Khouri H, Kitamoto K, Kobayashi T, Konzack S, Kulkarni R, Kumagai T, Lafon A, Latgé J-P, Li W, Lord A, Lu C, Majoros WH, May GS, Miller BL, Mohamoud Y, Molina M, Monod M, Mouyna I, Mulligan S, Murphy L, O'Neil S, Paulsen I, Peñalva MA, Pertea M, Price C, Pritchard BL, Quail MA, Rabbinowitsch E, Rawlins N, Rajandream M-A, Reichard U, Renauld H, Robson GD, Rodriguez de Córdoba S, Rodríguez-Peña JM, Ronning CM, Rutter S, Salzberg SL, Sanchez M, Sánchez-Ferrero JC, Saunders D, Seeger K, Squares R, Squares S, Takeuchi M, Tekaia F, Turner G, Vazquez de Aldana CR, Weidman J, White O, Woodward J, Yu J-H, Fraser C, Galagan JE, Asai K, Machida M, Hall N, Barrell B, Denning DW (2005) Genomic sequence of the pathogenic and allergenic filamentous fungus *Aspergillus fumigatus*. Nature 438:1151–1156

Nimis PL, Pinna D, Salvadori O (1992) Licheni e Conservatione dei Monumenti. Cooperativa Libraria Universitaria Editrice Bologna, Bologna

Nirenberg HI (1990) Recent advances in in the taxonomy of *Fusarium*. Stud Mycol 32:91–101

Noguerol-Seoane Á, Rifón-Lastra A (1997a) Aportación al conocimiento de la ficoflora epilítica en monumentos del Noroeste de España. Estudio del Monasterio de Samos (Lugo). Annales Jardín Botánico de Madrid 54:37–42

Noguerol-Seoane Á, Rifón-Lastra A (1997b) Epilithic phycoflora on monuments. A survey of San Esteban de Ribas de Sil monastery (Ourense, NW Spain). Cryptogamie, Algol 18:351–361

Norton TA, Melkonian M, Andersen RA (1996) Algal biodiversity. Phycologia 35(4): 308–326

Nováková S (2002) Algal flora of subalpine bog pools in the Krkonoše Mts. Preslia Praha 74:45–56

Nowicka-Krawczyk P, Żelazna-Wieszorek J, Koźlecki T (2017) Silver nanoparticles as a control agent against facades coated by aerial algae – A model study of *Apatococcus lobatus* (green algae). PlosOne. https://doi.org/10.1371/journal.pone.0183276

Nultsch W (1961) Der Einfluss des Lichtes auf die Bewegung der Cyanophyceen. I. Phototaxis von *Phormidium autumnale*. Planta 56:632–647

Nultsch W (1962) Der Einfluss des Lichtes auf die Bewegung der Cyanophyceen. II. Mitteilung. Photokinesis bei Phormidium autumnale. Planta 57:613–623

OENORM EN 1097-6 (2006) Prüfverfahren für mechanische und physikalische Eigenschaften von Gesteinskörnungen – Teil 6: Bestimmung der Rohdichte und der Wasseraufnahme. Ausgabe Juli 2006. Beuth, Berlin, Wien, Zürich

OENORM L1082 (1989) Bestimmung von Gesamtstickstoff (Determination of total nitrogen). Österreichisches Normungsinstitut, Wien

Ohtani S, Suyama K, Yamamoto H, Aridomi Y, Itoh R, Fukuoka Y (2000) Distribution of soil algae at the monitoring sites in the vicinity of Syova Station between austral summers of 1992/1993 and 1997/1998. Polar Biosci 13:113–132

Ong B-L, Lim M, Wee Y-C (1992) Effects of desiccation and illumination on photosynthesis and pigmentation of an edaphic population of *Trentepohlia odorata* (Chlorophyta). J Phycol 28:768–772

Ortega-Calvo JJ, Ariño X, Hernandez-Mariné M, Saiz-Jimenez C (1995) Biodeterioration of building materials by cyanobacteria and algae. Int Biodeterioration 28:165–185

Ortega-Calvo JJ, Ariño X, Hernandez-Mariné M, Saiz-Jimenez C (1991) Factors affecting the weathering and colonization of monuments by phototrophic microorganisms. Sci Total Environ 167:165–185

Ortega-Calvo JJ, Ariño X, Hernandez-Mariné M, Saiz-Jimenez C (1993) Isolation and characterization of epilithic Chlorophytes and Cyanobacteria from two Spanish cathedrals (Salamanca and Toledo)Isolation and characterization of epilithic Chlorophytes and Cyanobacteria from two Spanish cathedrals (Salamanca and Toledo). Nova Hedwigia 57:239–253

Ortega-Morales O, Guezennec J, Hernández-Duque G, Gaylarde CC, Gaylarde PM (2000) Phototrophic biofilms on ancient mayan buildings in Yucatan, Mexico. Curr Microbiol 40:81–85

Orynbaev SO, Ermekova BD (1973) Effect of pH of the medium on the development of *Helminthosporium sativum* and *Alternaria tenuis*. Mikol Fitopat 7:539–541

van Overeem MA (1937) On green organisms occuring in the lower atmosphere. Trav Bot Neerl 34:388–442

Pajdak-Stós A, Fiałkowska E, Fyda J (2001) *Phormidium autumnale* (Cyanobacteria) defense against three ciliate grazer species. Aquatic Microb Ecol 23:237–244

Palmer RJ, Friedmann EI (1990) Water relations and photosynthesis in the cryptoendolithic microbial habitat of hot and cold deserts. Microb Ecol 19:111–118

Palmer RJ, Hirsch P (1991) Photosynthesis-based microbial communities on two churches in Northern Germany: weathering of granite and glazed brick. Geomicrobiol J 9:103–118

Palmer RJ, Siebert J, Hirsch P (1991) Biomass and organic acids in sandstone of a weathering building: production by bacterial and fungal isolates. Microb Ecol 21:253–266

Panasenko VT (1967) Ecology of microfungi. Bot Rev 33:189–215

Papen H, Zumbusch E, Riedl G, Möller U, Kießl K (2001) Bauphysikalische Schädigungswirkungen an Außenputzen und Verblendmauerwerk durch Mikroorganismenbefall und Überprüfung von Schutzmittelansätzen. Bau- und Wohnforschung F 2231. Fraunhofer IRB Verlag, 41 pp

Parker P (1945) The Corrosion of Concrete: I. The Isolation of a species of bacterium associates with the corrosion of concrete exposed to atmospheres containing hydrogen sulfide. Aust J Exp Biol Med Sci 23:81

Parle JN, di Menna ME (1978) The ecology of *Pithomyces chartarum* and the control of facial eczema. In: Loutit MW, Miles JAR (eds) Microbial ecology. Springer, Berlin, pp 219–224

Parsons WM, Schlichting HE, Stewart KE (1966) Inflight transport of algae and protozoa by selected odonata. Trans Am Microsc Soc 85:520–527

Paulus W (1993) Microbicides for the protection of materials. Chapman & Hall, London

Paulus W (ed) (2005) Directory of microbicides for the protection of materials and processes. Kluwer Academic Publishers, Doordrecht, p 787

Pearson RC, Hall DH (1975) Factors affecting the occurrence and severity of blackmold of ripe tomato fruit caused by *Alternaria alternata*. Phytopathology 65:1352–1359

von Pechmann H (1966) Der Einfluss der Temperatur auf das Wachstum von Bläuepilzen. In: Becker G, Liese W (eds) Holz und Organismen. Beih. Mater. Org. Bd. 1, S 237–250

Pel HJ, de Winde JH, Archer DB, Dyer PS, Hofmann G, Schaap PJ, Turner G, de Vries RP, Albang R, Albermann K, Andersen MR, Bendtsen JD, Benen JA, van den Berg M, Breestraat S, Caddick MX, Contreras R, Cornell M, Coutinho PM, Danchin EG, Debets AJ, Dekker P, van Dijck PW, van Dijk A, Dijkhuizen L, Driessen AJ, d'Enfert C, Geysens S, Goosen C, Groot GS, de Groot PW, Guillemette T, Henrissat B, Herweijer M, van den Hombergh JP, van den Hondel CA, van der Heijden RT, van der Kaaij RM, Klis FM, Kools HJ, Kubicek CP, van Kuyk PA, Lauber J, Lu X, van der Maarel MJ, Meulenberg R, Menke H, Mortimer MA, Nielsen J, Oliver SG, Olsthoorn M, Pal K, van Peij NN, Ram AF, Rinas U, Roubos JA, Sagt CM, Schmoll M, Sun J, Ussery D, Varga J, Vervecken W, van de Vondervoort PJ, Wedler H, Wösten HA, Zeng A-P, van Ooyen AJ, Visser J, Stam H (2007) Genome sequencing and analysis of the versatile cell factory *Aspergillus niger* CBS 513.88. Nat Biotechnol 25:221–231

Pelhate J (1968) Recherche des besoins eu eau chez quelques moisissures des grains. Mycopath Mycol Appl 36:117–128

Peraza Zurita Y, Cultrone G, Sánchez Castillo P, Sebastián E, Bolivar FC (2005) Microalgae associated with deteriorated stonework of the fountain of Bibataín in Granada, Spain. Int Biodeterior Biodegradation 55:55–61

Perisic M, Stojanovic D (1967) A study of the biology of *Stemphylium botryosum* Wallr. – *Pleospora herbarum* (Pers.) Rabenh. – the cause of brown spots on alfalfa leafes. Zastita Bilja 18:37–44

Perrissol C, Roux M, Lepetit J (1993) Succession of bacteria attached to evergreen oak leaf surfaces. Eur J Soil 29(3–4):167–176

Peršoh D, Beck A, Rambold G (2004) The distribution of ascus types and photobiontal selection in Lecanoromycetes (Ascomycota) against the background of a revised SSU nrDNA phylogeny. Mycol Progr 3(2):103–121

Phillipson J (1935) Some algae of Victorian soils. Proc R Soc Victoria NS 47:262–287

Pickett-Heaps JD, Marchant HJ (1972) The phylogeny of the green algae: A new proposal. Cytobios 6:255–264

Pidoplichko NM (1953) Gribnaya Flora Grubykh Kormov. Izvo Akad. Nauk Ukr. SSR, Kiev

Piecková E (2003) *In vitro* toxicity of indoor *Chaetomium* Kunze ex Fr. Ann Agric Environ Med 10:9–14

Pierson CF (1966) Effects of temperature on the growth of *Rhizopus stolonifer* on peaches and on agar. Phytopathology 56:276–278

Piervittori R, Laccisaglia A (1993) Lichens as biodeterioration agents and biomonitors. Aerobiologia 9:181–186

Piñar G, Gurtner C, Lubitz W, Rölleke S (2001a) Identification of Archaea in objects of art by DGGE analysis and shot gun cloning. Methods Enzymol 336:356–366

Piñar G, Gurtner C, Ramos C, Lubitz W, Rölleke S (2001b) Identification of Archaea in deteriorated ancient wall paintings by DGGE and FISH analysis. In: Galan E, Zezza F (eds) Protection and conservation of the cultural heritage of the mediterranean cities. Swets and Zeitlinger Publishers, Balkema

Piñar G, Saiz-Jimenez C, Schabenreiter-Gurtner C, Blanco-Varela MT, Lubitz W, Rölleke S (2001c) Archaeal communities in two disparate deteriorated ancient wall paintings: detection, identification and temporal monitoring by denaturing gradient gel electrophoresis. FEMS Microbiol Ecol 37:45–54

Pitcairn CER, Fowler D (1995) Deposition of fixed atmospheric nitrogen and foliar nitrogen content of bryophytes and *Calluna vulgaris* (L.) Hull. Environ Pollut 88:193–205

Pitcairn CER, Fowler D, Leith I, Sheppard L, Tang S, Sutton M, Famulari D (2006) Diagnostic indicators of elevated nitrogen deposition. Environ Pollut 144:941–950

Pitt JI (1973) An appraisal of identification methods for *Penicillium* species – novel taxonomic criteria based on temperature and water relations. Mycologia 65:1135–1157

Pitt JI (1979) The genus *Penicillium* and its teleomorphic states *Eupenicillium* and *Talaromyces*. Academic Press, London, p 634

Pitt JI (2000) A laboratory guide to common *Penicillium* species. Third edition. Commonw Scient Industry Res Organisation. North Ryde, Australia, 197 pp

Pitt JI, Spotts RA, Holmes RJ, Cruickshank RH (1991) *Penicillium solitum* revived, and its role as a pathogen of pomaceous fruit. Phytopathology 81:1108–1112

Pocock MA (1937) Studies in South African Volvocales. 1. A new *Sphaerella* (*Haematococcus*). Proc Linn Soc Lond 149:55–58

Pocock MA (1960) *Haematococcus* in southern Africa. Trans R Soc South Africa 36(1):5–59

Poelt J (1969, Ergänzungsband I: 1977, II: 1981) Bestimmungsschlüssel europäischer Flechten. Verlag J Cramer, Lehre

Pohl W, Schneider J (2005) Geochemische Einflüsse endolithischer Mikroorganismen auf Gesteinsoberflächen. Z dt Ges Geowiss 156(1):81–92

Pombert J-F, Otis C, Lemieux C, Turmel M (2004) The complete mitochondrial DNA sequence of the green alga *Pseudendoclonium akinetum* (Ulvophyceae) highlights distinctive evolutionary trends in the Chlorophyta and suggests a sister-group relationship between the Ulvophyceae and Chlorophyceae. Mol Biol Evol 21:922–935

Pore RS, Larsh HW (1967) Aleuriospore formation in four related *Aspergillus* species. Mycologia 59:318–325

Porta D, Rippka R, Hernández-Mariné M (2000) Unusual ultrastructural features in three strains of *Cyanothece* (Cyanobacteria). Arch Microbiol 173:154–163

Potts M (1994) Desiccation tolerance of procaryotes. Microbiol Rev 58(4):755–805

Potts M, Friedmann EI (1981) Effects of water stress on cryptoendolithic Cyanobacteria from hot desert rocks. Arch Microbiol 130:267–271

Poulíčková A, Dřímalová D, Novotný R, Válová P, Havránek P (2001) Morphological responses to varying environmental conditions in *Klebsormidium flaccidum* isolated from Moroccan desert soils. Nova Hedwigia Beihefte 123:397–407

Preisig HR, Andersen RA (2004) Historical review of algal culturing techniques. In: Andersen RA (ed) Algal culturing techniques. Phycological Society of America. Elsevier Academic Press, San Diego, pp 1–12

Pringle A, Baker DM, Platt JL, Wares JP, Latgé JP, Taylor JW (2005) Cryptic speciation in the cosmopolitan and clonal human pathogenic fungus *Aspergillus fumigatus*. Evolution 59(9):1886–1899

Pringsheim EG (1924) Algenkultur. In: Abderhalden E (ed) Handbuch der biologischen Arbeitsmethoden, Abt. XI (2/1). Urban und Schwarzenberg, Berlin, pp 377–406

Pringsheim EG (1954) Algenreinkulturen, ihre Herstellung und Erhaltung. Gustav Fischer, Jena, p 109

Printz H (1939) Vorarbeiten zu einer Monographie der Trentepohliaceae. Nytt Mag Naturvidensk 80:137–210

Printz H (1964) Die Chaetophoralen der Binnengewässer. Hydrobiologia 24:1–376

Proctor VW (1966) Dispersal of desmids by waterbirds. Phycologia 5:227–232

Pröschold T (2005) Taxonomy of monadoid and coccoid green algae: conflict of classic and modern approaches. XVII international botanical congress - Abstracts, Wien, p 118

Pröschold T & Darienko T (2020) The green puzzle *Stichococcus* (Trebouxiophyceae, Chlorophyta): New generic and species concept among this widely distributed genus. Phytotaxa 441(2):113–142

Pröschold T, Marin B, Schlösser UG, Melkonian M (2001) Molecular phylogeny and taxonomic revision of *Chlamydomonas* (Chlorophyceae). Protist 152:265–300

Proteau PJ, Gerwick WH, Garcie-Pichel F, Castenholz R (1993) The structure of scytonemin, an ultraviolet sunscreen pigment from the sheaths of cyanobacteria. Experientia 49:825–829

Pryor BM, Bigelow DM (2003) Molecular characterization of *Embellisia* and *Nimbya* species and their relationship to *Alternaria, Ulocladium* and *Stemphylium*. Mycologia 95:1141–1154

Pulimood TB, Corden JM, Bryden C, Sharples L, Nasser SM (2007) Epidemic asthma and the role of the fungal mold *Alternaria alternata*. J Allergy Clin Immunol 120(3):610–617

Purvis OW, Coppins BJ, Hawksworth DL, James PW, Moore DM (1992) The Lichen Flora of Great Britain and Ireland. London, 710 pp

de Puymaly A (1924) Recherches sur les algues vertes aériennes. Thés Fac Sci Univ de Paris, Ser A, No. 991, Bordeaux

Raabová L, Elster J, Kováčik Ľ (2016) Phototrophic microflora colonizing substrates of man-made origin in Billefjorden Region, Central Svalbard. Czech Polar Rep 6(1):21–30

Rabie CJ, Lübben A, Steyn M (1976) Production of Sterigmatocystin by *Aspergillus versicolor* and *Bipolaris sorokiniana* on semisynthetic liquid media. Appl Environ Microbiol 32:206–208

Rai JN, Agarwal SC (1974) Increased osmotic tolerance of some Aspergilli isolated from usar (alkaline) soils. A possible indication of ecological specialization. Mycopath Mycol Appl 52:299–305

Rambold G, Friedl T, Beck A (1998) Phycobionts in Lichens: Possible indicators of phylogenetic relationships? Bryologist 101(3):392–397

Ranali G, Sorlini C (2003) Application of microorganisms for the deteriorated surface recovery. Coalition 6(2):2–4

Ranilla MJ, Carro MD (2003) Diet and procedures used to detach particle-associated microbes from ruminal digesta influence chemical composition of microbes and estimation of microbial growth in Rusitec fermenters. J Anim Sci 81:537–544

Raper KB, Fennell DI (1965) The genus *Aspergillus*. Williams and Wilkins, Baltimore

Rauch A, Schagerl M (2003) Ein Beitrag zur Algenflora des Schwingrasens am Lunzer Obersee (Österreich). Verh Zool-Bot Ges Österreich 140:67–92

Raven PH, Evert RE, Eichhorn SE (2005) Biology of plants. 7th ed. W. H. Freeman & Company, 944 pp

Reddy KJ, Haskell B, Sherman DM, Sherman LA (1993) Unicellular, aerobic nitrogen-fixing cyanobacteria of the genus *Cyanothece*. J Bacteriol 175(5):1284–1292

Redhead SA, Vilgalys R, Moncalvo J-M, Johnson J, Hopple JS Jr (2001) *Coprinus* Pers. and the Disposition of *Coprinus* Species sensu latu. Taxob 50(1):203–241

Reglinski T, Elmer PAG, Taylor JT, Parry FJ, Marsden R, Wood PN (2005) Suppression of *Botrytis* bunch rot in Chardonnay grapevines by induction of host resistance and fungal antagonism. Austr Plant Pathol 34(4):481–488

Renner O (1932) Zur Kenntnis des Wasserhaushaltes javanischer Kleinepiphyten. Planta (Berl.) 18:215–287

Renzl A (2006) Mykologische Charakterisierung von Prüfkörpern verschiedener Außenwandbekleidungen nach einem Jahr Freibewitterung und Analyse des potentiellen Nährstoffgehaltes. Diplomarbeit. Leopold-Franzens-Universität Innsbruck, 100 pp

de Reviers B (2002) Biologie et phylogénie des algues. Tome 1. Éditions Belin, Paris, 352 pp de Reviers B (2003) Biologie et phylogénie des algues. Tome 2. Éditions Belin, Paris, 255 pp

Reynolds ES (1950) *Pullularia* as a cause of deterioration of paint and plastic surfaces in south Florida. Mycologia 42:432–448

Richardson DHS (1975) The vanishing Lichens. Their history, biology and importance. David & Charles, Newton Abbot, London, Vancouver

Richardson MJ, Watling R (1997) Keys to fungi on dung, Revised edn. British Mycological Society, Stourbridge

Richtlinie 98/8/EG des Europäischen Parlamentes und des Rates vom 16. Februar 1998 über das Inverkehrbringen von Biozid-Produkten. ABl. L 123 vom 24.4.98, S 1–63.

Riedl H, Hubacek H (1988) Bauwerkzerstörung und -sanierung durch Mikroorganismen. Bundesministerium für Wissenschaft und Forschung, Wien, p 84

Rieth A (1966) Zur Kenntnis der Lebensbedingungen von *Porphyridium cruentum* (Ag.) Naeg. V. Faktoren, die die Kälteresistenz modifizieren. Genet Res Crop Evol 14(1):61–82

Rifón-Lastra A, Noguerol-Seoane A (2001) Green algae associated with the granite walls of monuments in Galicia (NW Spain). Cryptogamie-Algologie 22:305–326

Rifón-Lastra A, Noguerol-Seoane A (2002) Tres nuevas Citas de Chroococcales (Cyanophyta) para la Península Ibérica: *Hormothece cylindrocellulare* (Geitler) Komárek & Anagnostidis, *Gloeocapsopsis pleurocapsoides* (Novácek) Komárek & Anagnostidis y *Chroococcus montanus* Hansgirg. Ann Jardín Bot Madrid 59(2):332–334

Rindi F (2005) First report of a species of *Prasiola* (Chlorophyta: Prasiolaceae) from the Mediterranean Sea (Lagoon of Venice). Sci Mar 69(3):343–346

Rindi F (2007) Diversity, distribution and ecology of green algae and cyanobacteria in urban habitats. In: Seckbach J (ed) Algae and cyanobacteria in extreme environments. Cellular origin, life in extreme habitats and astrobiology, Vol 11. Springer, pp 619–638

Rindi F, Guiry MD (2002) Diversity, life history and ecology of *Trentepohlia* and *Printzina* (Trentepohliales, Chlorophyta) in urban habitats in Western Ireland. J Phycol 38:39–54

Rindi F, Guiry MD (2003) Composition and distribution of subaerial algal assemblages in Galway City, western Ireland. Cryptogamie Algol 24(3):245–267

Rindi F, Guiry MD (2004) Composition and spatial variability of terrestrial algal assemblages occuring at the bases of urban walls in Europe. Phycologia 43(3):225–235

Rindi F, Guiry MD, Barbiero RP, Cinelly F (1999) The marine and terrestrial Prasiolales (Chlorophyta) of Galway city, Ireland: a morphological and ecological study. J Phycol 35:469–482

Rindi F, Guiry MD, Critchley AT, Gall EA (2003) The distribution of some species of Trentepohliaceae in France. Cryptogamie Algol 24:133–144

Rindi F, Guiry MD, López-Bautista JM (2006) New records of Trentepohliales (Ulvophyceae, Chlorophyta) from Africa. Nova Hedwigia 83(3–4):431–449

Rindi F, Guiry MD, López-Bautista JM (2008) Distribution, morphology and phylogeny of *Klebsormidium* (Klebsormidiales, Charophyceae) in urban environments in Europe. J Phycol 44:1529–1540

Rindi F, McIvor L, Guiry MD (2004) The Prasiolales (Chlorophyta) of atlantic Europe: an assessment based on morphological, molecular, and ecological Data, Including the characterization of *Rosenvingiella radicans* (Kützing) comb. nov. J Phycol 40:977–997

Rindi F, McIvor L, Sherwood AR, Friedl T, Guiry MD, Sheath RG (2007) Molecular phylogeny of the green algal order Prasiolales (Trebouxiophyceae, Chlorophyta). J Phycol 43:811–822

Rindi F, Mikhailyuk T, Sluiman HJ, Friedl T, López-Bautista JM (2011) Phylogenetic relationships in *Interfilum* and *Klebsormidium* (Klebsormidiophyceae, Streptophyta). Mol Phylogent Evol 58:218–231

Rindi F, Sherwood AR, Guiry MD (2005) Taxonomy and distribution of *Trentepohlia* and *Printzina* (Trentepohliales, Chlorophyta) in the Hawaian Islands. Phycologia 44:270–284

Rippka R, Castenholz RW, Herdman M (2001a) Subsection IV. (Formerly Nostocales Castenholz 1989b sensu Rippka, Deruelles, Waterbury, Herdman and Stanier 1979). In: Boone DR, Castenholz RW, Garrity GM (eds) Bergey's Manual® of systematic bacteriology, vol One. The archaea and the deeply branching phototrophic bacteria. Springer, New York, Berlin, Heidelberg, pp 562–566

Rippka R, Castenholz RW, Herdman M (2001b) Form-genus I. *Calothrix* Agardh 1824. In: Boone DR, Castenholz RW, Garrity GM (eds) Bergey's Manual® of systematic bacteriology, vol One. The archaea and the deeply branching phototrophic bacteria. Springer, New York, Berlin, Heidelberg, pp 582–586

Rippka R, Castenholz RW, Herdman M (2001c) Form-genus II. Chroococcus Nägeli 1849. In: Boone DR, Castenholz RW, Garrity GM (eds) Bergey's Manual® of systematic bacteriology, vol One. The archaea and the deeply branching phototrophic bacteria. Springer, New York, Berlin, Heidelberg, pp 496–497

Rippka R, Castenholz RW, Herdman M (2001d) Form-genus V. Cyanothece. In: Boone DR, Castenholz RW, Garrity GM (eds) Bergey's Manual® of systematic bacteriology, vol One. The archaea and the deeply branching phototrophic bacteria. Springer, New York, Berlin, Heidelberg, pp 499–501

Rippon JW (1988) Medical Mycology, 3rd edn. W. B. Saunders Co., Philadelphia, USA

Roldán M, Clavero E, Hernández-Mariné M (2002) 3-D Biofilm Structure of Cyanobacteria in Catacombs. Coalition 5(2):6–8

Roldán M, Clavero E, Hernández-Mariné M (2004a) Biofilms fluorescence and image analysis in hypogaean monuments research. Arch Hydrobiol Algol Stud 111:127–143

Roldán M, Thomas F, Castel S, Quesada A, Hernández-Mariné M (2004b) Non invasive pigment identification in living phototrophic biofilms by confocal imaging spectrophotometry. Appl Environ Microbiol 70:3745–3750

Rölleke S, Muyzer G, Wawer C, Wanner G, Lubitz W (1996) Identification of bacteria in a biodegraded wall painting by denaturing gradient gel electrophoresis of PCR-amplified gene fragments coding for 16S rRNA. Appl Environ Microbiol 62(6):2059–2065

Rölleke S, Witte A, Wanner G, Lubitz W (1998) Medieval wall paintings – a habitat for archaea: identification of archaea by denaturing gradient gel electrophoresis (DGGE) of PCR-amplified gene fragments coding 16S rRNA in a medieval wall painting. Int Biodeterior Biodegradation 41:85–92

Rosas I, Roy-Ocotla G, Mosiño P (1989) Meteorological effects on variation of airborne algae in Mexico. Int J Biometeorol 33:173–179

Roscher JP (1967) Alga dispersal by muskrat intestinal contents. Trans Am Microsc Soc 86:497–498

Roux C (1986) *Leptosphaerulina chartarum* sp. nov., the teleomorph of *Pithomyces chartarum*. Trans Am Microsc Soc 86(2):319–323

Rudi K, Skulberg OM, Larsen F, Jakobsen KS (1997) Strain characterization and classification of oxyphotobacteria in clone cultures on the basis of 16S rRNA sequences from the variable regions V6, V7, and V8. Appl Environ Microbiol 63(7):2593–2599

Runa F, Park MS, Pryor BM (2009) *Ulocladium* systematics revisited: Phylogeny and taxonomic status. Mycol Progress 8:35–47

Russel RC (1958) Longevity studies with wheat seeds and certain seedborne fungi. Ca J Pl Sci 38:29–33

Russomanno OMR, Portugal MASC, Coutinho LN, Calil EMB, Figueiredo MB (2003) *Leptosphaerulina chartarum* (=*Pithomyces chartarum*) e seu envolvimento no eczema facial. Arq Inst Biol São Paulo 70(3):385–390

Ryass T, Borut S (1958) Contribution to the knowledge of soil fungi in Israel. Mycopath Mycol Appl 10:142–174

Ryšánek D, Holzinger A, Škaloud P (2016) Influence of substrate and pH on the diversity of the aeroterrestrial alga *Klebsormidium* (Klebsormidiales, Streptophyta): a potentially important factor for sympatric speciation. Phycologia 55:347–358

Saad S, Hagedorn DJ (1970) Growth and nutrition of an *Alternaria* pathogenic to snapbeans. Phytopathology 60:903–906

Sabbioni C (1995) Contribution of atmospheric deposition to the formation of damage layers. Sci Total Environ 167:49–55

Sabbioni C, Zappia G (1991) Oxalate Patinas on ancient monuments: the biological hypothesis. Aerobiologia 7:31–37

Saiz-Jimenez C (1993) Deposition of airborne organic pollutants on historic buildings. Atmos Environ 27B(1):77–85

Saiz-Jimenez C (1995) Deposition of anthropogenic compounds on monuments and their effect on airborne microorganisms. Aerobiologia 11:161–175

Saiz-Jimenez C (1997) Biodeterioration vs biodegradation: the role of microorganisms in the removal of pollutants deposited on historic buildings. Int Biodeterior Biodegradation 40(2–4):225–232

Saiz-Jimenez C (1999) Biogeochemistry of weathering processes in monuments. Geomicrobiol J 16(1):27–37

Saiz-Jimenez C, Garcia-Rowe J, Garcia Del Cura MA, Ortega-Calvo JJ, Roekens E, Van Grieken R (1990) Endolithic cyanobacteria in Maastricht limestone. Sci Total Environ 94:209–220

Sajjaphan K, Shapir N, Judd AK, Wackett LP, Sadowsky MJ (2002) Novel *psbA1* gene from a naturally occurring atrazine-resistant cyanobacterial isolate. Appl Environ Microbiol 68(3):1358–1366

Samson RA, Hocking AD, Pitt JI, King AD (eds) (1992) Modern methods in food mycology. Elsevier, Amsterdam

Samson RA, Hoekstra ES, Frisvad JC, Filtenborg O (2002a) Chapter 1: Identification of the common food-borne fungi. Introduction. Zygomycetes. Ascomycetes. Deuteromycetes. *Aspergillus*. *Cladosporium*. In: Samson RA, Hoekstra ES, Frisvad JC, Filtenborg O (eds) Introduction to food- and airborne fungi, 6th edn. Centraalbureau voor Schimmelcultures, Utrecht, pp 1–115

Samson RA, Hoekstra ES, Lund F, Filtenborg O, Frisvad JC (2002b) Methods for the detection, isolation and characterisation of food-borne fungi. In: Samson RA, Hoekstra ES, Frisvad JC, Filtenborg O (eds) Introduction to food- and airborne fungi, 6th edn. Centraalbureau voor Schimmelcultures, Utrecht, pp 283–297

Samson RA, Houbraken J, Summerbell RC, Flannigan B, Miller JD (2001) Common and important species of fungi and actinomycetes in indoor environments. In: Flannigan B, Samson RA, Miller JD (eds) Microorganisms in home and indoor work environments. Taylor and Francis, New York, pp 287–292

Samson RA, Houbraken JAMP, Kuijpers AFA, Frank JM, Frisvad JC (2004) New ochratoxin A or sclerotium producing species in *Aspergillus* section *Nigri*. Stud Mycol 50:45–61

Samson RA, Houbraken J, Varga J, Frisvad JC (2009) Polyphasic taxonomy of the heat resistant ascomycete genus *Byssochlamys* and its *Paecilomyces* anamorphs. Persoonia 22:14–27

Samson RA, Mouchacca J (1975) Additional notes on species of *Aspergillus*, *Eurotium* and *Emericella* from egyptian desert soil. Antonie von Leeuwenhoek 41:343–351

Samson RA, Stolk AC, Hadlock R (1976) Revision of the subsection *Fasciculata* of *Penicillium* and some allied species. Stud Mycol Baarn 12:1–47

Samson RA, Visagie CM, Houbraken J, Hong S-B, Hubka V, Klaassen CHW, Perrone G, Seifert KA, Susca A, Tanney JB, Varga J, Kocsubé S, Szigeti G, Yaguchi T, Frisvad JC (2014) Phylogeny, identification and nomenclature of the genus *Aspergillus*. Stud Mycol 78:141–173

Samuels GJ (1976) Perfect states of *Acremonium*. The genera *Nectria, Actiniopsis, Ijuhya, Neohenningsia, Ophiodictyon*, and *Peristomialis*. NZ J Bot 14:231–260

Sánchez-Antón F, Asencio Martínez AD (2007) Participation of Cyanophyceae in the biodeterioration of the stones of the Santo Domingo College in Orihuela, Alicante (SE Spain). Algol Stud 124:95–108

Sand W (1995) Mineralische Werkstoffe. In: Mineralische Werkstoffe H (ed) Mikrobielle Materialzerstörung und Materialschutz. Schädigungsmechanismen und Schutzmaßnahmen. Gustav Fischer, Jena, Stuttgart, pp 78–110

Sandler R, Tallman CB, Keamy DG, Irving WR (1971) Successfully treated rhinocerebral phycomycosis in well controlled diabetes. New Engl J Med 18:1180–1182

Sarma YSRK (1960) Some recent advances in the nuclear cytology of Chlorophyceae. In: Kachroo P (ed) Proceedings of the symposium on algology. Indian Council of Agricultural Research, New Delhi, pp 46–63

Sauer M (2001) 6. *Hypnum* Hedw. Schlafmoos. In: Nebel M, Philippi G (eds) Die Moose Baden-Württembergs. Band 2: Spezieller Teil (Bryophytina II, Schistostegales bis Hypnobryales). Ulmer, Stuttgart, pp 473–495

Sauer M, Philippi G (2001) Hypnaceae. Schlafmoose. In: Nebel M, Philippi G (eds) Die Moose Baden-Württembergs. Band 2: Spezieller Teil (Bryophytina II, Schistostegales bis Hypnobryales). Ulmer, Stuttgart, pp 461–463

Saxena VK (1983) Evidence of the biogenic nuclei involvement in Antarctic coastal clouds. J Phys Chem 87:4130–4134

Schaefer M (Hrsg) (1984) Paul Bromer. Fauna von Deutschland. Ein Bestimmungsbuch unserer heimischen Tierwelt. 16. Aufl, Quelle & Meyer, Heidelberg, 583 pp

Schäfer-Verwimp A (2001) 3. *Orthotrichum* Hedw. Goldhaarmoos. In: Nebel M, Philippi G (eds) Die Moose Baden-Württembergs. Band 2: Spezieller Teil (Bryophytina II, Schistostegales bis Hypnobryales). Ulmer, Stuttgart, pp 170–197

Schell WA, Perfect JR (1996) Fatal, disseminated *Acremonium strictum* infection in a neutropenic host. J Clin Microbiol 34:1333–1336

Scherer S (1991) Mikroorganismen in extremen Lebensräumen. Cyanobakterien in Wüstengebieten. Anpassungen an UV-B und Trockenstreß. Biologie in unserer Zeit 21(4): 220–222

Scherer S (1994) Anpassungen von Cyanobakterien in Wüsten. In: Hausmann K, Kremer BP (eds) Extremophile. Mikroorganismen in ausgefallenen Lebensräumen. VCH, Weinheim, New York, Basel, Cambridge, Tokyo, pp 179–193

Scherer S, Chen TW, Böger P (1988) A new UV-A/B protecting pigment in the Terrestrial Cyanobacterium *Nostoc commune*. Plant Physiol 88:1055–1057

Schipper MAA (1970) Two species of *Mucor* with oval- and spherical-spored strains. Antonie von Leeuvenhoek 36:475–488

Schipper MAA (1976) On *Mucor circinelloides, Mucor racemosus* and related species. Stud Mycol (Baarn) 12:40

Schipper MAA (1978) On certain species of *Mucor* with a key to all accepted species. Stud Mycol (Baarn) 17:1–52

Schipper MAA (1984) A revision of the genus *Rhizopus*. I. The *Rhizopus stolonifer* - group and *Rhizopus oryzae*. Stud Mycol (Baarn) 25:1–19

Schipper MAA, Samson RA (1994) Miscellaneous notes on Mucoraceae. Mycotaxon 50:475–491

Schipper MAA, Samson RA, Stalpers JA (1975) Zygospore ornamentation in the genera *Mucor* and *Zygorrhynchus*. Persoonia 8:321–328

Schipper MAA, Stalpers JA (1984) A revision of the genus *Rhizopus*. II. The *Rhizopus microsporus* - group. Stud Mycol (Baarn) 25:20–34

Schlensog M (2000) Zur Photosynthese antarktischer Kryptogamen unter besonderer Berücksichtigung von Photoinhibition. Dissertation Christian-Albrechts-Universität, Kiel.

Schlichting HE (1960) The role of waterfowl in the dispersal of algae. Trans Am Mocrosc Soc 79:160–166

Schlichting HE (1961) Viable species of algae and protozoa in the atmosphere. Lloydia 24:81–88

Schlichting HE, Speziale BJ, Zink RM (1978) Dispersal of algae and protozoa by antarctic flying birds. Antarct J US 13:147–149

Schlösser UG (1994) SAG-Sammlung von Algenkulturen at the university of Göttingen. Catalogue of Strains 1994. Bot Acta 107:111–186

Schmalfuß J, Matthes B, Mayer P, Böger P (1888) Chloroacetamide mode of action, I: inhibition of very long chain fatty acid synthesis in *Scenedesmus acutus*. Z Naturforsch 53c:995–1003

Schmidle W (1895) Beiträge zur alpinen Algenflora. Öst Bot Z 45(7):249–253

Schmitz F (1879) Untersuchungen über die Zellkerne der Thallophyten. Sitzungsber. Niederrrheinisch. Ges Nat u Heilk Bonn, S 345–376

Schneegurt MA, Sherman DM, Nayar S, Sherman LA (1994) Oscillating behavior of carbohydrate granule formation and dinitrogen fixation in the cyanobacterium *Cyanothece* sp. strain ATCC 51142. J Bacteriol 176:1586–1597

Schneegurt MA, Tucker DL, Ondr JK, Sherman DM, Sherman LA (2000) Metabolic rhytms of a diazotrophic cyanobacterium, *Cyanothece* sp. strain ATCC 51142, heterotrophically grown in continuous dark. J Phycol 36(1):107–117

Schneider R, Crueger G (1976) Eine neue Blattfleckenkrankheit an Gewächshaustomaten in der Bundesrepublik Deutschland, verursacht durch eine spezialisierte Form von *Stemphylium botryosum*. Phytopath Z 87:264–273

Schnug E, Ernst WHO, Kratz S, Knolle F, Haneklaus S (2004) Aspects of ecotoxicology of sulphur in the Harz region – a guided excursion. Landbauforschung Völkenrode 54(3):129–143

Schofield E, Ahmadjian V (1972) Field observations and laboratory studies of some Antarctic cold desert cryptogams. Antarctic Res Ser 20:97–142

Scholer HJ, Müller E, Schipper MAA (1983) Mucorales. In: Howard DH (ed) Fungi pathogenic for humans and animals, A. Marcel Dekker, New York, pp 9–59

Schol-Schwarz MB (1959) The genus *Epicoccum*. Trans Br Mycol Soc 42:149–173

Schopf JW (1993) Microfossils of the early archean apex chert: new evidence of the antiquity of life. Science 260:640–646

Schumann R, Häubner N, Klausch S, Karsten U (2005) Chlorophyll extraction methods for the quantification of green microalgae colonizing building facades. Int Biodeterior Biodegradation 55:213–232

Schwabe GH, Hickel B (1978) The morphologic variability of a *Synechococcus* clone from postvolcanic substrate on the island of Surtsey (Iceland). Surtsey Res Progr Rep 7:25–27

Schwarz K (1975) Bodenalgen aus küstennahen Gebieten der Adria. Diss Univ Innsbruck

Schwarz K (1979) Neue Bodenalgen aus Dalmatien. Pl Syst Evol 131:193–209

Schwerd R (2011) Verweilverhalten biozider Wirkstoffe in Außenbeschichtungen im mehrjährigen Freilandversuch. Forschungsergebnisse aus der Bauphysik 5, pp 221

Seckbach J, Oren A (2004) Introduction to the extremophiles. In: Seckbach J (ed) Cellular Origin, Life in Extreme Habitats and Astrobiology. Origins. Kluyver Academic Publishers, Netherlands, pp 373–393

Sedlbauer K (2001) Vorhersage von Schimmelpilzbildung auf und in Bauteilen. Dissertation Universität Stuttgart.

Sedlbauer K (2002) Was so alles blüht an und in Gebäuden. Tagungsbeitrag in Deutsch für die Rosenheimer Fenstertage. 17.–19. Okt. 2002, Rosenheim, p 193–204

Sedlbauer K, Krus M (2001) Schimmelpilzbildung auf WDVS infolge "Baufehlern". IBP-Mitteilungen 28 (391)

Sekita S, Yoshihira K, Natori S (1973) Structures of Chaetoglobosin-A and -B, cytotoxic metabolites of *Chaetomium globosum*. Tetrahedron Lett 1976/23:2109–2112

Sekita S, Yoshihira K, Natori S, Kuwano H (1976) Structures of Chaetoglobosins-C. -D, -E and -F, cytotoxic Indol-3-yl-(13)-cytochalasans of *Chaetomium globosum*. Tetrahedron Lett 1976:1351

Shariat C, Collard HR (2007) Acute lung injury after exposure to *Stachybotrys chartarum*. Respir Med Extra 3(2):74–75

Sharma NK, Singh S, Rai AK (2006) Diversity and seasonal variation of viable algal particles in the atmosphere of a subtropical city in India. Environ Res 102(3):252–259

Sharp RF (1975) Some observation on the penetration sequences, effect of pH and humus formation by wood microfungi. Mycopathologia 55:41–46

Shihira I, Krauss RW (1965) *Chlorella*. Physiology and taxonomy of forty-one isolates. Univ Maryland Publ, 97 pp

Shirakawa MA, Gaylarde CC, Gaylarde PM, John V, Gambale W (2002) Fungal colonization and succession on newly painted buildings and the effect of biocide. FEMS Microbiol Ecol 39:165–173

Shirakawa MA, John VM, Gaylarde CC, Gaylarde P, Gambale W (2004) Mould and phototroph growth on masonry façades after repainting. Mater Struct 37:472–479

Shirkey B, McMaster NJ, Smith SC, Wright DJ, Rodriguez H, Jaruga P, Birincioglu M, Helm RF, Potts M (2003) Genomic DNA of *Nostoc commune* (Cyanobacteria) becomes covalently modified during long-term (decades) desiccation but is protected from oxidative damage and degradation. Nucleic Acids Res 31(12):2995–3005

Shoemaker RA (1959) Nomenclature of *Drechslera* and *Bipolaris*, grass parasites segregated from *Helminthosporium*. Can J Bot 37:879–887

Shubert LE, Gärtner G (2015) Nonmotile coccoid and colonial green algae. In: Wehr JD, Sheath RG, Kociolek JP (eds) Freshwater algae of North America: ecology and classification, 2nd edn. Academic Press, Amsterdam, pp 313–372

Shubert LE, Starks TL (1978) Algal succession on orphaned coal mine spoils. In: Wali MK (ed) Ecology and coal resource development. Pergamon Press, New York, pp 652–659

Simmons EG (2007) *Alternaria*: an identification manual. CBS biodiversity series 6. Centraalbureau voor Schimmelcultures, Utrecht, p 775

Simpson PD, Van Valkenburg SD (1978) The ultrastructure of *Mychonastes ruminatus* gen. et sp. nov., a new member of the Chlorophyceae isolated from brackish water. Br Phycol J 13(2):117–130

Sinclair NA, Herring CM (1975) Isolation of *Penicillium corylophilum* Dierckx from acid mine water and its optimal growth on hydrocarbons at acid pH. Mycopathologia 57:19–22

Singh P (1976) Some fungi in the forest soils of Newfoundland. Mycologia 68:881–890

Sinha RP, Ambasht NK, Sinha JP, Häder D-P (2003) Wavelength-dependent induction of a mycosporine-like amino acid in a rice-field cyanobacterium, *Nostoc commune*: role of inhibitors and salt stress. Photochem Photobiol Sci 2:171–176

Sitte P, Weiler EW, Kadereit JW, Bresinsky A, Körner C (2002) Lehrbuch der Botanik für Hochschulen. Begründet von E. Strasburger, F. Noll, H. Schenk, A.F.W. Schimper. 32. Auflage/ neubearbeitet von Peter Sitte, Elmar W. Weiler, Joachim W. Kadereit, Andreas Bresinsky, Christian Körner. Spektrum Akademischer Verlag, Berlin, 1123 pp

Škaloud P, Rindi F (2013) Ecological differentiation of cryptic species within an asexual protist morphospecies: a case study of filamentous green alga *Klebsormidium* (Streptophyta). J Eucar Microbiol 60:350–362

Škaloud P (2006) Variation and taxonomic significance of some morphological features in Europaean strains of *Klebsormidium* (Klebsormidiophyceae, Streptophyta). Nova Hedwigia 83(3–4):533–550

Škaloud P, Neustupa J, Radochová B, Kubínová L (2005) Confocal microscopy of chloroplast morphology and ontogeny in three strains of *Dictyochloropsis* (Trebouxiophyceae, Chlorophyta). Phycologia 44(3):261–269

Škaloud P, Neustupa J, Škaloudová M (2008) 3.4 Species composition and diversity of algae on anthropogenic substrata. Novitates Bot Univ Carol 19:33–37

Škaloud P, Peksa O (2010) Evolutionary inferences based on ITS rDNA and actin sequences reveal extensive diversity of the common lichen alga Asterochloris (Trebouxiophyceae, Chlorophyta). Molecular Phylogenetics and Evolution 54: 36–46

Skinner FA, Jones PCT, Mollison JE (1952) A comparison of a direct and plate counting technique for quantitative estimation of soil microorganisms. J Gen Microbiol 6:261–271

Smith BL, Towers NR (1984) Pithomycotoxicosis (facial eczema) in New Zealand and the use of zinc salts for its prevention. In: Seawright AA, Hegarty MP, James LF, Keeler RF (eds) Australia – USA poisonous plants symposium, Brisbane. Quensland, Poisonous Plant Committee, pp 70–79

Smith T, Olson R (2007) A taxonomic survey of lamp flora (Algae and Cyanobacteria) in electrically lit passages within Mammouth Cava National Park, Kentucky. Int J Speleol 36(2):105–114

Søchting U (1997) Epiphyllic cover on spruce needles in Denmark. Ann Bot Fennici 34:157–164

Soltis PS, Soltis DE, Wolf PG, Nickrent DL, Chaw S-M, Chapman RL (1999) The phylogeny of land plants inferred from 18S rDNA sequences: pushing the limits of rDNA signal? Mol Biol Evol 16(12):1774–1784

Soper BW, Reddy KJ (1991) Characterization of plasmids in marine unicellular, aerobic nitrogen-fixing *Cyanothece* spp. Abst. VII Int Symp photosynthetic procaryotes, Amherst, Mass, p 176

Souza-Egipsy V, Wierzchos J, Sancho C, Belmonte A, Ascaso C (2004) Role of biological soil crust cover in bioweathering and protection of sandstones in a semi-arid landscape (Torrollones de Gabarda, Spain). Earth Surf Proc Land 29:1651–1661

Spieck E, Meincke M, Bock E (1992) Taxonomic diversity of Nitrosovibrio strains isolated from building sandstones. FEMS Microbiol Ecol 102:21–26

Spiller H, Shanmugam KT (1987) Physiological conditions for nitrogen fixation in a unicellular marine cyanobacterium, *Synechococcus* sp. strain SF1. J Bacteriol 169:5379–5384

Srivastava A, Strasser RJ (1995) Govindjee Polyphasic rise of chlorophyll a fluorescence intensity and quantum yield of photosystem II of herbicide-resistant D1 mutants of *Chlamydomonas reinhardtii*. Photosynth Res 43:131–141

Stapper NJ, Kricke R (2004) Epiphytische Moose und Flechten als Bioindikatoren von städtischer Überwärmung, Standorteutrophierung und verkehrsbedingten Immissionen. Limprichtia 24:187–208

Starks TL, Shubert LE (1978) Algal colonization on a reclaimes surface-mined area in western North Dakota. In: Wali MK (ed) Ecology and coal resource development. Pergamon Press, New York, pp 652–660

Starmach K (1972) Chlorophyta III. In: Starmach K (ed) Flora slodkowodna Polski. T. 10. PWN, Warszawa & Krakow, 750 pp

Starmach K (1977) Phaeophyta, Rhodophyta. In: Starmach K (ed) Flora slodkowodna Polski, T. 14. PWN, Warszawa & Krakow, 445 pp

Starr RC, Zeikus JA (1993) UTEX – The culture collection of algae at the University of Texas at Austin. J Phycol 23(Suppl):1–106

Sterflinger K, Hain M (1999) *In situ* hybridization with tRNA targeted probes as a new tool for the detection of black yeasts and meristematic fungi. Stud Mycol 43:23–30

Sterflinger K, Krumbein WE (1997) Dematiaceous fungi as a major agent for biopitting on Mediterranean marbles and limestones. Geomicrobiol J 14(3):219–230

Sterflinger K, Prillinger H (2001) Molecular taxonomy and biodiversity of rock fungal communities in an urban environment (Vienna, Austria). Antonie van Leeuvenhoek 80:275–286

Stielow JB, Lévesque CA, Seifert KA, Meyer W, Iriny L- Smits D, Renfurm R, Verkley GJ, Groenewald M, Chaduli D, Lomascolo A, Welti S, Lesage-Meessen L, Favel A, Al-Hatmi AM, Damm U, Yilmaz N, Houbraken J, Lombard L, Quaedvlieg W, Binder M, Vaas LA, Vu D, Yurkov A, Begerow D, Roehl O, Guerreiro M, Fonseca A, Samerpitak K14, van Diepeningen AD1, Dolatabadi S15, Moreno LF16, Casaregola S17, Mallet S17, Jacques N, Roscini L, Egidi E, Bizet C, Garcia-Hermoso D, Martín MP, Deng S, Groenewald JZ, Boekhout T, Beer de ZW, Barnes I, Duong TA, Wingfield MJ, Hoog de GS, Crous PW, Lewis CT, Hambleton S, Moussa TA, Al-Zahrani HS, Almaghrabi OA, Louis-Seize G, Assabgui R, McCormick W, Omer G, Dukik K, Cardinali G, Eberhardt U, Vries de M, Robert V (2015) One fungus, which genes? Development and assessement of universal primers for potential secondary fungal DNA barcodes. Persoonia 35:242–263

Stoyneva-Gärtner M, Uzunov B, Gärtner G, Radkova M, Atanassov I, Atanasova R, Borisova C, Draganova P, Stoikova P (2019) Review on the biotechnological and nanotechnological potential of the streptophyte genus *Klebsormidium* with pilot data on its phycoprospecting and polyphasic identification in Bulgaria. Biotechnol Biotechnol Equip 33 (1): 559–578

Strasburger E (1880) Zellbildung und Zellteilung. 3. Aufl, Jena, 392 pp

Strøm KM (1926) Norwegian mountain algae. Skr Norske Vid Akad Oslo 1:1–263

Subramanian CV, Jain BL (1966) A revision of some graminicolous Helminthosporia. Curr Sci 35:352–355

Sudakin DL (2000) *Stachybotrys chartarum*: current knowledge of its role in disease. MedGenMed 2(1):7

Surek B, Beemelmanns U, Melkonian M, Bhattacharya D (1994) Ribosomal RNA sequence comparison demonstrate an evolutionary relationship between Zygnematales and Charophytes. Pl Syst Evol 191:171–181

Sutton BC, Carmichael JW (1973) 11. Coelomycetes. In: Ainsworth GC, Sparrow FK, Sussman AS (eds) The fungi. An advanced treatise, vol IVA. A taxonomic review with keys: ascomycetes and fungi imperfecti. Academic Press, New York, London, pp 513–582

Sutton BC (1980) The Coelomycetes. Fungi imperfecti with pycnidia, acervuli and stromata. Commonw Myc Inst Kew, Surrey, p 696

Swanson CP (1957) Cytology and cytogenetics. Prentice-Hall Inc., New Jersey, USA, p 596

Taber RA, Pettit RE, Taber WA, Dollahite JW (1968) Isolation of *Pithomyces chartarum* in Texas. Mycologia 60:727–730

Tan CK, Lee YK, Ho KK (1993) Effect of light intensity and ammonium-N on carotenogenesis of *Trentepohlia odorata* and *Dunaliella bardawil*. J Appl Phycol 5:547–549

Tardieux P, Perea-Dallos M, Falcou J, Hervecottu T (1972) Preliminary study on the effect of gamma irradiation on a fungal population. Agrochimica 16:83–98

Temraleevaa AD, Moskalenkoa SV, Bachura YM (2017) Morphology, Ecology, and 18S rDNA Phylogeny of the Green Microalgal Order Protosiphonales (Chlorophyceae, Chlorophyta). Microbiology 86(2): 159–169

Teneva I, Dzhambazov B, Mladenov R, Schirmer K (2005) Molecular and phylogenetic characterization of *Phormidium* species (Cyanoprokaryota) using the CPCB-IGS-CPCA locus. J Phycol 41(1):188–194

The R (1972) Über den Einfluss der Temperatur auf die Merkmalsausbildung einiger Mucoraceen und Thamnidiaceen. Diss Univ Saarbrücken

Thornbush M, Viles H (2003) Integrated digital photography and image processing for the quantification of colouration on soiled limestone surfaces in Oxford, England. J Cult Heritage 5(3):285–290

Thornbush M, Viles H (2007) Photo-based decay mapping of replaced stone blocks on the boundary wall of Worcester College, Oxford. Geological Society, London, Special Publications, vol 271, pp 69-75

Thornton RH, Percival JC (1959) A hepatotoxin from *Sporodesmium bakeri* capable of producing facial eczema diseases in sheep. Nature 183:63

Tiano P (2003) Stone reinforcement by biomediated calcite treatment. Coalition 6(2):5–8

Timberlake HG (1903) The nature and function of the pyrenoid. Science NS, XVII, p 460

Tormo R, Recio D, Silva I, Muñoz AF (2001) A quantitative investigation of airborne algae and lichen soredia obtained from pollen traps in south-west Spain. Eur J Phycol 36:385–390

Tran TH, Govin A, Guyonnet R, Grosseau P, Lors C, Garcia-Diaz Eric, Damidot D, Deves O, Ruot B (2012) Influence of the intrinsic characteristics of mortars on biofouling by *Klebsormidium flaccidum*. Int Biodeter Biodegradation 70:31–39

Trenkwalder H (1975) Neue Bodenalgen aus Föhrenwäldern im Raum von Brixen (Südtirol, Italien). Ber nat-med Ver Innsbruck 62:7–19

Tresner HD, Hayes JA (1971) Sodium chloride tolerance of terrestrial fungi. Appl Microbiol 22:210–213

Tribe HT (1960) Decomposition of buried cellulose film, with special reference to the ecology of certain soil fungi. In: Parkinson D, Waid SJ (eds) Ecology of soil fungi. Liverpool Univ. Press, pp 246–256

Trinkaus P (2001) Wiederbesiedlung weiter Bereiche des Grazer Stadtgebietes durch *Xanthoria parietina* (L.) Th.Fr. Joannea Bot 2:5–11

Trique B (1970) Croissance et sporulation de l′*Aspergillus versicolor* (Vuill.) Tiraboschi et du *Penicillium cyclopium* Westl. en culture stationnaire ou agitee. Revue Mycol 34:365–375

Tschaikner A, Gärtner G, Kofler W (2008) *Coelastrella aeroterrestrica* n. sp. (Chlorophyta, Scenedesmoideae) – a new, obviously often overlooked species. Algol Stud 128:11–20

Tschaikner A, Ingolić E, Gärtner G (2007) Observations in a New Isolate of *Coelastrella terrestris* (Reisigl) Hegewald & Hanagata (Chlorophyta, Scenedesmaceae) from Alpine Soil (Tyrol, Austria). Phyton 46(2):237–245

Tschermak-Woess E (1948) Über wenig bekannte und neue Flechtengonidien I. *Chlorella ellipsoidea* Gerneck, als neue Flechtenalge. Österr Bot Zeitschr 95(3):341–343

Tschermak-Woess E (1979) Über Plastidenstapel bei *Botrydiopsis alpina* sowie Anlage und Vermehrung der Stigmen bei dieser und *Heterococcus* (Xanthophyceae). Pl Syst Evol 131:179–192

Tschermak-Woess E (1980a) *Chaenothecopsis consociata* – kein parasitischer oder parasymbiontischer Pilz, sondern lichenisiert mit *Dictyochloropsis symbiontica*, spec. nova. Pl Syst Evol 136:287–306

Tschermak-Woess E (1980b) *Elliptochloris bilobata*, gen. et spec. nov., der Phycobiont von *Catolechia wahlenbergii*. Pl Syst Evol 136:63–72

Tschermak-Woess E (1983) Das Haustoriensystem von Dictyonema kennzeichnend für die Gattung. Pl Syst Evol 143:109–115

Tschermak-Woess E (1984) Über die weite Verbreitung lichenisierter Sippen von *Dictyochloropsis* und die systematische Stellung von *Myrmecia reticulata* (Chlorophyta). Pl Syst Evol 147:299–322

Tschermak-Woess E (1988a) II.B. The algal partner. In: Galun M (ed) CRC handbook of lichenology, vol I. CRC Press, Boca Raton, Florida, pp 39–92

Tschermak-Woess E (1988b) New and known taxa of *Chlorella* (Chlorophyceae): occurrence as lichen phycobionts and observations on living dictyosomes. Pl Syst Evol 159:123–139

Türk R, Gärtner G (2001) Biological soil crusts of the subalpine, alpine and nival areas in the Alps. In: Belnarp J, Lange OL (eds) Biological soil crusts: structure, function, and management. Springer, Berlin, Heidelberg, pp 67–73

Turland NJ, Wiersema JH, Barrie FR, Greuter W, Hawksworth DL, Herendeen PS, Knapp S, Kusber W-H, Li D-Z, Marhold K, May TW, McNeill J, Monro AM, Prado J, Price MJ, Smith GF (eds) (2018) International Code of Nomenclature for algae, fungi, and plants (Shenzhen Code) adopted by the Nineteenth International Botanical Congress, Shenzhen, China, July 2017. Regnum Vegetabile, vol 159. Koeltz Botanical Books, Glashuetten, pp 1–254

Turmel M, Ehara M, Otis C, Lemieux C (2002) Phylogenetic relationships among streptophytes as inferred from chloroplast small and large subunit rRNA gene sequences. J Phycol 38:364–375

Turner S, Huang T-C, Chaw S-M (2001) Molecular phylogeny of nitrogen-fixing unicellular cyanobacteria. Bot. Bull. Acad. Sin. 42:181–186

Ueda K (1961) Structure of plant cells with special reference to lower plants. VI. Structure of chloroplasts in algae. Cytologia 26:344–358

Ueno K, Giam CS, Taber WA (1974) Absence of sporidesmin in a Texas isolate of *Pithomyces chartarum*. Mycologia 66:360–362

Uher B, Kováčik L (2002) Epilithic cyanobacteria of subaerial habitats in National Park Slovak Paradise (1998–2000). Bull Slov Bot Spoločn 24:25–29

Uher B, Aboal M, Kovacik L (2005a) Epilithic and chasmoendolithic phycoflora of monuments and buildings in South-Eastern Spain. Cryptogamie Algol 26(3):275–308

Uher B, Kováčik L, Kučera P, Hindáková A, Pivko D (2005b) Cyanobacteria and algae on stone of cultural heritage in Bratislava. Bull Slov Spoločn 27:11–16

UKNCC (2001) Catalogue of the UK National Culture Collection: list of algae and protozoa. UKNCC Collection, Oban, p 231

Umeda M, Yamashita T, Saito M, Sekita S, Takahashi C, Yoshihira K, Natori S, Kurata H, Udagawa S (1974) Chemical and cytotoxicity survey on the metabolites of toxic fungi, Japan. J Exp Med 44:83–96

Uzunov BA, Stoyneva MP, Gärtner G, Kofler W (2008) First record of *Coelastrella* species (Chlorophyta: Scenedesmaceae) in Bulgaria. Ber. Nat.-med Verein Innsbruck 95:27–34

van der Aa HA, van Kesteren HA (1971) The identity of Phyllosticta destructiva Desm. and similar Phoma-like fungi described from Malvaceae and Lycium halimifolium. Acta Bot Neerl 20:552–563

VDL-RL 06: Pilzbefall. Richtlinie zur Bestimmung der Beständigkeit von Beschichtungsstoffen gegen Pilzbefall. Ausgabe: November 1998. Verband der Deutschen Lackindustrie

VDL-RL 07: Algenbefall. Richtlinie zur Bestimmung der Beständigkeit von Beschichtungsstoffen gegen Algenbefall. Ausgabe: November 1998. Verband der Deutschen Lackindustrie

Vesonder RF, Horn BW (1985) Sterigmatocystin in dairy cattle feed contaminated with *Aspergillus versicolor*. Appl Environ Microbiol 49(1):234–235

Vězda A (1978) Neue oder wenig bekannte Flechten in der Tschechoslowakei II. Folia Geobot. Phytotax. 13:397–420

Vinatzer G (1975) Untersuchungen über die Bodenalgen in der alpinen Stufe des Pitschberges (2300 m), Südtirol. Dissertation, Leopold Franzens Universität Innsbruck, 142 pp

Vincent WF (1988) Microbial ecosystems of Antarctica. Studies in polar research. Cambridge University Press, Cambridge, New York, New Rochelle, Melbourne, Sydney, p 304

Visagie CM, Houbraken J, Frisvad JC, Hong S-B, Klaassen CHW, Perrone G, Seifert KA, Varga J, Yaguchi T, Samson RA (2014) Identification and nomenclature of the genus *Penicillium*. Stud Mycol 78:343–371

Vischer W (1927) Zur Biologie von *Coelastrum proboscideum* und einigen andern Grünalgen. Verh Naturf Ges Basel 38:386–416

Vischer W (1936) Über Heterokonten und Heterokonten-ähnliche Grünalgen (*Bumilleriopsis, Heterothrix, Heterococcus* und *Dictyococcus, Muriella*). Ber Schweiz Bot Ges 45:372–410

Vischer W (1945) Heterokonten aus alpinen Böden, speziell dem schweizerischen Nationalpark. Ergebn. wiss. Unters. Schweiz. Nationalpark, N. F. vol 1, pp 477–512

Vischer W (1953) Über primitivste Landpflanzen. Ber Schweiz Bot Ges 63:169–193

Vischer W (1960) Reproduktion und systematische Stellung einiger Rinden- und Bodenalgen. Schweiz Z Hydrol 22(1):330–349

von Arx JA (1981) The genera of fungi sporulating in pure culture. J Cramer Verlag 424 pp

von Arx JA (1984) On Mucoraceae s. str. and other families of the Mucorales. Sydowia 35:10–26

von Arx JA, Guarro J, Figueras MJ (1986) The Ascomycete genus *Chaetomium*. Beih Nova Hedwigia 84:162

von Arx JA, Storm PK (1967) Über einige aus dem Erdboden isolierte, zu *Sporormia, Preussia* und *Westerdykiella* gehörende Ascomyceten. Persoonia 4:407–415

Wagner E, Schwarz W (1965) Geomikrobiologische Untersuchungen. IV. Untersuchungen über die mikrobielle Verwitterung von Kalkstein im Karst. Z Allg Mikrobiol 5(1):52–76

Wainwright M, Ali TA, Barakah F (1993) A review of the role of oligotrophic microorganisms in biodeterioration. Int Biodeterior Biodegradation 31:1–13

Wakefield RD, Jones MS, Wilson MJ, Young ME, Nicholson K, Urquhart DCM (1996) Investigations of decayed sandstone colonised by a species of *Trentepohlia*. Aerobiologia 12:19–25

Wakefield RD, Jones MS (1998) An introduction to stone colonizing micro-organisms and biodeterioration of building stone. Q J Eng Geol 31:301–313

Walsh JH, Stewart CS (1971) Effect of temperature, oxygen and carbon dioxide on cellulolytic activity of some fungi. Trans Br Mycol Soc 57:75–84

Walter H (1990) Vegetation und Klimazonen. Grundriss der globalen Ökologie. 6. Aufl. Ulmer, Stuttgart

Wang XW, Houbraken J, Groenewald JZ, Meijer M, Andersen B, Nielsen KF, Crous PW, Samson RA (2016) Diversity and taxonomy of *Chaetomium* and chaetomium-like fungi from indoor environments. Stud Mycol 84:145–224

Wang XW, Yang FY, Meijer M, Kraak B, Sun BD, Jiang YL, Wu YM, Bai FY, Seifert KA, Crous PW, Samson RA, Houbraken J (2019) Redefining *Humicola* sensu stricto and related genera in the Chaetomiaceae. Stud Mycol 93:65–153

Ward EWB, Henry AW (1961) Comparative response of two saprophytic and two plant parasitic soil fungi to temperature, hydrogen ion concentration, and nutritional factors. Can J Bot 39:65–79

Warnock DW (1971) Assay of fungal mycelium in grains of barley, including the use of the fluorescent antibody technique for individual fungal species. J. Gen. Microbiol. 67:197–205

Warnock DW (1973) Use of immunofluorescence to detect the mycelium of *Alternaria, Aspergillus* and *Penicillium* in Barley grains. Trans Br Mycol Soc 61:547–552

Warscheid T, Oelting M, Krumbein WE (1991) Physico-chemical aspects of biodeterioration processes on rocks with special regard to organic pollution. Int Biodet 28:37–48

Watanabe S (1977) The genus *Chlorella* (Chlorococcales) from Japanese soils (2). J Jpn Bot 52:129–136

Watanabe S (1979) Green soil algae from Papua New Guinea. In: Kurokawa S (ed) Studies on cryptogams of Papua New Guinea. Academic Scientific Book Inc., Tokyo, pp 87–102

Webster J, Dix NJ (1960) Succession of fungi on decaying cocksfoot culms. Trans Br Mycol Soc 43:85–99

Wee YC, Lee KB (1980) Proliferation of algae on surfaces of buildings in Singapore. Int Biodeter Bull 16:113–117

Wee YC (1988) Growth of Algae on exterior painted masonry surfaces. Int Biodeter 24(4/5):367–371

Weitzman I, McGough DA, Rinaldi MG, Della-Lata P (1996) *Rhizopus schipperae*, sp. nov., a new agent of zygomycosis. Mycotaxon 59:217–225

Wheeler KA, Hocking AD (1988) Water relations of *Paecilomyces variotii*, *Eurotium amstelodami*, *Aspergillus candidus* and *Aspergillus sydowii*, xerophilic fungi isolated from Indonesian dried fish. Int J Food Microbiol 7(1):73–78

Whittle AM (1977) Mycoflora of cones and seeds of *Pinus sylvestris*. Trans Br Mycol Soc 69:47–57

Whittle SJ, Casselton PJ (1975a) The chloroplast pigments of the algal classes Eustigmatophyceae and Xanthophyceae. I. Eustigmatophyceae. Br. Phycol. J. 10:179–191

Whittle SJ, Casselton PJ (1975b) The chloroplast pigments of the algal classes Eustigmatophyceae and Xanthophyceae. II. Xanthophyceae. Br Phycol J 10:192–204

Wilcox L, Lewis LA, Fuerst PA, Floyd GL (1992) Assessing the relationships of autosporic and zoosporic chlorococcalean green algae with 18S rRNA sequence data. J Phycol 28:381–386

Wilhelm C, Krämer P, Wiedemann I (1987) Die Lichtsammelkomplexe der verschiedenen Algenstämme. Biologie in unserer zeit 17(5):138–143

Wilkins CK, Larsen ST, Hammer M, Poulsen OM, Wolkoff P, Nielsen G (1998) Respiratory effects in mice exposed to airborne emmissions from *Stachybotrys chartarum* and implications for risk assessment. Pharmacol, Toxicol, p 83

Wille N (1923) Karplanter og ferskwandsalger fra øerne Husøy, Ona og Rosholmen paa Nordmør. Nyt Mag Naturv 61:53–89

Wirth V (1995a) Die Flechten Baden-Württembergs. Teil 1 und 2. 2. Auflage, Ulmer, Stuttgart

Wirth V (1995b) Flechtenflora. Bestimmung und ökologische Kennzeichnung der Flechten Südwestdeutschlands und angrenzender Gebiete. 2., neubearbeitete und ergänzte Auflage. Eugen Ulmer, Stuttgart

Wirth V, Hauck M, Schultz M (2013) Die Flechten Deutschlands, 1 u. 2, Ulmer, Stuttgart, 1244 pp

Wolf M, Buchheim M, Hegewald E, Krienitz L, Hepperle D (2002) Phylogenetic position of the Sphaeropleaceae (Chlorophyta). Plant Syst Evol 230:161–171

Wolf M, Hepperle D, Krienitz L (2003) On the phylogeny of *Radiococcus*, *Planktosphaeria* and *Schizochlamydella* (Radiococcaceae, Chlorophyta). Biol Bratisl 58(4):759–765

Wollum AG (1982) Cultural methods for soil microorganisms. In: Page AL, Miller RH, Keeney DR (eds) Methods of soil analysis, part II. Chemical and microbiological properties. American Society of Agronomy, pp 781–802

Wolseley PA, James PW, Theobald MR, Sutton MA (2006) Detecting changes in epiphytic lichen communities at sites affected by atmospheric ammonia from agricultural sources. Lichenologist 38:161–176

Woudenberg JHC, Groenewald JZ, Binder M, Crous PW (2013) *Alternaria* redefined. Stud Mycol 75:171–212

Woudenberg JHC, Seidl MF, Groenewald JZ, de Vries M, Stielow JB, Thomma BPHJ, Crous PW (2015) *Alternaria* section *Alternaria*: species, formae speciales or pathotypes? Stud Mycol 82:1–21

Woudenberg JHC, Hanse B, van Leeuwen GCM, Groenewald JZ, Crous PW (2017) *Stemphylium* revisited. Stud Mycol 87:77–103

Wu M, Hanlin RT (1992) Ascomal development in *Leptosphaerulina crassiasca*. Mycologia 84(2):241–252

Xiong F, Kopecky J, Nedbal L (1999) The occurrence of UV-B absorbing mycosporine-like amino acids in freshwater and terrestrial microalgae (Chlorophyta). Aquat Bot 63:37–49

Xiong F, Lederer F, Kukavský J, Nedbal L (1996) Screening of freshwater algae (Chlorophyta, Chromophyta) for ultraviolet.B sensitivity of the photosynthetic apparatus. J Plant Physiol 148:42–48

Yadav AS, Madelin MF (1968) Experimental studies on microfungi from decaying stems of *Heracleum sphondylium* and *Urtica dioica*. Trans Br Mycol Soc 51:261–267

Yaguchi T, Horie Y, Tanaka R, Matsuzawa T, Ito J, Nishimura K (2007) Molecular phylogenetics of multiple genes on *Aspergillus* section *Fumigati* isolated from clinical specimens in Japan. Jpn J Med Mycol 48:37–46

Yano S, Koyabashi K, Kato K (2003) Case report. Intrabronchial lesion due to *Cladosporium sphaerospermum* in a healthy, non-asthmatic woman. Mycoses 46(8): 330–332

Yeghen T, Fenelon L, Campbell CK, Warnock DW, Hoffbrand AV, Prentice HG, Kibbler CC (1996) *Chaetomium* pneumonia in a patient with acute myeloid leukaemia. J Clin Pathol 49:184–186

Young TWK (1969) Electron and phase-contrast microscopy of spores in two species of the genus *Mycotypha* (Mucorales). J Gen Microbiol 55:243–249

Zaas AK, Moyce M, Schell W, Alexander Lodge B, Miller JL, Perfect JR (2003) Risk of fungemia due to *Rhodotorula* and antifungal susceptibility testing of *Rhodotorula* isolates. J Clin Microbiol 41(11):5233–5235

Zalar P, Gostinčar C, de Hoog GS, Uršič V, Sudhadham M, Gunde-Cimerman N (2008) Redefinition of *Aureobasidium pullulans* and its varieties. Stud Mycol 61:21–38

Zanardini E, Abbruscato P, Realini M, Brusetti L, Sorlini C (2002) Molecular techniques applied to the study of microbial communities colonising different lithotypes. Coalition 5(2):8–10

Zagulyayeva ZA (1971) Physiology of cellulose decomposing micromycetes. Mikol Fitopat 5:385–389

Zahradníková M, Andersen HL, Tønsberg T, Beck A (2017) Molecular Evidence of Apatococcus, including *A. fuscideae* sp. nov., as Photobiont in the Genus *Fuscidea*. Protist 168:425–438

Zeuch L (1934) Untersuchungen zum Wasserhaushalt von *Pleurococcus vulgaris*. Planta 22(5):614–643

Zheng R-Y, Cheng G-Q, Huang H, Liu X-Y (2007) A monograph of *Rhizopus*. Sydowia 59(2):273–372

Zycha H, Siepmann R, Linnemann G (1969) Mucorales. Eine Beschreibung aller Gattungen und Arten dieser Pilzgruppe, J Cramer Lehre, p 355

Index

© Springer-Verlag GmbH Deutschland, ein Teil von Springer Nature 2020

W. K. Hofbauer and G. Gärtner, *Microbial life on Façades,*

https://doi.org/10.1007/978-3-662-54833-2